Wiring Regulations in Brief

This newly updated edition of *Wiring Regulations in Brief* provides a user-friendly guide to the newest amendments to BS 7671 and the IET Wiring Regulations. Topic-based chapters link areas of working practice – such as earthing, cables, installations, testing and inspection, and special locations – with the specifics of the Regulations themselves. This allows quick and easy identification of the official requirements relating to the situation in front of you. The requirements of the regulations, and of related standards, are presented in an informal, easy-to-read style to remove confusion. Packed with useful hints and tips, and highlighting the most important or mandatory requirements, this book is a concise reference on all aspects of the eighteenth edition of the IET Wiring Regulations. This handy guide provides an on-the-job reference source for electricians, designers, service engineers, inspectors, builders, and students.

Ray Tricker (MSc, IEng, CQP-FCQI, FIET, FCMI, FIRSE) is a Senior Consultant with over 50 years' continuous service in Quality, Safety and Environmental Management; Project Management; Communication Electronics; Railway Command, Control and Signalling Systems; Information Technology; and the development of Molecular Nanotechnology.

He served with the Royal Corps of Signals (for a total of 37 years) during which time he held various managerial posts culminating in being appointed as the Chief Engineer of NATO's Communication Security Agency (ACE COMSEC). Most of Ray's work since leaving the Services has centred on the European Railways. He has held a number of posts with the Union Internationale des Chemins de fer (UIC) (e.g. Quality Manager of the European Train Control System (ETCS)) and with the European Union (EU) Commission (e.g. European Rail Traffic Management System (ERTMS) Users Group Project Coordinator, HEROE Project Coordinator and T500 Review Team Leader). He was also a UKAS Assessor (for the assessment of certification bodies for the harmonisation of the Trans-European, High Speed, Railway Network) and recently he was appointed as the Quality, Safety and Environmental Manager for the Project Management Consultancy responsible for overseeing the multi-billion-dollar Trinidad Rapid Rail System.

Currently, as well as writing books on diverse subjects such as ISO 9001:2015, Building, Wiring and Water Regulations, he is busy assisting Small Businesses from around the world (usually on a no-cost basis) to produce their own auditable Quality and/or Integrated Management Systems to meet the requirements of ISO 9001, ISO 14001 and OHSAS 18001 etc.

Wiring Regulations in Brief

4th Edition

Ray Tricker

Routledge
Taylor & Francis Group

LONDON AND NEW YORK

Fourth edition published 2021
by Routledge
2 Park Square, Milton Park, Abingdon, Oxon, OX14 4RN

and by Routledge
52 Vanderbilt Avenue, New York, NY 10017

Routledge is an imprint of the Taylor & Francis Group, an informa business

First edition published by Butterworth Heinemann, an imprint of Elsevier 2007

Second edition published by Butterworth Heinemann, an imprint of Elsevier 2008

Third edition published by Routledge 2015

British Library Cataloguing-in-Publication Data

A catalogue record for this book is available from the British Library

Library of Congress Cataloging-in-Publication Data
Names: Tricker, Ray, author.
Title: Wiring regulations in brief / Ray Tricker.
Description: Fourth edition. | Abingdon, Oxon ; New York : Routledge, 2021.
| Includes bibliographical references and index.
Identifiers: LCCN 2020022558 (print) | LCCN 2020022559 (ebook)
| ISBN
9780367432010 (hardback) | ISBN 9780367431983 (paperback) |
ISBN
9781003001829 (ebook)
Subjects: LCSH: Electric wiring, Interior--Handbooks, manuals, etc. |
Electric wiring, Interior--Standards.
Classification: LCC TK3271 .T75 2021 (print) | LCC TK3271
(ebook) | DDC
621.319/240941--dc23
LC record available at https://lccn.loc.gov/2020022558
LC ebook record available at https://lccn.loc.gov/2020022559

ISBN: 978-0-367-43201-0 (hbk)
ISBN: 978-0-367-43198-3 (pbk)
ISBN: 978-1-003-00182-9 (ebk)

Typeset in Goudy Oldstyle Std
by KnowledgeWorks Global Ltd.

Contents

List of figures

List of tables

Preface

The Industrial Revolution during the 1800s was responsible for causing poor living and working conditions in ever-expanding, densely populated urban areas. Outbreaks of cholera and other serious diseases (through poor sanitation, damp conditions and lack of ventilation) forced the UK Government to take action.

Building control took on the greater role of health and safety through the first Public Health Act in 1875 which eventually led to the first set of national building standards (i.e. *The Building Regulations*).

As is the case with most official documents, as soon as they were published, they were virtually out of date and consequently required frequent revision. So it wasn't too much of a surprise to learn that the committee responsible for writing the *Public Health Act* of 1875 had overlooked the increased use of electric power for street lighting and for domestic purposes. Electricity was beginning to become more popular, but as there were no rules and regulations governing their installation at that time, the companies, or people responsible, simply dug up the roads and laid the cables as and where they felt like it!

From a health and safety point of view, the Government of the day expressed extreme concern at this exceedingly dangerous situation and so in 1882 *The Electric Lighting Clauses Act* (modelled on the previous 1847 *Gas Act*) was passed by Parliament. This legislation was implemented by *Rules and Regulations for the prevention of Fire Risks Arising from Electric Lighting*, and it is this document that is the forerunner of today's IET Wiring Regulations. Since then, this document has seen a succession of amendments, new editions and new titles, and we now have the new 18th Edition of these Regulations, which has been written in alignment with the latest editions of the CENELEC, IEC, and EN Harmonised Documents for electrical safety, and is now republished as an officially recognised British Standard.

By alignment with these other regulations and standards, BS 7671:2018 (*Requirements for Electrical Installations*) has virtually become a European document which is widely used in many EU countries as well as within the UK. So we now have an official regulation which came into effect on 1 January 2019, and this categorically states that:

> *"All new installations from this point MUST fully comply with BS 7671:2018."*

 Although the Regulations primarily apply to the design, erection and verification of **new** electrical installations, they **also** apply to **additions and alterations** to existing installations.

The new Wiring Regulations have been thoroughly revised and, in some places, restructured so as to reflect current building requirements etc. The entire standard has also been thoroughly copy-edited in order to make it a bit more 'readable', and quite a lot of the sections have either been modified, moved to other sections or completely rewritten. The most significant amendments now shown in BS 7671:2018 concern:

Chapter 41 Protection against electric shock

The wording and structure of this chapter has been changed in accordance with other associated standards. It also contains a number of significant changes such as:

- a.c. final circuits supplying luminaires in a domestic property **must** now be protected by an RCD;
- metallic pipes entering a building with an insulating section at their point of entry do **not** need to be connected to protective equipotential bonding;
- switching or isolating devices shall **not** be inserted in a PEN conductor.

 The regulations concerning IT systems have also been revised and restructured.

Chapter 42 Protection against thermal effects

This chapter now recommends installing of Arc Fault Detection Devices (AFDDs) in order to reduce the risk of fire in a.c. final circuits of a fixed installation, and cables (especially those supplying safety circuits) must now satisfy the fire protection requirements of the Construction Products Regulation (CPR).

Chapter 44 Protection against voltage disturbances and electromagnetic disturbances

Chapter 44 (which deals with protection against overvoltages of atmospheric origin (e.g. lightning) or due to switching) has been redrafted – and in some cases a risk assessment is now required.

Chapter 46 Devices for isolation and switching

This is a completely new chapter which deals with non-automatic (local and remote) isolation and switching devices that are used for electrical installations or electrically powered equipment.

Chapter 52 Selection and erection of wiring systems

This chapter now includes the following requirements:

- all cables **must** be supported against premature collapse in the event of a fire;

 Note: This applies throughout the installation, not just in escape routes as previously required.

- cables must now satisfy the fire protection requirements of the CPR.

Chapter 53 Protection, isolation, switching, control and monitoring

Chapter 53 has been totally revised and now deals with the general requirements for protection, isolation, switching, control and monitoring, together with the selection and erection of the devices used to control these operations.

Chapter 54 Earthing arrangements and protective conductors

New regulations in this Chapter include earth electrodes and the insertion of a switching device in a protective conductor (particularly where an installation is supplied from more than one source of energy).

Chapter 55 Other equipment

This has been revised with respect to the selection and erection of ground-recessed luminaires.

Part 6 Inspection and testing

This Part has been completely restructured and now

- includes the regulation being renumbered so as to align with the CENELEC standard;
- Chapters 61, 62 and 63 (having been reorganised and modified) now form two new Chapters 64 and 65.

Part 7 Special installations or locations

A number of small changes and modifications to the Regulations have been made to this Part, including the consideration of:

- construction and demolition site installations;
- extra-low-voltage lighting installations;
- electrical installations in caravans and motor caravans;
- electric vehicle charging installations;
- floor and ceiling heating systems;
- medical locations;
- onshore units of electrical shore connections for inland navigation vessels (this is an entirely new section).

The Appendices

The majority of the existing appendices now include minor manuscript amendments. However, a new appendix (Appendix 17) has now been written which provides recommendations and instructions for the design and erection of electrical installations primarily intended for industrial purposes.

Author's note

*Please note, however, that this is only the author's impression of the most important aspects of the Wiring Regulations and their association with the Building Regulations. It should, therefore, only be treated as an aide-memoire to the Regulations and electricians should **always** consult the requirements of BS 7671 to satisfy compliance.*

The current legislation for **all** building control is the Building Act 1984 (implemented by the Building Regulations 2010). Together they set minimum requirements for the design and construction of buildings and are intended to ensure the health, safety and welfare of people in and around those buildings.

The legislation includes requirements to ensure that fuel and power are conserved and that facilities are provided for people (including those with disabilities) to get into and move around inside buildings, with ease. They are basic performance standards which are supported by a series of documents called Approved Documents which contain practical and technical guidance on ways in which the requirements of the Building Act 1984 can be met.

Since the introduction of the Public Health Act in 1875 there has always, therefore, been a direct link between electrical installations and building control, and parts of all of the Approved Documents have effect on these sorts of installations.

 If you log onto https://www.gov.uk/government/collections/approved-documents you can download free copies of these Approved Documents.

From an electrical safety point of view, the publication of the revised Approved Document P for Electrical Safety in 2013 (Figure P1) linked the design, installation, inspection and testing of electrical installations to building control and

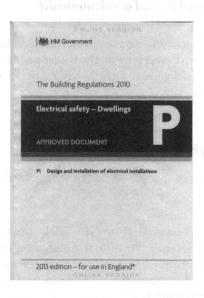

Figure P1 Approved Document P: Electrical Safety – Dwellings. (2013 Edition – for use in England)

the aim of this book is to draw all of the various requirements of this huge library of information together and to make life a little bit easier for the reader to understand the *'officialese'*.

Over the past 120 years there have been literally hundreds of books written on the subject of electrical installations, but the aim of my 4th edition of *Wiring Regulations in Brief* is not just to become another book on the library shelf to be occasionally looked at. The intention is that it will provide professional engineers, students and (to a lesser degree) the unqualified DIY fraternity with an easy-to-read reference source to the official requirements of BS 7671:2018 for electrical safety and electrical installations.

Although BS 7671:2018 is well structured and has separate sections for all the main topics (e.g. safety protection, selection and erection of equipment and so on), it is not the easiest of standards to get to grips with for a particular situation. Occasionally it can be very confusing, requiring the reader to constantly flick backwards and forwards through the book to find what it is all about.

For example, Regulation 411.4.202 states that:

> *Where a circuit-breaker is used to satisfy the requirements of Regulation 411.3.2.2 or Regulation 411.3.2.3, the maximum value of earth fault loop impedance (Zs) shall be determined by the formula in Regulation 411.4.4. Alternatively, for a nominal voltage (Uo) of 230 V and a disconnection time of 0.4 s in accordance with Regulation 411.3.2.2 or 5 s in accordance with Regulation 411.3.2.3, the values specified in Table 41.3 for the types and ratings of overcurrent devices listed may be used instead of calculation[!]*

The intention of *Wiring Regulations in Brief*, therefore, is to peel away some of this confusion and jargon and provide the reader with an on-the-job reference source that can be quickly used without having to delve backwards and forwards though the standard.

Content of this book

To reflect these changes, this 4th edition of *Wiring Regulations in Brief* is structured as follows:

Chapter 1 –   The background to BS 7671, what it contains and a
Introduction   description of its unique numbering system, objectives and legal status. The effect that the Wiring Regulations have on other statuary instruments and how BS 7671 can be implemented.

Chapter 2 –
Building
regulations

For many years the Wiring Regulations have not fully supported the requirements of the Building Regulations. However,
owing to the growing number of electrical accidents occurring in domestic and non-domestic buildings, in 2005 the
Government decided that in future:

*'All new electrical wiring and components (within England and
Wales) must be designed and installed in accordance with the
electrical safety requirements contained in Part P of the Building
Regulations.'*

This chapter, therefore, provides:

- an overview of the structure and contents of the Building
  Regulations and their Approved Documents (or Parts);
- responsibilities for electrical safety;
- the need to oversee and control extensions (material
  alterations and material changes of use);
- electricity distributors' responsibilities (earthing, electrical
  installation work, types of wiring and wiring systems etc.);
- inspection and testing; and (most importantly)
- who and what a competent firm or competent person is,
  what they are responsible for, and how a building can be
  self-certified.

 Note: While the requirements from the Wiring Regulations are normally pref-
aced by the word *'shall'* (meaning that this section **is** a mandatory requirement), you will notice that the Building Regulations use the words *'should'* (i.e.
recommended), *'may'* (i.e. permitted) or *'can'* (i.e. possible).

The reason for this is that Approved Documents reproduce the actual
requirements contained in the Building Regulations relevant to a particular subject area. This is then followed by *Practical and Technical Guidance* (together
with examples) showing how the requirements can be met in some of the more
common building situations. There may, however, be alternative ways of complying with the Building Regulations 2010's requirements to those shown in
the Approved Documents, and you are, therefore, under no obligation to adopt
any particular solution in an Approved Document – if your prefer to meet the
requirement(s) in some other way – but, you **must** meet the requirement!

Chapter 3 –
Earthing

This chapter reminds the reader about the different types of
earthing systems and earthing arrangements. It then lists the
main requirements from the various Approved Documents
and Regulations for safety protection (direct and indirect
contact), protective conductors and protective equipment,
before briefly touching on the test requirements for earthing.

Chapter 4 –
Safety
protection

This chapter lists the mandatory and fundamental requirements for safety protection contained in both the Wiring Regulations and the Building Regulations 2010. It also includes information concerning the basic protection against electric shock, fault protection, protection against direct and indirect contact, protective conductors and protective equipment, and then lists the test requirements for safety protection.

Chapter 5 –
Electrical
equipment,
components,
accessories
and supplies

The amount of different types of equipment, components, accessories and supplies for electrical installations currently available is enormous and any attempt to cover every type, model and/or manufacture would prove an impossible task for a book such as this.

The intention of this chapter, therefore, is to provide a catalogue of all the different types identified and referred to in the Wiring Regulations (e.g. luminaires, RCDs, plugs and sockets) and then make a list of the specific requirements that are sprinkled throughout the Regulations.

Chapter 6 –
Cables,
conductors
and conduits

Within BS 7671:2018 there is frequent reference to different types of cables (e.g. single-core, multicore, fixed, flexible) conductors (e.g. live supply, protective, bonding) and conduits, cable ducting, cable trunking and so on. Unfortunately, as is the case for equipment and components, the requirements for these items are liberally sprinkled throughout the Standard.

The aim of this chapter, therefore, is to provide a catalogue of all the different types identified and referred to in the Wiring Regulations, under two main headings (i.e. 'cables', and 'conductors and conduits'), and then make a list of their essential requirements.

Chapter 7 –
Special
installations
and locations

Whilst the Regulations apply to electrical installations in all buildings, there are also some indoor and out-of-doors special installations and locations that are subject to additional requirements due to the extra dangers they pose. This chapter considers the requirements for these special locations and installations and, while perhaps not providing a complete list, represents the most important requirements.

Chapter 8 –
External
influences

Although Chapter 51 of BS 7671:2018) still contains the requirements for external influences, the 18th edition of the Wiring Regulations has now been further developed in accordance with new ISO standards and the BS EN 60721 and BS EN 61000 series on environmental conditions.

This chapter provides a brief overview (together with extracts from the current Regulations) on how the environment in which an electrical product, service or piece of equipment is situated will effect it meeting the quality and safety requirements of BS 7671:2018.

Chapter 9 – Installation, maintenance and repair	While this is a comparatively small chapter(compared with some of the others in this book!), it is nevertheless an extremely important one as it provides some guidance on the requirements for installation, maintenance and repair. It also lists the Regulations' requirements for maintenance etc. with respect to electrical installations. However, as per the previous chapters which contain similar lists, please remember that this is only the author's impression of the most important aspects of the Wiring Regulations, and electricians should always consult the latest edition of BS 7671 to satisfy compliance. Finally, in an annex, there is a complete set of checklists for the quality, safety and environmental control of electrical equipment and installations.
Chapter 10 – Inspection and testing	In accordance with the Regulations, every installation (or alteration to an existing installation) during erection and on completion before being put into, or back into, service, has to be inspected and tested to verify, so far as is reasonably practicable, that the requirements of the Regulations have been met.

Part 6 of BS 7671:2018 has been completely restructured and now aligns with the CENELEC (European Committee for Electrotechnical Standardization) Standard in order to meet the latest requirements for the initial and periodic inspection and testing.

This final chapter provides some guidance on the requirements for the installation, maintenance, inspection, certification and repair of electrical installations. It lists the Regulations' primary requirements for these activities with respect to electrical installations and (in an annex) provides an example Stage Audit Checklist for designers and engineers to use.

The ten chapters are then supported by the following annexes:

Annex A – Symbols used in electrical installations
Annex B – List of electrical and electromechanical symbols
Annex C – SI units for existing technology
Annex D – The IP code
Annex E – Acronyms and abbreviations
Annex F – Standards
Annex G – Other books associated with the Wiring Regulations

plus a full Index

The following symbols will help you get the most out of this book. In the margins you will find:

 An important requirement or point.

 A good idea or suggestion.

and within the text:

 Notes: are used to provide further amplification or information.

For your convenience (and to save you having to look backwards and forwards through the book for the correct requirement), quite of a lot of these requirements have been shown more than once (i.e. in different chapters and/or sections of the book) as have a few of the figures and tables.

 Note: If any reader has any thoughts about the contents of this book (such as areas where perhaps they feel I have not given sufficiently coverage, omissions and/or mistakes etc.), then please let know by e-mailing me at **ray@herne.org.uk**, and I will make suitable amendments in the next edition of this book.

I hope that you find this book useful!!

Ray Tricker

Foreword

BS 7671 *Requirements for Electrical Installations – IEE Wiring Regulations* has been an integral part of most of my professional life as an electrical engineer, and previously as an apprentice, only slipping out of sight during my time overseas. It is a document that has to guide many different people on safe and acceptable practice, including professional engineers, designers, technicians, and electricians. It also has to remain up to date, taking cognisance of the evolution of legislation, technological developments, and changes in custom and practice. It is thus a difficult task to make the requirements readily accessible to such a diverse readership. Commentaries and further explanations are, therefore, often helpful and are sometimes essential reading to gain sufficient understanding to carry out work associated with electrical installations.

I am, therefore, pleased to contribute in a small way to Ray's book, which sheds another useful light on the Wiring Regulations. Significantly for me, and perhaps many other older hands, it places these Regulations within their context in Part P of the Building Regulations and cross-references requirements. To have the Wiring Regulations virtually side by side with the legal requirements of the Building Regulations and other relevant background information in one handy volume should be particularly useful. Ray's book is a welcome contribution to working with the Wiring Regulations and Part P of the Building Regulations.

I like the simple approach, not delving too deeply into the fine detail and making information easy to find by grouping it and laying it out logically. This is what most readers need much of the time. For me, this book is a must for work associated with electrical installations complying with the Building Regulations and the Wiring Regulations. This book can save the cost of buying parts of the Building Regulations, making it a cost-effective option. Some readers may find that they do not need to buy the Wiring Regulations as well.

We live every day with the multitude of benefits provided by electricity. Modern life would be impossible without it. Yet electricity is potentially dangerous to property, humans, and many living creatures. Only the application of human ingenuity can reduce the risks to levels that are deemed acceptable. That we can live, work, and play surrounded by live electrical equipment and installations is testimony to the dedication, understanding, and skills of many people. It is also testimony to the success of the Wiring Regulations, in its many revisions and editions over the years, and books such as Ray's commentary, which helps to make the Regulations more accessible to every user.

Nigel Moore BSc (Hons), MBA, CEng, MIET, MCIM, HNC

BS7671 Colour Arrangements

The interface between old and new cable colours

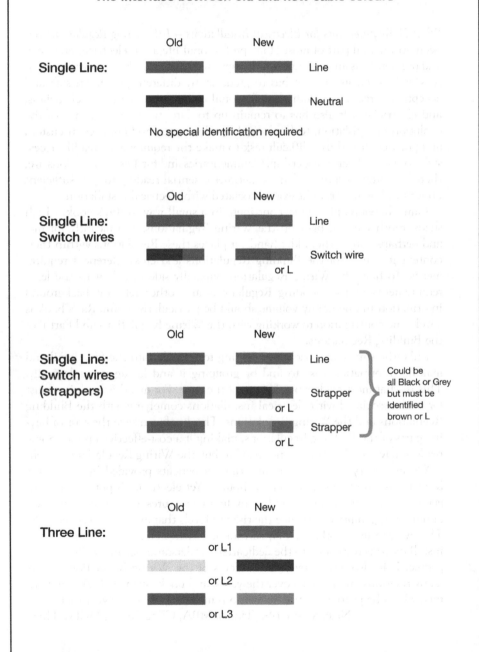

Single Line:

Old   New

Line

Neutral

No special identification required

**Single Line:
Switch wires**

Old   New

Line

Switch wire
or L

**Single Line:
Switch wires
(strappers)**

Old   New

Line

Strapper
or L

Strapper
or L

Could be
all Black or Grey
but must be
identified
brown or L

Three Line:

Old   New

or L1

or L2

or L3

1
Introduction

Author's start note

This initial chapter contains the background to British Standard 7671, what it contains, a description of its unique numbering system, and its objectives and legal status. It also discusses the effect that the Wiring Regulations have on other statuary instruments and how this British Standard can be implemented.

1.1 Introduction

The latest edition of the IET Wiring Regulations has now grown to a massive 560-page document that defines the way in which all electrical installation work must be carried out. It does not matter whether the work is carried out by a professional electrician or an unqualified DIY enthusiast, the installation **must** always comply with the Wiring Regulations.

The current edition of the Regulations is BS 7671:2018 and is entitled *Requirements for Electrical Installations* – IET Wiring Regulations (Eighteenth Edition). This is a bit of a mouthful to remember (!), and so it is normally referred to as *The Wiring Regulations, The Blue Book, The 18th Edition* or simply BS 7671:2018.

This British Standard is published with the full support of the BEC (the British Electrotechnical Committee – the UK's national body responsible for formal standardisation within the electrotechnical sector) in partnership with the BSI (the British Standards Institution – which has ultimate responsibility for all British Standards produced within this sector) and the Institution of Engineering and Technology (IET) – which, with more than 168,000 members worldwide in 150 countries, is Europe's largest grouping of professional engineers involved in power, engineering, communications, electronics, computing, software, control, informatics and manufacturing.

The technical authority for BS 7671:2018 is the Joint IET/BSI Technical Committee (JPEL/64) and BS7 671:2018 came into effect on 1 January 2019; from that date, **all** new installations (as well as additions and alterations to existing installations) **must** fully comply with BS 7671:2018.

Note: All references made in this book to the '*Wiring Regulations*' or the '*Regulation(s)*' – where not otherwise specifically identified – refer to BS 7671:2018 '*Requirements for Electrical Installations*' (IET Wiring Regulations – Eighteenth Edition).

1.2 Historical background

The first public electricity supply in the UK was at Godalming in Surrey, in November 1881 and mainly provided street lighting. At that time, there were no existing rules and regulations available to control the installation and so the electricity company just dug up the roads and laid the cables in the gutters. This particular electricity supply was discontinued in 1884.

On 12 January 1882, the steam-powered Holborn Viaduct Power Station officially opened and this facility supplied 110 V d.c. for both private consumption and street lighting. Once more, there was no one in authority to stipulate how the cables should be laid, and their positioning was, therefore, dependent on the electrician responsible for that particular section of the work.

Later in 1882 *The Electric Lighting Clauses Act* (modelled on the previous 1847 *Gas Act*) was passed by Parliament and this enabled the Board of Trade to authorise the supply of electricity in any area by a local authority, company or person and to grant powers to install this electrical supply (including breaking up the streets) through the use of the 1882 *Rules and Regulations for the prevention of Fire Risks Arising from Electric Lighting*.

 This document was the forerunner of today's IET Wiring Regulations.

Historically, since 1882 there has been a succession of amendments and new editions of the Regulations, as shown in Table 1.1.

 By aligning with all existing and new CENELEC, IEC and EN Harmonised Documents for electrical safety, in essence, BS 7671: 2008 has now virtually become a European Regulation its own right and is widely used many EU countries as shown in Figure 1.1.

Table 1.1 BS 7671:2018 – publication details

Date	Cover colour	Edition	Comments
1882		First edition	Entitled *Rules and Regulations for the prevention of Fire Risks Arising from Electric Lighting*
1888		Second edition	Entitled *Wiring Rules and Regulations in Buildings*
1897		Third edition	Entitled *General Rules recommended for Wiring for the Supply of Electrical Energy*
1903		Fourth edition	Issued as IEE Wiring Regulations, called *Wiring Rules*

Table 1.1 *(continued)*

Date	Cover colour	Edition	Comments
1907		Fifth edition	Issued as IEE Wiring Regulations
1911		Sixth edition	Issued as IEE Wiring Regulations
1916		Seventh edition	Issued as IEE Wiring Regulations
1924		Eighth edition	Issued as IEE Wiring Regulations, called *Regulations for the Electrical Equipment of Buildings*
1927		Ninth edition	Issued as IEE Wiring Regulations
1934		Tenth edition	Issued as IEE Wiring Regulations
1939		Eleventh edition	Issued as IEE Wiring Regulations Revised issue (1943), reprinted with minor amendments (1945), supplement issued (1946), revised Section 8 (1948)
1950		Twelfth edition	Issued as IEE Wiring Regulations Supplement issued (1954)
1955		Thirteenth edition	Issued as IEE Wiring Regulations Reprinted, 1958, 1961, 1962 and 1964
1966		Fourteenth edition	Issued as IEE Wiring Regulations Reprinted, 1968, 1969, 1970 (in metric units), 1972, 1973, 1974 and 1976

By now this continual updating was seen as a bit of a problem, particularly for designers and installers, who had to ensure that they were always working in compliance with the latest regulations. With the publication of the 15th edition, therefore, it was decided that in future all reprints of the same edition would be contained in one of five different coloured covers (i.e. red, green, yellow, blue and brown), and a new edition would be published when the brown-covered reprint required updating.

Date	Cover colour	Edition	Comments
1981	Red	Fifteenth edition	Issued as IEE Wiring Regulations Entitled *Regulations for Electrical Installations* [closely corresponding to the international standard IEC 60364]
1983	Green		Reprinted incorporating amendments
1984	Yellow		Reprinted incorporating amendments
1986	Blue		Reprinted incorporating amendments
1987	Brown		Reprinted incorporating amendments
1988	Brown		Reprinted with minor corrections
1991	Red	Sixteenth edition	Issued as IEE Wiring Regulations in 1991 with amendments Amendment No. 1 issued December 1994

Table 1.1 (*continued*)

Date	Cover colour	Edition	Comments
1992			The British Standards Institute adopts the IEE Wiring Regulations as the basis for BS 7671:1992 *Requirements for Electrical Installations*
1994	Green		Reprinted incorporating Amendment No 1
1997	Yellow		Reprinted incorporating Amendment No 2 Amendment No 3 issued Apr 2000
2001	Blue		BS7671:2001 published June 2001 Amendment No 1 issued February 2002 Amendment No 2 [implementation of the cable core colours introduced in the revision of CENELEC Harmonization Document HD 308 S2: 2001]
2004	Brown		Reprinted incorporating Amendments No 1 and No 2 [This edition was officially adopted as mandatory requirements for the new Building Regulations Part P '*Electrical Safety*'
2008	Red	Seventeenth edition	BS 7671:2008 published January 2008 This as a new edition aligned with old, existing and new CENELEC, IEC and EN Harmonised Documents
2011	Green		Reprinted incorporating Amendment No 1 [new material on electromagnetic compatibility (EMC), harmonics, surge protective devices, and new special locations including medical locations] Amendment No 2 issued 2013 (includes requirements for Electric Vehicle Charging Installations)
2015	Light brown		New edition issued August 2015 including new Amendment N0 3 (includes requirements for CCTV systems)
2018	Light blue	Eighteenth edition	BS7671:2018 published June 2018 [New edition following a thorough overhaul and changes to Fire Safety, Earth Fault Loop Impedance, RCD protection of Socket Outlets and a revised Inspection and Testing Documentations]

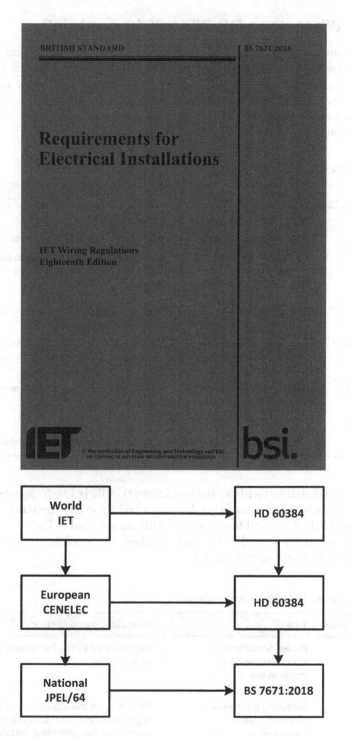

Figure 1.1 Installation standards at world, European and national levels.

1.3 What does BS 7671:2018 contain?

The Standard "*contains the rules for the design and erection of electrical installations so as to provide for safety and proper functioning for the intended use*". It is based on the plan agreed internationally (i.e. through the Comité Européen de Normalisation Électrotechnique (CENELEC)) for the "*arrangement of safety rules for electrical installations*".

The structure of BS 7671:2018 is as shown in Table 1.2.

Table 1.2 BS 7671:2018 – structure

Part	Description
Part 1	Sets out the scope, object and fundamental principles
Part 2	Defines certain terms that are used throughout the Regulations
Part 3	Identifies the characteristics of an installation that will need to be taken into account in choosing and applying the requirements of the subsequent parts of the Regulations
	These characteristics may vary from one part of an installation to another and will, therefore, need careful assessment
Part 4	Describes the basic measures that are available for the protection of persons, property and livestock, and against the hazards that may arise from the use of electricity
Part 5	Describes precautions that need to be taken in the selection and erection of the equipment of an installation
Part 6	Covers inspection and testing
Part 7	Identifies particular requirements for special installations or locations

Any intended departure from the requirements of Parts 1 to 6 requires special consideration by the installation designer – and these changes **must** be documented in the Electrical Installation Certificate specified in Part 6.

The seven parts of the standard are then supported by 17 informative Appendices as shown in Table 1.3:

Table 1.3 BS 7671:2018 – appendices

Appendix	Title	Description and remarks
1	British Standards to which reference is made in the Regulations	Reproduced in the Reference section of this book
2	Statutory regulations and associated memoranda	Details of all the statutory regulations, legislation and EU Harmonised Directives that electrical installations are required to comply with

Table 1.3 (*continued*)

Appendix	Title	Description and remarks
3	Time/current characteristics of overcurrent protective devices and RCDs	Details of time/current characteristics for • circuit breakers • fuses • RCDs
4	Current-carrying capacity and voltage drop for cables	Information and schedules for • circuit parameters • methods of installing cables overhead and underground, together with their rating factors and overload protection • tables of voltage drop in consumers' installations • determination of the size and type of cable to be used • effect of harmonic currents • relationship of current-carrying capacity to other parameters
5	Classification of external influences (extracted from HD 60364-5-51)	Lists and schedules of external influences having an influence on electrical installations
6	Model forms of certification and reporting	Information and tables concerning the requirements for and the completion of • the Electrical Installation Certificate • the Minor Electrical Works Certificate • the Electrical Installation Condition Report
7	Harmonised cable core colours	Current details of the harmonised (i.e. harmonised with CENELEC Standard HD 384-5-514) cable core marking and colours that are to be used in all installations. (For details see inside front and back covers of this book.)
8	Current-carrying capacity and voltage drop for busbar trunking and power track systems	Information concerning: • effective current-carrying capacity • protection against overload current • rating factors for current-carrying capacity • the basis of current-carrying capacity • voltage drop

Table 1.3 (*continued*)

Appendix	Title	Description and remarks
9	Definitions – multiple source, d.c. and other systems	Diagrammatic explanations of TNS supplies, where the T signifies Terre (or Earth), N is Neutral and S stands for Separate. The most common arrangement is a two-core cable (Live and Neutral) with a lead outer sheath (Earth)
10	Protection of conductors in parallel against overcurrent	Information concerning: • overload protection of conductors in parallel • short-circuit protection of conductors in parallel
11	Effect of harmonic currents.	Moved to Appendix 4
12	Voltage drop in consumers' installations	Moved to Appendix 4
13	Methods for measuring the insulation resistance/impedance of floors and walls to earth or to the protective conductor system	Information concerning • test electrode 1 • test electrode 2 • test methods for testing the impedance of floors and walls with a.c. voltage
14	Determination of prospective fault current	Informative concerning the requirement for the prospective fault current to be determined at every point of an installation
15	Ring and radial final circuit arrangements	Information concerning the co-ordination between conductor and overload protective device
16	Devices for protection against overvoltage	Information concerning the design and installation of surge protection devices
17	Energy efficiency	A new appendix which provides recommendations for the design and erection of electrical installations

1.3.1 What about the standard's numbering system?

The numbering system used to identify specific requirements in BS 7671:2018 is as follows:

- the first digit signifies a Part;
- the second digit a Chapter;
- the third digit a Section.

The subsequent digits that follow this third digit are the actual Standard's Regulation numbers. For example, Section number **413** is made up as follows:

- Part **4** – Protection for Safety
- Chapter **41** (first chapter of Part 4) – Protection against Electric Shock
- Section **413** (third section of Chapter 41) – Protective Measure: Electrical Separation

1.4 What are the objectives of the Wiring Regulations?

Current legal requirements for everyone involved in certain electrical activities – even simply choosing the size of cable or fuse – requires them to be aware of the regulative requirements associated with such work. BS 7671:2018 is the traditionally approved Code of Practice for those who are involved in (or supervise) electrical work such as electrical maintenance, control and/or instrumentation.

The stated intention of wiring safety codes is to "*provide technical, performance and material standards that will allow sufficient distribution of electrical energy and communication signals, at the same time protecting persons in the building from electric shock and preventing fire and explosion*". In other words:

> *To ensure the protection of people and livestock from fire, shock or burns from any installation that complies with their requirements.*

The Regulations form the basis of safe working practice throughout the electrical industry.

1.5 What is the legal status of the original Wiring Regulations?

Although the original IET Wiring Regulations were always held in high esteem throughout Europe, they had **no** legal status and did not require overseas contractors who were carrying out installation work in the UK to abide by them. This problem was overcome in October 1992 when the IET Wiring Regulations became British Standard 7671 – thus providing them with a national/international status.

Note: Although the Regulations are *non-statutory regulations*, they may be used as evidence in a court of law to claim compliance with a statutory requirement.

1.6 What do the Wiring Regulations cover?

The Wiring Regulations cover both electrical installations and electrical equipment.

1.6.1 Electrical installation

> **Definition:** For the purpose of the Regulations:
>
> *"Electrical installations (or Installation) means an assembly of associated electrical equipment having certain co-ordinated characteristics."*

The Regulations apply to the design, erection and verification of electrical installations such as those for:

- agricultural and horticultural premises;
- caravans and motor caravans;
- caravans, caravan parks and similar sites;
- commercial premises;
- conducting locations with restricted movement;
- construction and demolition sites;
- electric vehicle charging;
- exhibitions, shows and stands;
- extra-low-voltage lighting;
- floor and ceiling heating systems;
- highway equipment and street furniture;
- locations containing a bath or shower;
- low-voltage generating sets;
- marinas and similar locations;
- medical locations;
- mobile or transportable units;
- offshore units of electrical shore connections for inland navigation vessels;
- operating and maintenance gangways;
- outdoor lighting;
- prefabricated buildings;
- public premises;
- residential premises;
- rooms and cabins containing sauna heaters;
- solar photovoltaic (PV) power supply systems;
- swimming pools and other basins;
- temporary installations for structures, amusement devices and booths at fairgrounds, amusement parks and circuses, including professional stage and broadcast applications.

The Regulations also include requirements for:

- consumer installations external to buildings;
- circuits supplied at nominal voltages up to and including 1000 V a.c. or 1500 V d.c;
- circuits (but not apparatus and/or equipment internal wiring) that are operating at voltages greater than 1000 V and derived from an installation having a voltage not exceeding 1000 V a.c. (e.g. discharge lighting, electrostatic precipitators);
- fixed wiring for communication and information technology, signalling, command and control etc. (but not the actual apparatus and/or the equipment's internal wiring);
- the addition to (or alteration of) installations and parts of existing installations;
- wiring systems and cables not specifically covered by the standards for appliances.

 Notes:

1. The standard nominal supply voltage for domestic single-phase 50 Hz installations in the UK has been 230V a.c. (rms) since 1 January 1995. Previously it was 240V.
2. Although the preferred frequencies are 50 Hz, 60 Hz and 400 Hz, the use of other frequencies for special purposes is not excluded.
3. 'Premises' covers the land and all facilities (including buildings) belonging to it.

Although the Regulations are intended as a standard for electrical installations, in certain cases they may need to be supplemented by the requirements and/or recommendations from other British, European and International Standards or by the requirements of the person ordering the work. Such cases could include (amongst others) the following:

- electrical apparatus for use in the presence of combustible dust (covered by BS EN 50281 and BS EN 61241);
- electrical installations for open-cast mines and quarries (covered by BS 69070);
- electric signs and high-voltage luminous discharge tube installations (covered by BS 559 and BS EN 50107);
- electric surface heating systems (covered by BS EN 60335-2-96);
- emergency lighting (covered by BS 5266 and BS EN 1838);
- explosive atmospheres (covered by BS EN 600790);
- fire-detection and alarm systems in buildings (covered by BS 5839);
- life safety and firefighting applications (covered by BS 8519 and BS 9999);
- telecommunications systems (covered by BS 6701);
- temporary electrical systems for entertainment and related purposes (covered by BS 7909).

1.6.1.1 Exclusions

The Regulations do **not** apply to the following installations:

- aircraft equipment;
- electric fences (covered by BS EN 60335-2-76);
- electrical equipment of machines (covered by BS EN 60204);
- escalator or moving walk installations (specifically covered by relevant parts of BS 5655 and BS EN 115);
- lift installations (specifically covered by relevant parts of BS 5655 and BS EN 81-1);
- lightning protection systems (for buildings and structures covered by BS EN 62305);
- mines specifically covered by statutory regulations;
- mobile and fixed offshore installation equipment;
- motor vehicle equipment (except where Regulations concerning caravans or mobile units are applicable);
- onboard ship equipment (covered by BS 8450, BS EN 60092-507 or BS EN 13297);
- radio interference suppression equipment (except so far as it affects safety of the electrical installation);
- railway traction, rolling stock and signalling equipment;
- systems for the distribution of electricity to the public;
- the d.c. side of cathode protection systems.

1.6.2 Electrical equipment – definition

Electrical equipment (or Equipment) means any item used for the generation, conversion, transmission, distribution or utilisation of electrical energy, such as machines, transformers, apparatus, measuring instruments, protective devices, accessories and appliances, and luminaires alternators, generators and batteries.

The Regulations are only applicable to the actual selection and application of items of electrical equipment **in** an electrical installation.

The Regulations do **not** deal with requirements for the construction of assemblies of electrical equipment, which are required to comply with the appropriate standards.

1.7 What affect does using the Regulations have on other Statutory Instruments?

The requirements of the Wiring Regulations also have an effect on the implementation of other Statutory Instruments, such as:

- the Fire Precautions (Workplace) Regulations 1997;
- the Building Act 1984;

- the Disability and Equality Act 2010;
- the Electricity at Work Regulations 1989;
- the Health and Safety at Work Act 2015;
- the Sustainable and Secure Building Act 2004.

1.7.1 What is the Building Act 1984?

The Building Act 1984 (as implemented by the Building Regulations 2010) is the enabling Act under which all Building Regulations have been made. The Secretary of State (under the power given in the Building Act 1984) is required to:

- secure the health, safety, welfare and convenience of persons in or about buildings and of others who may be affected by buildings or matters connected with buildings;
- further the conservation of fuel and power;
- prevent waste, undue consumption, misuse or contamination of water.

and may make regulations with respect to the design and construction of buildings and the provision of services, fittings and equipment in (or in connection with) buildings.

 Note: A copy of the current regulations governing the Building Regulations 2010 (see front cover at Figure 1.2) can be downloaded from: http://www. legislation.gov.uk/uksi/2010/2214/contents/made.

For many years, the UK has managed to maintain a relatively high standard of electrical safety within buildings (domestic and non-domestic) based on voluntary controls centred around the Wiring Regulations. With the growing number of electrical accidents occurring in the 'home', the government was forced to implement a legal requirement for safety in **all** electrical installation work in dwellings – so:

 in future **all** new electrical wiring or electrical components for domestic premises (or small commercial premises linked to domestic accommodation) **must** now to be designed and installed in accordance with the Building Regulations, Approved Document P. This Approved Document is based on the fundamental principles set out in Part 7 of BS 7671:2018 and also introduced the requirement for the standardisation of cable core colours of all a.c. power circuits.

In addition, **all** fixed electrical installations (i.e. wiring and appliance fixed to the building fabric – such as socket outlets, switches, consumer units and ceiling fittings) **must** now be designed, installed, inspected, tested and certified to BS 7671:2018.

 Note: Part P only applies to fixed electrical installations that are intended to operate at low voltage or extra-low voltage which are not controlled by the

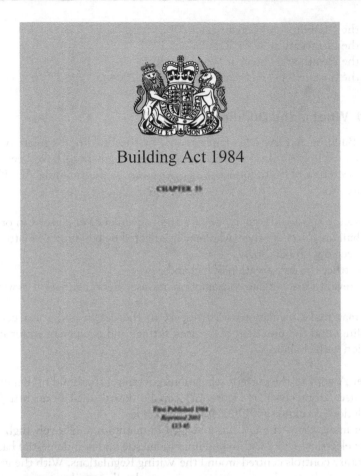

Figure 1.2 The Building Act 1984.

Electricity (Standards of Performance) Regulations 2015 as amended, or the Electricity at Work Regulations 1989 (as amended).

1.7.1.1 Competent Persons Scheme

Under Part P of the Building Regulations, all domestic installation work **must** now be inspected by Local Authority Building Control officers **unless** the work has been completed out by a *competent person*, who is able to self-certify the work.

For more details about the Building Regulations, visit https://www.gov.uk/building-regulations-approval, see '*Building Regulations in Brief*' 9th edition https://www.routledge.com/Building-Regulations-in-Brief/Tricker-Alford/p/book/9781138285163 or get a copy of the '*Building Regulations Pocket Book*' https://www.routledge.com/Building-Regulations-Pocket-Book/Tricker-Alford/p/book/9780815368380.

1.7.2 What is the effect of the Wiring Regulations on the Disability and Equality Act 2010?

The Equality Act (see cover shown in Figure 1.3) replaced most of the Disability Discrimination Act (DDA) in October 2010.

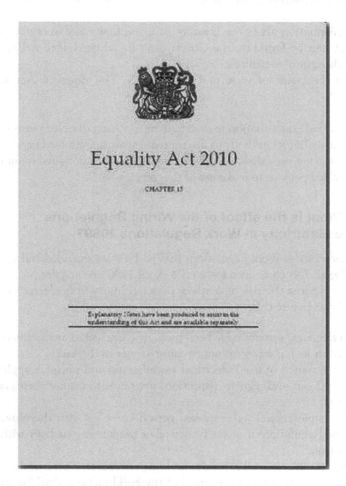

Figure 1.3 The Equality Act 2010.

The Equality Act 2010 aims to protect disabled people and prevent disability discrimination. It provides legal rights for disabled people in the areas of:

- access to goods, services and facilities including larger private clubs and land-based transport services;
- buying and renting land or property;
- education;
- employment;
- functions of public bodies, for example the issuing of licences.

The Equality Act also provides rights for people not to be directly discriminated against or harassed because they have an association with a disabled person. This can apply to a carer or parent of a disabled person. In addition, people must not be directly discriminated against or harassed because they are wrongly perceived to be disabled.

More information about the Equality Act, and how you can obtain copies of the Act, can be found on the Government Equalities Office website: http://homeoffice.gov.uk/equalities.

From the point of view of BS 7671:2018, The Equality Act makes it unlawful:

- for a trade organisation to discriminate against a disabled person;
- for a qualifications body to discriminate against a disabled person;
- for service providers to make it impossible or unreasonably difficult for disabled persons to make use of that service.

1.7.3 What is the effect of the Wiring Regulations on the Electricity at Work Regulations 1989?

The Electricity at Work Regulations (EWR) 1989 (as amended and as depicted in in Figure 1.4) came into force on 1 April 1990 and require precautions to be taken against the risk of death or personal injury from electricity in work activities to ensure that:

- all electrical systems have been properly constructed, maintained and are used in such a way so as not to cause danger to the user;
- maintenance of fixed electrical installations and portable appliances is carried out and regular inspections are made to ensure their continued safety;
- the employer and self-employed person complies with the provisions of these Regulations in so far as they relate to matters which are within their control;
- the person responsible for the building ensures that electrical test certificates are in place to confirm that the building's installations and appliances have been appropriately tested.

The Electricity at Work Regulations 1989 also state that where an accident occurs and it is found that the systems are not covered by a valid test certificate, then the Health & Safety Executive (HSE) will be involved in prosecutions resulting from electrocution or death within the workplace. Reducing the risk of such an accident **is** a legal requirement.

Overall, the Regulations require that:

- all electrical equipment and installations are maintained in a safe condition;

Figure 1.4 The Electricity at Work Regulations 1989.

- all electrical systems shall be constructed and maintained to prevent danger;
- all people working with electricity are competent to do the job.

Complicated tasks (i.e. equipment repairs, alterations, installation work and testing) may require a suitably qualified electrician, in which case it is required that:

- all staff are aware of an organisation's electrical safety procedure;
- all work activities are carried out so as not to give rise to danger;
- equipment and procedures are safe and suitable for the working environment;
- equipment is switched off and/or unplugged before making adjustments. 'Live working' must be eliminated from work practices.

Electricity is recognised as a major hazard; not only can it kill (research has shown that the majority of electric shock fatalities occur at voltages up to 240V), but it can also cause fires and explosions. Even non-fatal shocks can cause severe and permanent injury. Most of the electrical risks can be controlled by using suitable equipment, following safe procedures when carrying out electrical work and/or ensuring that all electrical equipment and installations are properly maintained.

Additional precautions will be required for harsh and particular conditions (e.g. wet surroundings, cramped spaces, work out of doors or near live parts of equipment). For this reason, the Electricity at Work Regulations 1989 is used to impose health and safety requirements for all electricity used at work.

Whilst the majority of the Regulations concern hardware requirements, others are more generalised. For example:

- installations shall be of proper construction;
- conductors shall be insulated;
- means of cutting off the power (i.e. for electrical isolation) shall be available.

In brief, Regulations (as shown in Table 1.4) concern:

Table 1.4 Contents of the Regulations

Regulation	Activity	Effect
Regulation 4	Systems, work activities and protective equipment	All systems shall at all times be so constructed to prevent, insofar as is reasonably practicable, danger
Regulation 5	Strength and capability of electrical equipment	No electrical equipment is to be used where its strength and capability may be exceeded so as to give rise to danger
Regulation 6	Adverse or hazardous environments	Electrical equipment sited in adverse or hazardous environments must be suitable for those conditions
Regulation 7	Insulation, protection and placing of conductors	Permanent safeguarding or suitable positioning of live conductors is required
Regulation 8	Earthing and other suitable precautions	Equipment must be earthed or other suitable precautions must be taken (e.g. the use of RCDs, double-insulated equipment, reduced-voltage equipment etc.)
Regulation 9	Integrity of reference conductors	Nothing is to be placed in an earthed circuit conductor which might, without suitable precautions, give rise to danger by breaking the electrical continuity or by introducing a high impedance

Table 1.4 (*continued*)

Regulation	Activity	Effect
Regulation 10	Connections	All joints and connections in systems must be mechanically and electrical suitable for use
Regulation 11	Means for protecting from excess current	Suitable protective devices should be installed in each system to ensure all parts of the system and users of the system are safeguarded from the effects of fault conditions

Note: Regulations 4 to 11 in effect, therefore, place a duty on the designer, the installer and the end user to ensure the suitability and protection of all electrical equipment.

Regulation 12	Means of cutting off the supply and for isolation	Where necessary to prevent danger, suitable means shall be available for cutting off the electrical supply to any electrical equipment and isolating that particular equipment
Regulation 13	Precautions for work on equipment made dead	Adequate precautions must be taken to prevent electrical equipment, which has been made dead in order to prevent danger, from becoming live while any work is carried out
Regulation 14	Work on or near live conductors	No work can be carried out on live electrical equipment unless this can be properly justified
		Which means that risk assessments are required
		If such work is to be carried out, suitable precautions must be taken to prevent injury
Regulation 15	Working space, access and lighting	Adequate working space, adequate means of access and adequate lighting shall be provided at all electrical equipment on which or near which work is being done in circumstances that may give rise to danger
Regulation 16	Persons to be competent to prevent danger and injury	No person shall engage in work that requires professional technical knowledge or experience to prevent danger or injury, unless he has that knowledge or experience, or is under appropriate supervision

 For more information about the Electricity at Work Regulations 1989, contact the Health and Safety Executive (HSE) www.hse.gov.uk/lau/lacs/19-3.htm.

Or to download a copy of the Electricity at Work Regulations 1989 (Statutory Instrument 1989 No. 635) go to: http://www.opsi.gov.uk/si/si1989/Uksi_19890635_en_1.htm .

1.7.4 What is the effect of the Wiring Regulations on the Fire Precautions (Workplace) Regulations

The Fire Precautions (Workplace) Regulations 1997 (as depicted in Figure 1.5 and as amended by the Fire Precautions (Workplace) (Amendment) Regulations 1999) stipulate that:

> *"All offices, shops, railway premises and factories which have more than 20 persons employed in the building (or more than 10 person*

Figure 1.5 The Fire Precautions (Workplace) Regulations 1997.

employed anywhere other than on the ground floor) require a Fire Certificate."

"Any hotel or boarding house provided sleeping accommodation for more than six persons (guests or staff) or where this sleeping accommodation is above the first floor or below the ground floor, require a Fire Certificate."

When a Fire Certificate is issued the owner or occupier is required to provide and maintain:

- a means of escape;
- means of fighting fire;
- means of providing warning in case of fire;
- other means for ensuring that the means of escape can be safely and effectively used at all material times.

These requirements are reflected in the electrical installation.

 For further information about the Fire Precautions (Workplace) (Amendment) Regulations 1999 (Statutory Instrument 1999 No. 1877) visit: http://www. odpm.gov.uk/stellent/groups/odpm_fire/documents/sectionhomepage/odpm_fire_page.hcsp.

Or for a copy of the Act, use the following link: http://www.legislation.gov. uk/uksi/1999/1877/contents/made.

Figure 1.6 The Health and Safety at Work Act 1974.

1.7.5 What is the effect of the Wiring Regulations on the Health and Safety at Work Act 1974?

Any company with more than five employees is legally obliged to possess a comprehensive health and safety policy.

Over the years, the IET Wiring Regulations have been regularly used by the HSE in their guidance and installation notices, and installations which conform to BS 7671 (as amended) are regarded by the HSE as likely to achieve conformity with the relevant parts of the Electricity at Work Regulations 1989. In certain instances where the Regulations have been used, they may also be accompanied by Codes of Practice approved under Section 16 of the Health and Safety at Work Act 1974.

Although some existing installations may have been designed and installed to conform to the standards set by earlier editions of the Wiring Regulations, this does **not** necessarily mean that they will fail to achieve conformity with the relevant parts of the Electricity at Work Regulations 1989.

For further information about the Health and Safety at Work Act 1974 visit: http://www.legislation.gov.uk/ukpga/1974/37/contents.

1.8 How are the IET Wiring Regulations implemented?

Although the IET Wiring Regulations rely (primarily) on British Standards for their implementation (see Annex F 'Reference section' for details) they do, however, include the policy decisions made in a number of Statutory Instruments and by the Council of European Communities in the relative EU Harmonised Directives.

1.8.1 Statutory Instruments

In Great Britain the following classes of electrical installations are required to comply with the Statutory Regulations as described in Table 1.5.

The full texts of **all** Statutory Instruments that have been published since 1987 are now available from the Office of Public Sector Information (OPSI) via their website: http://www.legislation.gov.uk/uksi.

With effect from July 1999, Statutory Instruments which have also been made by the National Assembly for Wales have been published via the Wales Legislation section: http://www.legislation.gov.uk/wsi.

And the series of Scottish Statutory Instruments have been published via the Scottish Legislation: http://www.legislation.gov.uk/ssi.

Table 1.5 Statutory Instruments affecting electrical installations

Type of electrical installation	Statutory Instrument
Building generally (subject to certain exemptions)	Building Regulations 2010 (as amended) (for England and Wales): • SI 2010 No 2214 (as amended by SI 2011/1515) Building (Scotland) Amendment Regulations 2011(as amended): • Scottish SI 2011 No 120 Building Regulations (Northern Ireland) 2000 (as amended): • Statutory Rule 2000 No 398
Cinematograph installations	Cinematograph (Safety) Regulations 1955 (as amended under the Cinematograph Act, 1909, and/or Cinematograph Act, 1952): • SI 1982 No 1856
Distributors' installations generally (subject to certain exemptions)	Electricity Safety, Quality and Continuity Regulations 2002: • SI 2002 No 2665 • SI 2006 No 1521
High-voltage luminous tube	Conditions of licence under: • in England and Wales – Local Government (Miscellaneous provisions) Act 1982 • in Scotland – Civic Government (Scotland) Act 1982
Machinery	The Supply of Machinery (Safety) Regulations 1992 as amended • SI 1992 No 3073 • SI 1994 No 2063
Theatres and other places licensed for public entertainment, music, dancing etc.	Conditions of licence under: • in England and Wales – Local Government (Miscellaneous provisions) Act 1982 • in Scotland – Civic Government (Scotland) Act 1982
Work activity Places of work Non-domestic installations	The Electricity at Work Regulations 1989 as amended: • SI 1989 No 635 • SI 1996 No 192 • SI 1997 No 1993 • SI 1999 No 2024 The Electricity at Work Regulations (Northern Ireland) 1991: • Statutory Rule No 13

1.8.2 British Standards, International Standards and Harmonised Documents

As well as British and International Standards, the Wiring Regulations also take account of the technical substance of agreements reached in CENELEC.

For your convenience, a list of the current (i.e. at the time of writing this book) Standards and Directives relevant to the BS 7671:2018 have been listed in Annex F to this book as a follows:

- British Standards currently used with the wiring regulations:
  - o by standard; and
  - o by subject.
- Other standards to which reference is made in the regulations:
  - o IEC and ISO;
  - o CENELEC Harmonised Directives:
    - – by subject;
    - – by directive.

 Note: By the time you read this edition of *Wiring Regulations in Brief*, it is quite possible that some of the Standards and Directives listed in this book and the BS 7671:2018 will have been reviewed and updated and, although the ones listed in Annex F are still relevant, the latest edition of these documents should always be taken into account. For this reason, the reader is recommended to always have a quick check via https://shop.bsigroup.com/, or via Google (or some other search engine), to make sure that they are using the most up-to-date regulation and/or recommendation.

Author's end note

In Chapter 2 we shall now take a look at the structure and contents of the Building Regulations: the responsibilities for electrical safety; the need to oversee and control extensions (material alterations and material changes of use); the electricity distributor's responsibilities (earthing, electrical installation work, types of wiring and wiring systems etc.); and inspection, testing and self-certification.

2

Building regulations

Author's start note

For many years the Wiring Regulations have not fully supported the requirements of the Building Regulations. However, owing to the growing number of electrical accidents occurring in domestic buildings, in 2005 the Government decided that in future:

> **All new electrical wiring and components (within England and Wales) must be designed and installed in accordance with the electrical safety requirements contained in Part P of the Building Regulations.**

Chapter 2, therefore, provides:

- *an overview of the structure and contents of the Building Regulations and their 'Approved Documents' (or 'Parts');*
- *responsibilities for electrical safety;*
- *the need to oversee and control extensions (material alterations and material changes of use);*
- *electricity distributors' responsibilities (earthing, electrical installation work, types of wiring and wiring systems etc.);*
- *inspection and testing; plus (most importantly);*
- *who and what a competent firm or competent person is responsible for, and how a building can be self-certified.*

With a mandatory requirement that all electrical installation work should comply with the statement made in Figure 2.1:

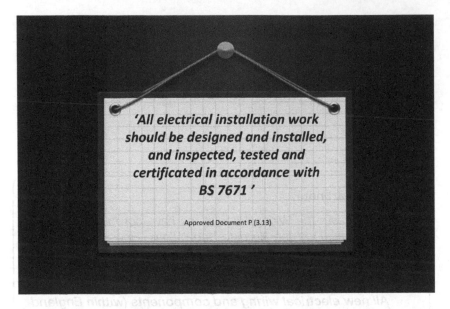

'All electrical installation work
should be designed and installed,
and inspected, tested and
certificated in accordance with
BS 7671'

Approved Document P (3.13)

Figure 2.1 Mandatory requirements for domestic and non-domestic buildings.

2.1 The Building Act 1984

The Building Act 1984 is the mechanism by which the Secretary of State ensures that the health, welfare and convenience of persons living in or working in or near buildings is secured.

As well as health and safety, the primary purpose of the Building Act 1984 is to assist in the conservation of fuel and power, and to prevent waste, undue consumption, misuse and contamination of water. The Building Act 1984 imposes a set of requirements on owners and occupiers of buildings which cover the design and construction of buildings, and the provision of services, fittings and equipment used in (or in connection with) buildings. These involve and cover:

- a method of controlling (inspecting and reporting) buildings;
- how services, fittings and equipment may be used; and
- the inspection and maintenance of any service, fitting or equipment used.

The Building Act 1984 does **not** apply to Scotland or to Northern Ireland who under devolved government have their own legislation– see Figure 2.2 Section 2.3.1 for details.

2.1.1 Who polices the Building Act?

Under the terms of the Building Act 1984, **local authorities** are responsible for ensuring that any building work that takes place in their area conforms

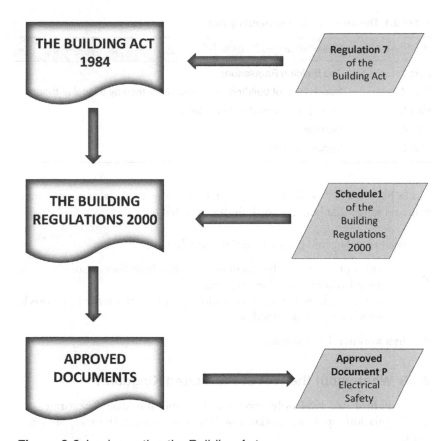

Figure 2.2 Implementing the Building Act.

to the requirements of the associated Building Regulations. They have the authority to:

- make you complete alterations so that your work complies with the Building Regulations; or
- make you take down and remove or rebuild anything that contravenes a regulation; or
- employ a third party (and then send you the bill!) to take down and rebuild non-conforming buildings or parts of buildings.

 You can be prosecuted or ordered to carry out remedial work on a property whether you are the owner of the property or merely the occupier. They can, in certain circumstances, even take you to court and have you fined – especially if you fail to remove or rebuild non-conforming work.

2.1.2 What does the Building Act 1984 contain?

The Building Act 1984 is made up of five parts, as shown in Table 2.1.

Table 2.1 The structure of the Building Act

Building Act	Description
Part 1	The Building Regulations
Part 2	Supervision of building work, etc. other than by a local authority
Part 3	Other provisions about buildings
Part 4	General
Part 5	Supplementary

Note: Regulation 7 of the Building Act 1984 covers materials and workmanship and states that building work shall be carried out

- with adequate and proper materials; which:
  - o   are appropriate for the circumstances in which they are used;
  - o   are adequately mixed or prepared; and
  - o   are applied, used or fixed so as adequately to perform the functions for which they are designed; and

- in a workmanlike manner.

2.1.3 What about the rest of the United Kingdom?

The Building Act 1984 **only** applies to England and Wales. Separate Acts and Regulations apply in Scotland or Northern Ireland; these are shown in Table 2.2:

Table 2.2 Building legislation within the United Kingdom

Nation	Act	Regulations	Implementation
England and Wales	Building Act 1984	Building Regulations 2010	Approved Documents
Scotland	Building (Scotland) Act 2003	Building (Scotland) Regulations 2004 (as amended)	Technical Handbooks
Northern Ireland	Building Regulations (Northern Ireland) Order 1979 (as amended)	Building Regulations (Northern Ireland) 2012 (as amended)	Technical Booklets

2.1.3.1 Scotland

In Scotland, the requirements for buildings are controlled by the Building (Scotland) Regulations 2004 as amended by the Building (Scotland) Amendment Regulations 2010. The methods for implementing these requirements are similar to those used in England and Wales, except that the guidance documents for achieving compliance are contained in two Technical Handbooks (one for domestic work, and the other for non-domestic work) and other associated guidance.

The main procedural difference between the Scottish system and the others is that a building warrant is **still** required before work can start.

2.1.3.2 Northern Ireland

In Northern Ireland the primary legislation for buildings is the Building Regulations (Northern Ireland) Order 1979 under which Building Regulations (Northern Ireland) 2000 are made. These regulations revoked and replaced with amendments the previous 1994 Building Regulations, and they have since been amended by the Building (Amendment) Regulations (Northern Ireland) 2010. The Principle Regulations comprise 15 Parts and are supported by Technical Booklets which are then used to ensure that the requirements are implemented.

 Note: Table 2.3 provides a summary of the titles of the sections of Building Regulations that apply in the United Kingdom.

Table 2.3 Building Regulations for the United Kingdom

England and Wales		Scotland		Northern Ireland
Part A	Structure	Section 1 – Structure	Technical Handbooks 2017	Part D – Structure
Part B	Fire safety	Section 2 – Fire	Technical Handbooks 2017	Part E – Fire safety
Part C	Site preparation and resistance to contaminants and water	Section 3 – Environment	Technical Handbooks 2017	Part C – Preparation of site and resistance to moisture
Part D	Toxic substances	Section 3 – Environment	Technical Handbooks 2017	Part B – Materials and workmanship
Part E	Resistance to the passage of sound	Section 5 – Noise	Technical Handbooks 2017	Part G – Sound insulation of dwellings
Part F	Ventilation	Section 3 – Environment	Technical Handbooks 2017	Part K – Ventilation
Part G	Sanitation, hot water safety and water efficiency	Section 3 – Environment	Technical Handbooks 2017	Part P – Sanitary appliances and unvented hot water storage systems
Part H	Drainage and waste disposal	Section 3 – Environment	Technical Handbooks 2017	Part N – Drainage

Table 2.3 (*continued*)

England and Wales	Scotland		Northern Ireland
Part J Combustion appliances and fuel storage	Section 3 – Environment Section 4 – Safety	Technical Handbooks 2017	Part L – Heat-producing appliances and liquefied petroleum gas installations
Part K Protection from falling, collision and impact	Section 4 – Safety	Technical Handbooks 2017	Part H – Stairs, ramps and protection from impact
Part L Conservation of fuel and power	Section 6 – Energy	Technical Handbooks 2017	Part F – Conservation of fuel and power
Part M Access and facilities for disabled people	Section 4 – Safety	Technical Handbooks 2017	Part R – Access and facilities for disabled people
Part P Electrical safety	Section 4 – Safety	Technical Handbooks 2017	
Part Q Security in dwellings			
Part R High speed electronic communications networks			Technical Booklet M

For full details of the Building Act 1984, Building Regulations 2010 and all of their associated Approved Documents, please see the latest Edition (i.e. 9th) of *Building Regulations in Brief* by Ray Tricker and Samantha Alford (Routledge).

2.2 What are the supplementary regulations?

The Supplementary Regulations that make up Part 5 of the Building Act comprise seven schedules, the function of which is to list the principal areas requiring regulation and to show how the Building Regulations are to be controlled by local authorities. These schedules are listed in Table 2.4:

Note: Schedule 1 is the most important Schedule from the point of view of builders and electricians. This is because it shows, in general terms, how the Building Regulations are to be administered by local authorities, the approved methods of construction and the approved types of materials that are to be used in (or in connection with) buildings.

Table 2.4 The seven schedules of the Building Act's Supplementary Regulations

Schedule	Title	Description
Schedule 1	Building Regulations	This describes the mandatory requirements for completing all building work
Schedule 2	Relaxation of Building Regulations for existing work	This provides guidance in connection with work that has been carried out prior to a local authority (under the Building Act 1984 Section 36) dispensing with or relaxing some of the requirements contained in the Building Regulations
		Schedule 2 is quite difficult to understand and if it affects you, then I would strongly advise that you discuss it with the local authority before proceeding any further
Schedule 3	Inner London	This applies to how Building Regulations are to be used in Inner London, as well as ruling which sections of the Act may be omitted It also details how by-laws concerning the relation to the demolition of buildings (in Inner London) may be made
Schedule 4	Provisions consequential upon Public Body's notice	This concerns the authority and ruling of Public Bodies' notices and certificates
Schedule 5	Transitional provisions	It lists the transitional effect of the Building Act 1984 concerning existing Acts of Parliament.
Schedule 6	Consequential amendments	It lists the consequential amendments that will have to be made to existing Acts of Parliament owing to the acceptance of the Building Act 1984
Schedule 7	Repeals	It lists the cancellation (repeal) of some sections of existing Acts of Parliament, owing to acceptance of the Building Act 1984

2.3 What are the Building Regulations?

The Building Regulations 2010 (SI 2010/2214) relate to Schedule 1 of the Building Act and are a set of minimum requirements for implementing the Building Act and basic performance standards designed to secure the health, safety and welfare of people in and around buildings and the conservation of fuel and energy in England and Wales.

They are legal requirements laid down by parliament and based on the Building Act 1984. The Building Regulations:

- are approved by parliament;
- are designed to ensure structural stability;
- contribute to meeting the needs of disabled people;

- deal with the minimum standards of design and building work for the construction of domestic, commercial and industrial buildings;
- ensure the health and safety of people in and around buildings (by providing functional requirements for building design and construction);
- promote energy efficiency in buildings;
- promote the use of suitable materials to provide adequate durability, fire and weather resistance, and the prevention of damp;
- set out the procedure for ensuring that building work meets the standards laid down;
- stipulate the minimum amount of ventilation and natural light to be provided for habitable rooms.

The level of safety and standards acceptable are set out as guidance in **Approved Documents** which are discussed below. Compliance with the detailed guidance of the Approved Documents is usually considered as evidence that the Building Regulations themselves have been fulfilled.

2.3.1 How is my building work evaluated for conformance with the Building Regulations?

Part of the local authority's duty is to make regular checks that all building work being completed is in conformance with the approved plan submitted and the Building Regulations themselves. These checks would normally be completed at certain stages of the work (e.g. the excavation of foundations) and tests will include:

- tests of any service, fitting or equipment that has been, is being or is proposed to be provided in, or in connection with, a building;
- tests of any material, component, or combination of components that has been, is being or is proposed to be used in the construction of a building;
- tests of the soil or sub-soil of the site of the building.

 The cost of carrying out these tests will normally be charged to the owner or occupier of the building. The Local Authority has the power to ask the person responsible for the building work to complete some of these tests on their behalf.

2.4 Approved Documents

The Secretary of State makes available a series of documents (called **Approved Documents**) which contain practical and technical guidance on ways in which the requirements of Schedule 1 and Regulation 7 of the Building Act 1984 can be met.

 More details concerning Approved Documents are available on the Planning Portal at http://www.planningportal.gov.uk/buildingregulations/approveddocuments).

Each Approved Document reproduces the actual *requirements* contained in the Building Regulations relevant to the subject area (e.g. Approved Document P deals with Electrical Safety). This is then followed by *practical and technical guidance* (together with examples) showing how the requirements can be met in some of the more common building situations.

There may, however, be alternative ways of complying with the requirements to those shown in the Approved Documents and you are, therefore, under no obligation to adopt any particular solution contained in an Approved Document if you prefer to meet the requirement(s) in some other way.

If you intend to carry out building work you should **always** check with your local authority (or an approved inspector) that your proposals comply with Building Regulations.

The current sets of Approved Documents are in 15 Parts, A to R (less 'I', 'O' and 'N') and consist of:

Approved Document A (Structure) contains design standards and guidance for the structural stability and safety of all buildings and provides direction on how not to affect the structural integrity of other buildings.

Approved Document B (Fire Safety) addresses fire safety precautions, which must be adhered to in order to ensure the safety of occupants, firefighters and those close to the building in the event of a fire. The document includes means of escape, the ability to internally isolate a blaze to prevent a fire from spreading, external fire spread, firefighter access to the building and facilities, fire detection and warning systems in place within a building. It also addresses the internal spread of a fire due to the structure or lining used within a building and safety measures related to this.

Approved Document C (Site preparation and resistance to contaminants and moisture) provides instruction on resistance to contaminants and moisture, including ensuring buildings are protected from both weather and water damage, from dangerous substances such as radon and methane, and that guidelines are followed when preparing a site for construction to take place.

Approved Document D (Toxic substances) provides guidance on controlling toxic substances used in building works. It focuses on health and safety guidance for using a barrier in a cavity wall to prevent harmful fumes from entering a property due to injecting urea formaldehyde into the cavity fill insulation system.

Approved Document E (Resistance to the passage of sound) provides guidance on soundproofing, including the transmission of sounds between walls, ceilings, windows and floors. It covers unwanted sound travel within different areas of a building, including common areas within schools and buildings containing flats, and in between, connecting buildings.

Approved Document F (Ventilation) provides guidance on building ventilation, including building air quality and preventing condensation in a domestic or non-domestic structure.

Approved Document G (Sanitation, hot water safety and water efficiency) provides guidance on the supply of water to a property, including water safety, hot water supply, sanitation and water efficiency (i.e. an easily accessible water supply that doesn't incur wastage.

Approved Document H (Drainage and waste disposal) offers guidance on drainage including foul and surface water and rainwater, and sanitary waste disposal, including sewage structures and their upkeep.

Approved Document J (Combustion appliances and fuel storage systems) provides guidance on the safe installation and usage of heat-producing appliances, including boilers, chimneys and flues, and offers advice on safe fuel storage installations, including solid fuel, liquid oil fuels, and gas-fired heating.

Approved Document K (Protection from falling, collision and impact) Protection from falling involves the fitting of safety measures on staircases, ramps and ladders, as well as advice about the positioning of balusters, vehicle barriers and windows to avoid injury. Guidance for avoiding collision and impact is detailed in the positioning of doors and windows within a property, ensuring that no injuries occur due to occupants colliding with open windows, skylights, ducts etc., that large panes of glass are marked to avoid accidental impact with them, and that doors and windows are not positioned in a way that could trap someone.

 Note: This document is particularly relevant to those who are building a loft conversion and contains guidance in relation to building a staircase for access to the loft.

Approved Document L (Conservation of fuel and power) guidance provided includes insulation regulations, boiler productivity, lighting, and storage techniques for hot water.

Approved Document M (Access to and use of Buildings) provides information about the ease of access to, and use of, buildings, including facilities for disabled visitors or occupants, and the ability to move through a building easily including to toilets and bathrooms.

Approved Document P (Electrical safety) contains guidance on electrical safety in **dwellings**, including detailed information about what procedures need to be in place and who may carry these out, such as when a professional electrician must be hired. Those undertaking electrical works must be considered '*competent*', having a complete and proficient understanding of electrical fittings and being able to check safety circuits.

 Note: This document details electrical safety to avoid injuries and fires caused by electrical installations, including the design, installation, inspection and testing of any electrical works made within a dwelling.

Approved Document Q (Security – Dwellings) contains guidance on security in new dwellings, including measures taken to avoid any unauthorised entrance to dwellings and flats within a building.

Approved Document R (Electronic communications) contains guidance on high-speed electronic communications networks, i.e. the use of physical infrastructures within a building to ensure it may be connected to a broadband network.

Regulation 7 (Materials and workmanship) provides guidance on materials and workmanship, i.e. the use of the appropriate materials for a construction and how those who are working on the building must behave in a workmanlike manner.

 Free pdf copies of these Approved Documents are available from https://www.gov.uk/government/collections/approved-documents or if you prefer, printed copies can be obtained via:

Email: sales@ribabookshops.com
Telephone: 0191 244 5557
Website: RIBA online bookshop – https://www.architecture.com/riba-books

2.5 Electrical safety

For many years, the UK has managed to maintain relatively high electrical safety standards with the support of voluntary controls based on BS 7671. However, with a growing number of electrical accidents occurring in the 'home', the government have been forced to consider the legal requirement for safety in electrical installation work in dwellings.

Since 1 Jan 2005, all new electrical wiring or electrical components for domestic premises (or small commercial premises linked to domestic accommodation) have had to be designed and installed in accordance with Approved Document P of the Building Regulations. Therefore, all fixed electrical installations (i.e. wiring and appliance fixed to the building fabric such as socket outlets, switches, and consumer units and ceiling fittings) must also be designed, installed, inspected, tested and certified to BS 7671:2018.

 Part P also introduced new requirements for cable core colours for a.c. power circuits and, with effect from 31 March 2006, **all** new installations or alterations to existing installations must use the new (harmonised) colour cables. (Further information, concerning cable identification colours for extra-low-voltage and d.c. power circuits is available in Appendix 7 of BS 7671:2018 and for ease of reference they are shown inside the front cover of this book and as listed in Table 2.5).

 For single-phase installations in domestic premises, the new colours are the same as those for flexible cables to appliances (namely green-and-yellow, blue and brown for the protective, neutral and phase conductors, respectively).

Table 2.5 Identification of conductors in a.c. power and lighting circuits

Conductor	Colour
Protective conductor	Green and Yellow
Neutral	Blue
Phase of single-phase circuit	Brown
Phase 1 of 3-phase circuit	Brown
Phase 2 of 3-phase circuit	Black
Phase 3 of 3-phase circuit	Grey

Note: Part P only applies to fixed electrical installations that are intended to operate at low voltage or extra low voltage which are **not** controlled by the Electricity Safety, Quality and Continuity Amendment Regulations 2009 (as amended) or the Electricity at Work Regulations 1989 (as amended).

Currently, electricians carrying out work in England and Wales **will** have to comply with Part P of the Building Regulations whereas in Scotland it is the Building Standards system.

At the present time Northern Ireland has no equivalent statutory requirement.

2.6 What is the aim of Approved Document P?

The aim of Part P is to increase the safety of householders by improving the design, installation, inspection and testing of electrical installations in dwellings when they (i.e. the installations) are being newly built, installed, extended or altered.

This current 2013 edition of Approved Document P (*Electrical safety – Dwellings*) has been updated and replaces all previous editions. There are some legal and technical changes in this edition, such as:

- reduction in the range of electrical installation work that is notifiable;
- installers who are not a registered competent person may now use a '*registered third-party certifier*' to endorse work as an alternative to using a Building Control Body.

The technical guidance throughout Approved Document P (2013) remains relevant to BS 7671:2018.

This Approved Document provides guidance for compliance with the Building Regulations for building work carried out in England. It also applies to building work carried out on excepted energy buildings in Wales as defined in the Welsh Ministers (Transfer of Functions) Order 2018.

The Government is currently in the process of introducing a mandatory scheme whereby domestic installations shall be checked at regular intervals (as well as when they are sold and/or purchased) to make sure that they comply with Approved Document P **and** BS 7671:2018. This will mean, of course, that if you had an installation which was not correctly certified, then your house insurance might well not be valid!

2.7 Who is responsible for electrical safety?

Basically, there are three people who are responsible for the electrical safety of (and within) buildings. These are:

> **The owner** – needs to determine whether the works carried out are minor or notifiable work. If the work is notifiable, then the owner needs to make sure that the person(s) carrying out the work is either registered under one of the self-certified schemes (see Table 2.6) or is able to certify their work under the local authority Building Control Approval route.

Table 2.6 Authorised competent person self-certification schemes for installers

Type of installation	Schemes
In dwellings – installation of fixed low- or extra-low-voltage electrical installations	BESCA, Blue Flame Certification, Certsure, NAPIT, OFTEC, Stroma
In dwellings – fixed low- or extra-low-voltage electrical installations as part of other work being carried out by the registered person	APHC, BESCA, Blue Flame Certification, Certsure, APIT, Stroma
Buildings other than dwellings – installation of lighting or electrical heating systems	BESCA, Blue Flame Certification, Certsure, NAPIT, Stroma

> **The designer** – needs to ensure that all electrical work is designed, constructed, inspected and tested in accordance with the BS 7671 (current issue) and falls under either a Competent Persons Scheme or the local authority Building Control Approval route.
>
> **The builder/developer** – needs to ensure that they have electricians who can self-certify their work or who are qualified/experienced enough to enable them to sign off under the Electrical Installation Certification form.

2.7.1 What are the statutory requirements?

All electrical installations need to:

- be designed and installed to protect against mechanical and thermal damage;
- be designed and installed so that they will not present an electrical shock and/or fire hazard;
- be tested and inspected to meet relevant equipment/installation standards;
- provide sufficient information so that persons wishing to operate, maintain or alter an electrical installation can do so with reasonable safety;
- comply with such requirements placed by the Building Regulations.

2.7.2 What does all this mean?

With a few exceptions, **any** electrical work undertaken in a home must be reported to the local authority Building Control for inspection if it includes the addition of a new electrical circuit, or involves work in the:

- kitchen;
- bathroom;
- garden area.

This statutory requirement includes any work undertaken professionally, or by you or another family member or by a friend.

The **only** exception is when the installer has been approved by a Competent Persons organisation such as those shown in Table 2.6.

2.7.3 Electrical installations

The following protective schemes (as listed in Table 2.6) are currently available to ensure the safety of electrical systems)

All electrical installations shall provide adequate protection for persons against the risks of electric shock, burn or fire injuries, and should be designed and installed (suitably enclosed and appropriately separated) to provide mechanical and thermal protection.

Electrical installations must be inspected and tested during installation, at the end of installation and before they are taken into service to verify that they:

- are safe to use, maintain and alter;
- comply with Part P (and any other relevant Parts) of the Building Regulations;
- meet the relevant equipment and installation standards;
- meet the requirements of the Building Regulations.

 Note: Any proposal for a new mains supply installation (or where significant alterations are going to be made to an existing mains supply) must be agreed with the electricity distributor.

2.7.4 What types of building does Approved Document P cover?

Part P applies to **all** electrical installations in (**and around**) buildings or parts of buildings comprising:

- dwelling houses and flats;
- dwellings and business premises that have a common supply;
- land associated with domestic buildings;
- fixed lighting and pond pumps in gardens;
- shops and public houses with a flat above;
- common access areas in blocks of flats such as corridors and stairways;
- shared amenities of blocks of flats such as laundries and gymnasiums.

Table 2.7 provides the details of works that are notifiable to local authority and/or must be completed by a company registered as a 'competent firm'.

Table 2.7 Notifiable work

Locations where work is being completed	Extensions and modifications to circuits	New circuits
Bathrooms	Yes	Yes
Bedrooms	Yes	Yes
Ceiling (overhead) heating	Yes	Yes
Communal area of flats	Yes	Yes
Computer cabling	No	No
Conservatories	No	Yes
Dining rooms	No	Yes
Garages (integral)	No	Yes
Garages (detached)	No	Yes*
Garden – lighting	Yes	Yes
Garden – power	Yes	Yes
Greenhouses	Yes	Yes
Halls	No	Yes
Hot air saunas	Yes	Yes
Kitchen	Yes	Yes
Kitchen diners	Yes	Yes
Landings	No	Yes
Lounge	No	Yes
Paddling pools	Yes	Yes
Remote buildings	Yes	Yes
Sheds	Yes	Yes
Shower rooms	Yes	Yes

Table 2.7 (*continued*)

Locations where work is being completed	Extensions and modifications to circuits	New circuits
Small-scale generators	Yes	Yes
Solar power systems	Yes	Yes
Stairways	No	Yes
Studies	No	Yes
Swimming pools	Yes	Yes
Telephone cabling	No	Yes
TV Rooms	No	Yes
Underfloor heating	Yes	Yes
Workshops (remote)	Yes	Yes

*   If the installation requires outdoor wiring.

2.7.5 What is a competent firm?

For the purposes of Part P, the Government has defined *competent firms* as electrical contractors:

- who work in conformance with the requirements to BS 7671:2018;
- whose standard of electrical work has been assessed by a third party;
- who are registered under the NICEIC (National Inspection Council for Electrical Installation Counselling) Approved Contractor scheme and the Electrotechnical Assessment Scheme.

2.7.6 What is a competent person responsible for?

When a competent person undertakes installation work, that person is responsible for:

- ensuring compliance with BS 7671:2018 and all relevant Building Regulations;
- providing the person ordering the work with a signed Building Regulations self-certification certificate;
- providing the relevant Building Control Body with an information copy of the certificate;
- providing the person ordering the work with a completed Electrical Installation Certificate.

2.7.7 Who is entitled to self-certify an installation?

Part P affects **every** electrical contractor carrying out fixed installation and/or alteration work in homes. **Only** registered installers are entitled to self-certify the electrical work, however, and they **must** be registered as a competent person under one of the schemes shown in Table 2.6.

TrustMark was established in 2005 in conjunction with the Government, industry bodies and consumer protection groups. Since this time, in response to the industry-led, Government-commissioned Each home Counts (EHC) review, the TrustMark remit has expanded to include all Repair, Maintenance and Improvement (RMI), Energy Efficiency and retrofit measures, providing a level playing field of quality for consumers having work carried out on or around their home.

TrustMark (see logo at Figure 2.3) operates within a Master Licence Agreement issued by the Government's Department for business, Energy and Industrial Strategy (BEIS).

Figure 2.3 The TrustMark Initiative. (Logo produced courtesy of TrustMark.)

The TrustMark replaced the Quality Mark scheme which closed on 31 December 2004 because too few firms joined. For more information about TrustMark, see their website at: http://www.trustmark.org.uk.

2.7.8 When do I have to inform the local authority Building Control Body?

All proposals to carry out electrical installation work **must** be notified to the local authority's Building Control Body before work begins, **unless** the proposed installation work is undertaken by a person who is a competent person registered under a government-approved Part P Self Certification Scheme or the work is agreed non-notifiable work, such as:

- connecting an electric gate or garage door to an existing isolator (but, be careful, the installation of the circuit up to the isolator **is** notifiable!);
- fitting and replacing cookers and electric showers (unless a new circuit is required);

- installing equipment (e.g. security lighting, air conditioning equipment and radon fans) that is attached to the outside wall of a house (unless there are exposed outdoor connections and/or the installation is a new circuit, or an extension of a circuit in a kitchen, or special location, or is associated with a special installation);
- installing fixed equipment where the final connection is via a 13 A plug and socket (unless it involves fixed wiring and the installation of a new circuit or the extension of a circuit in a kitchen or special location);
- installing prefabricated, 'modular' systems such as kitchen lighting systems and armoured garden cabling that are linked by plug and socket connectors (provided that products are CE-marked and that any final connections in kitchens and special locations are made to existing connection units or points, e.g. a 13 A socket outlet);
- installing or upgrading main or supplementary equipotential bonding (provided that the work complies with other applicable legislation, such as the Gas Safety (Installation and Use) Regulations);
- installing mechanical protection to existing fixed installations (provided that the circuit's protective measures and current-carrying capacity of conductors are unaffected by increased thermal insulation);
- re-fixing or replacing the enclosures of existing installation components;
- replacement, repair and maintenance jobs;
- replacing fixed electrical equipment (e.g. socket outlets, control switches and ceiling roses) which do not require the provision of any new fixed cabling;
- replacing the cable of a single circuit cable (where damaged, for example, by fire, rodent or impact – provided that the replacement cable has the same current-carrying capacity, follows the same route and does not serve more than one sub-circuit through a distribution board);
- work that is not in a kitchen or special location, which does not involve a special installation and which only consists of:

  o adding lighting points (light fittings and switches) to an existing circuit;
  o adding socket outlets and fused spurs to an existing ring or radial circuit (provided that the existing circuit protective device is suitable and supplies adequate protection for the modified circuit);

- work that is not in a special location and only concerns:

  o adding a telephone, extra-low-voltage wiring and equipment for communications, information technology, signalling, command, control and other similar purposes;
  o adding prefabricated equipment sets (and their associated flexible leads) with integral plug and socket connections.

All of this work can be completed by a DIY enthusiast (family member or friend) but still needs to be installed in accordance with manufacturers' instructions and done in such a way that it does not present a safety hazard. This work does not need to be notified to a local authority Building Control Body (unless it is installed in an area of high risk such as a kitchen or a bathroom etc.), but **all** DIY electrical work (unless completed by a qualified professional – who is responsible for issuing a Minor Electrical Installation Certificate) will still need to be checked, certified and tested by a competent electrician.

Any work that involves adding a new circuit to a dwelling will need to be either notified to the Building Control Body (who will then inspect the work) or needs to be carried out by a competent person who is registered under a government-approved Part P Self-Certification Scheme.

Work involving any of the following will also have to be notified:

- consumer unit replacements;
- electric floor or ceiling heating systems;
- extra-low-voltage lighting installations (other than pre-assembled, CE-marked lighting sets);
- garden lighting and/or power installations;
- installation of a socket outlet on an external wall;
- installation of outdoor lighting and/or power installations in the garden or that involves crossing the garden;
- installation of new central heating control wiring;
- solar photovoltaic (PV) power supply systems;
- Small-scale generators such as micro-CHP (Combined Heat and Power Generation) units.

 Note: Where a person who is **not** registered to self-certify intends to carry out the electrical installation, then a Building Regulation (i.e. a Building Notice or Full Plans) application will need to be submitted together with the appropriate fee, based on the estimated cost of the electrical installation. The Building Control Body will then arrange to have the electrical installation inspected at first-fix stage and tested upon completion.

In any event, the electrical work will still need to be certified under BS7671:2018 by a suitably competent person who will be responsible for the design, installation, inspection and testing of the system (on completion) and have the confidence of completing a certificate to say that the work is satisfactory and complies with current codes of practice.

The main things to remember are:

- is the work notifiable or non-notifiable?
- does the person undertaking the work need to be registered as a competent person?
- what records (if any) need to be kept of the installation?

2.7.9 What if the work is completed by a friend, a relative or me?

- You do **not** need to tell your local authority's Building Control Department about non-notifiable work such as:

o   repairs, replacements and maintenance work;
o   extra power points or lighting points or other alterations to existing circuits (unless these are in a kitchen, a bathroom or outdoors).

You **do**, however, need to tell them about most other work.

If you are not sure about this, or you have any questions, ask your local authority's Building Control Department.

2.7.10 What if the work is completed by a contractor or an installer?

If the work is of a notifiable nature then the installer(s) must be registered with one of the schemes shown in Table 2.6.

Figure 2.4 provides a quick guide to the requirements.

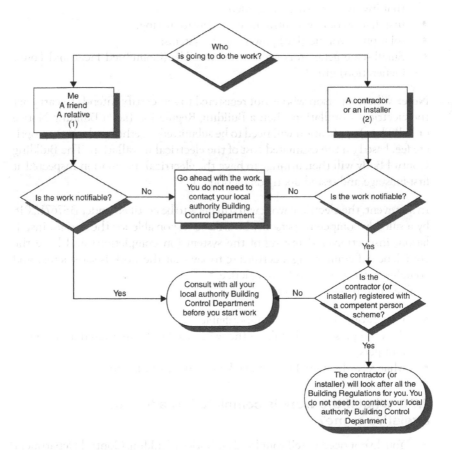

Figure 2.4 How to meet the new rules.

2.8 What inspections and tests will have to be completed and recorded?

As shown in Table 2.8 there are four types of electrical installation certificates and one Building Regulation compliance certificate that have to be completed.

Table 2.8 Types of installation

Type of inspection	When is it used?	What should it contain?	Remarks
Electrical Installation Certificate	For the initial certification of a new installation or for the alteration or addition to an existing installation where new circuits have been introduced	A schedule of inspections and test results as required by Part 6 (of BS 7671) A certificate, including guidance for recipients (standard form from Appendix 6 of BS 7671)	The original certificate shall be given to the person ordering the work and a duplicate retained by the contractor
Minor Electrical Installation Works Certificate	For additions and alterations to an installation such as an extra socket outlet or lighting point to an existing circuit, the relocation of a light switch etc.	Relevant provisions of Part 6 of BS 7671	This certificate may also be used for the replacement of equipment such as accessories or luminaires, but *not* for the replacement of distribution boards (or similar items) or the provision of a new circuit
Electrical Installation Report	For the inspection of an existing electrical installation	A schedule of inspections and a schedule of test results as required by Part 6 (of BS 7671)	For safety reasons, the electrical installation will need to be inspected at appropriate intervals by a competent person
Building Regulations Compliance Certificate	For confirmation that the work carried out complies with the Building Regulations	The basic details of the installation, the location, the completion date and the name of the installer	A purchaser's solicitor may request this document when you come to sell your property. Looking further ahead, it may be required as one of the documents that will make up your 'Home Information Pack'

Copies of these various certificates and forms are contained in Appendix 6 of BS 7671:2018.

2.8.1 What should be included in the records of the installation?

All *original* certificates should be retained in a safe place and be shown to any person inspecting or undertaking further work on the electrical installation in the future. If you later vacate the property, this certificate will demonstrate to the new owner that the electrical installation complied with the requirements of BS 7671 at the time the certificate was issued. The Construction (Design and Management) Regulations require that for a project covered by those Regulations, a copy of this certificate (together with schedules and test results) is included in the project health and safety documentation.

Figure 2.5 indicates how to choose what type of inspection is required.

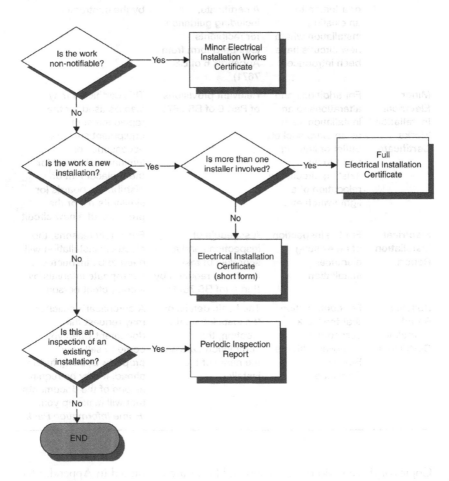

Figure 2.5 Choosing the correct inspection certificate.

2.8.2 Where can I get more information about the requirements of Part P?

Further guidance concerning the requirements of Part P (Electrical safety) is available from:

- the IET (Institution of Engineering Technology) at http://electrical.theiet. org/building-regulations/part-p/index.cfm;
- the NICEIC (National Inspection Council for Electrical Installation Contracting) at http://www.niceic.org.uk;
- the ECA (Electrical Contractors' Association) at http://www.eca.co.uk.

2.9 Requirements from the Approved Documents

Although

- Part E (Resistance to the passage of sound),
- Part J (Combustion appliances and fuel storage systems), and
- Part K (Protection from falling, collision and impact)

have a number of requirements concerning electrical safety and electrical installations (see Figure 2.6 for details), the main requirements are contained in Part P (Electrical safety) together with:

- Part M (Access and use of buildings); and
- Part L (Conservation of fuel and power);
- Part B (Fire safety).

Figure 2.6 Building Regulations.

 Note: Part L consists of four separate sub-parts. L1A, L1B (for domestic buildings) and L2A and L2B (for non-domestic buildings).

2.9.1 Part P – Electrical safety

Reasonable provision shall be made in the design, installation, inspection and testing of electrical installations in order to protect persons from fire or injury.

Sufficient information shall be provided so that persons wishing to operate, maintain or alter an electrical installation can do so with reasonable safety.

Note: The statutory requirements for electrical installations are different in England and Wales from those in Scotland and Northern Ireland.

Electricians carrying out work in England and Wales shall have to comply with Part P of the Building Regulations whereas in Scotland it is the Building Standards system.

At the present time Northern Ireland has no equivalent statutory requirement.

2.9.2 Part M – Access and facilities for disabled people

In addition to the requirements of the Disability and the Equality Act 2019, reasonable provision and precautions need to be taken to ensure that people, regardless of their disability, age or gender to gain access to and to make use of the facilities of the buildings.

2.9.3 Part L1 – Conservation of fuel and power

Energy efficiency measures shall be provided which:

- provide lighting systems that utilise energy-efficient lamps with manual switching controls, or, in the case of external lighting fixed to the building, automatic switching, manual switching or automatic switching controls;
- present information, in a suitably concise and understandable form (including results of performance tests carried out during the works), that shows building occupiers how the heating and hot water services can be operated and maintained.

Responsibility for achieving compliance with the requirements of Part L rests with the person carrying out the work. That *person* may be, for example, a developer, a main (or sub-) contractor, or a specialist firm directly engaged by a private client.

Note: The person responsible for achieving compliance should either themselves provide a certificate, or obtain a certificate from the sub-contractor, that commissioning has been successfully carried out. The certificate should be made available to the client and the Building Control Body.

2.9.4 Part B – Fire safety

The building needs to be designed and constructed so that:

- in the event of fire, its stability will be maintained for a reasonable period;

- there are appropriate provisions for the early warning of fire;
- there are appropriate means of escape in case of fire from the building to a place of safety outside the building;
- the unseen spread of fire and smoke within concealed spaces in its structure and fabric is inhibited.

A wall common to two or more buildings shall be designed and constructed so that it adequately resists the spread of fire between those buildings.

To inhibit the spread of fire within the building, the internal linings shall:

- adequately resist the spread of flame over their surfaces; and
- have, if ignited, a rate of heat release or a rate of fire growth which is reasonable in the circumstances.

2.9.5 Additional requirements and facilities for disabled people

During 2016, Approved Document M (*Access to and use of buildings*) was thoroughly overhauled and restructured in order to meet the future changed requirements of the Disability Discrimination Act 2017.

 On 1 October 2019, the Equality Act replaced most of the Disability Discrimination Act (DDA) and it was written into law that:

> 2019 Equality Act
>
> *All those who provide services to the public, irrespective of their size, are required to take reasonable steps to remove, alter or provide a reasonable means of avoiding a physical feature of their premises which makes it unreasonably difficult or impossible for disabled people to make use of their services.*

Forewarning of the changes to these Acts naturally resulted in Part M ('*Access and use of Buildings*') being completely overhauled and it now covers:

- the conversion of a building for use as a shop now being redefined as a '*material change of use*';
- amendments to omit specific references to (and a definition of) disabled people;
- expansion of the terms to include parents with children, elderly people and people with all types of disabilities (e.g. mobility, sight and hearing etc.);
- the use of a building to disabled people as residents, visitors, spectators, customers or employees, or participants in sports events, performances and conferences (which resulted in amendments being made to M1 (accessibility), M2 (sanitary accommodation) and M3 (audience and/or spectator seating)).

The current edition, therefore, no longer primarily concentrates on wheelchair users, but includes people using walking aids, people with impaired sight (and other mobility and sensory problems) and mothers with prams, as well as people with luggage etc.

Reasonable provision should, therefore, be made to make sure that dwellings (including any purpose-built student living accommodation, other than traditional halls of residence providing mainly bedrooms and not equipped as self-contained accommodation) provide sufficient access for disabled people.

2.9.6 Design

Electrical installations should be designed and installed (suitably enclosed and appropriately separated) so that they:

- are safe to use, maintain and alter;
- comply with the requirements of BS 7671:2018;
- comply with Part P (and any other relevant parts) of the Building Regulations;
- comply with the relevant equipment and installation standards;
- do not present an electric shock or fire hazard to people;
- provide adequate protection against mechanical and thermal damage;
- provide adequate protection for persons against the risks of electric shock, burn or fire injuries.

Note: See Appendix A of Part P to the Building Regulations for details of the types of electrical services normally found in dwellings, some of the ways that they can be connected, and the complexity of wiring and protective systems that can be used to supply them.

2.9.7 Earthing

Distributors are required to provide an earthing facility for all new connections and to ensure:

- all electrical installations are properly earthed;
- all lighting circuits include a circuit-protective conductor;
- all socket outlets which have a rating of 32 A or less and which may be used to supply portable equipment for use outdoors shall be protected by a residual current device (RCD).

Note: The most usual type of earthing is an electricity distributor's earthing terminal, which is provided for this purpose, near the electricity meter.

It is **not** permitted to use a gas, water or other metal service pipe as a means of earthing for an electrical installation (this does not rule out, however, equipotential bonding conductors being connected to these pipes).

In addition, distributors should note and ensure that:

- new or replacement, non-metallic light fittings, switches or other components (e.g. non-metallic varieties) do not require earthing unless new circuit-protective (earthing) conductors are provided;
- socket outlets that are capable of accepting unearthed (2-pin) plugs must **not** be used to supply equipment that needs to be earthed;
- where electrical installation work is classified as an extension, a material alteration or a material change of use, the work must consider and include that the earthing and bonding systems are satisfactory and meet the requirements.

 See Figure 2.7 for details of some earth and bonding conductors that might be part of an electrical installation.

 All accessible consumer units should be fitted with a childproof cover or installed in a lockable cupboard.

2.9.8 Extensions, material alterations and material changes of use

 Note: The whole of the existing installation does **not** have to be upgraded to current standards, but only to the extent necessary for the new work to meet the these standards, except where upgrading is required by the energy efficiency requirements of the Building Regulations.

Where any electrical installation work is classified as an extension, a material alteration or a material change of use, the work needs to consider and include:

- confirmation that the mains supply equipment is suitable to (and can) carry the additional loads envisaged;
- the amount of additions and alterations that will be required to the existing fixed electrical installation in the building;
- the earthing and bonding systems required being satisfactory and meeting the requirements;
- the necessary additions and alterations to the circuits which feed them;
- the rating and the condition of existing equipment (belonging to both the consumer and the electricity distributor) being sufficient;
- the protective measures required to meet the requirements.

 Note: Appendix C to Part P of the Building Regulations offers guidance on some of the older types of installations that might be encountered during alteration work, while Appendix D provides guidance on the application of the now harmonised European cable identification system.

 See Figure 2.8 for details of some of the types of electrical services normally found in dwellings, some of the ways they can be connected, and the complexity of wiring and protective systems that can be used to supply them.

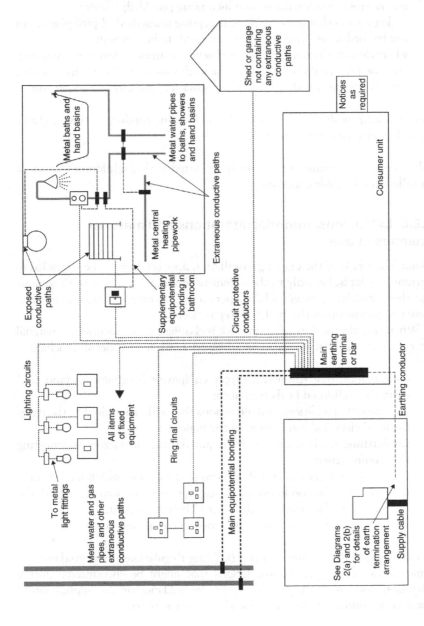

Figure 2.7 Typical earth and bonding conductors that might be part of the electrical installation consumer units.

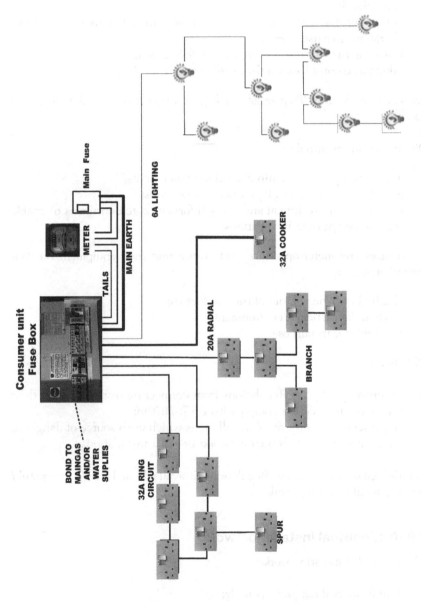

Figure 2.8 Typical fixed installations that might be encountered in new or upgraded existing dwellings.

2.9.9 Electricity distributors' responsibilities

The electricity distributor is responsible for:

- ensuring that the installation is mechanically protected and can be safely maintained;
- evaluating and agreeing proposals for new installations or significant alterations to existing ones;
- installing the cut-out and meter in a safe location;
- taking into consideration the possible risk of flooding.

 See the Gov.uk site's *'Prepare for flooding'* at https://www.gov.uk/prepare-for-flooding.

Distributors are required to:

- maintain the supply within defined tolerance limits;
- provide an earthing facility for new connections;
- provide certain technical and safety information to consumers to enable them to design their installations.

Distributors and meter operators must ensure that their equipment on consumers' premises:

- clearly shows the polarity of the conductors;
- is safe in its particular environment;
- is suitable for its purpose.

Distributors:

- are prevented by the Regulations from connecting installations to their networks which do not comply with BS 7671:2018;
- may disconnect consumers' installations which are a source of danger or cause interference with their networks or other installations.

 Detailed guidance on these Regulations is available at https://www.gov.uk/building-regulations-approval.

2.9.10 Electrical installation work

All electrical installation works:

- shall be carried out professionally;
- shall comply with the Electricity at Work Regulations 1989 (as amended);
- may only be carried out by persons that are competent to prevent danger and injury while doing it, or who are appropriately supervised.

Note: Persons installing domestic combined heat and power equipment must advise the local distributor of their intentions before (or at the time of) commissioning the source.

2.9.10.1 Types of wiring or wiring system

The wiring system chosen for electrical installations will depend on location and designed use. For example:

- cables concealed in floors and (in certain circumstances) walls are required to have an earthed metal covering, be enclosed in steel conduit, or have some form of additional mechanical protection (see BS 7671:2018 for more information);
- cables to an outside building (e.g. garage or shed) if run underground, should be:
  - o routed and positioned so as to give protection against electric shock and fire as a result of mechanical damage to a cable;
  - o have some form of underground coloured plastic warning tape installed directly above the service line, at a depth of 6–8 inches, to ensure the safety of someone excavating in that area.
- heat-resistant flexible cables are required for the final connections to certain equipment (see maker's instructions).

 PVC insulated and sheathed cables are likely to be suitable for much of the wiring in a typical dwelling.

2.9.10.2 Equipotential bonding conductors

- Main equipotential bonding conductors are required to water service pipes, gas installation pipes, oil supply pipes and certain other 'earthy' metalwork that may be present on the premises.
- The installation of supplementary equipotential bonding conductors is required for installations and locations where there is an increased risk of electric shock (e.g. bathrooms and shower rooms).
- The minimum size of supplementary equipotential bonding conductors (without mechanical protection) is 4 mm^2.

2.9.10.3 Electrical components and installations

 New or replacement, non-metallic light fittings, switches or other components do **not** require earthing **unless** new circuit protective (earthing) conductors are provided.

2.9.11 Inspection and test

Electrical installations need to be inspected and tested during and at the end of installation, and before they are taken into service, to verify that they:

- are safe to use;
- comply with BS 7671:2018;
- meet the relevant equipment and installation standards.

All electrical work should be inspected (during installation as well as on completion) to verify that the components have:

- been selected and installed in accordance with BS 7671:2018;
- been made in compliance with appropriate British Standards or harmonised European Standards;
- been evaluated against external influences (such as the presence of moisture);
- not been visibly damaged (or are defective) so as to be unsafe;
- been tested to check satisfactory performance with respect to continuity of conductors, insulation resistance, separation of circuits, polarity, earthing and bonding arrangements, earth fault loop impedance and functionality of all protective devices including residual current devices;
- been inspected and tested for compliance with the requirements of BS 7671:2018 using appropriate and accurate instruments.

 Note: Inspections and testing of DIY work should **also** meet the above requirements.

2.9.12 Alarm systems

2.9.12.1 Emergency alarms

Emergency alarm pull cords should be:

- coloured red;
- located as close to a wall as possible;
- have two red 50 mm diameter bangles.

Front plates should contrast visually with their backgrounds.

 The colours red and green should **not** be used in combination as indicators of 'ON' and 'OFF' for switches and controls.

2.9.12.2 Emergency assistance alarms

Emergency assistance alarm systems should have:

- visual and audible indicators to confirm that an emergency call has been received;
- a reset control reachable from a wheelchair, WC, or from a shower/changing seat;
- a signal that is distinguishable visually and audibly from the fire alarm.

2.9.12.3 Fire alarms

 Fire alarms should emit an audio and visual signal to warn occupants with hearing or visual impairments.

2.9.13 Controls and switches

The aim should be to ensure that all controls and switches should be easy to operate, visible and free from obstruction, and:

- should be located between 750 mm and 1200 mm above the floor;
- should not require the simultaneous use of both hands to operate (unless absolutely necessary for safety reasons);
- switched socket outlets should indicate whether they are ON;
- mains and circuit isolator switches should clearly indicate whether they ON or OFF;
- where possible, light switches with large push pads should be used in preference to pull cords.
- individual switches on panels and on multiple-socket outlets should be well separated;
- controls that need close vision (e.g. thermostats) should be located between 1200 mm and 1400 mm above the floor;
- front plates should contrast visually with their backgrounds;

2.9.14 Heat emitters

- Heat emitters should either be screened or have their exposed surfaces kept at a temperature below 43°C.

In toilets and bathrooms, heat emitters (if located) should **not** restrict:

- the minimum clear wheelchair manoeuvring space;
- the space beside a WC used to transfer from the wheelchair to the WC.

2.9.15 Lighting circuits

 All lighting circuits shall include a circuit protective conductor.

Light switches should be:

- aligned horizontally with door handles and have large push pads (in preference to pull cords);
- located between 900 and 1100 mm of the entrance door opening;
- located between 750 mm and 1200 mm above the floor.

The colours red and green should **not** be used in combination as indicators of ON and OFF for switches and controls.

2.9.15.1 Fixed lighting

In locations where lighting can be expected to have most use, fixed lighting (e.g. fluorescent tubes and compact fluorescent lamps – but **not** GLS tungsten lamps with bayonet-cap or Edison-screw bases) with a luminous efficacy greater than 40 lumens per circuit-watt should be available.

Note: Table 2.9 provides an indication of the recommended number of locations (excluding garages, lofts and outhouses) that need to be equipped with efficient lighting.

Table 2.9 Lighting requirements

Number of rooms created (hall, stairs and landing(s) count as one room, as does a conservatory)	Recommended minimum number of locations (fixed lighting)
1–3	1
4–6	2
7–9	3
10–12	4

External lighting (including lighting in porches, but not lighting in garages and carports) should:

- automatically extinguish when there is enough daylight, and when not required at night;
- have sockets that can only be used with lamps having an efficacy greater than 40 lumens per circuit Watt. (Such as a fluorescent or compact fluorescent lamp types, but not GLS tungsten lamps with bayonet cap or Edison screw bases).

2.9.16 Power operated doors

Doors to accessible entrances shall be provided with a power-operated door opening and closing system if a force greater than 20 N is required to open or shut a door.

Once open, all doors to accessible entrances should be wide enough to allow unrestricted passage for a variety of users, including wheelchair users, people carrying luggage, people with assistance dogs, and parents with pushchairs and small children.

The effective clear width through a single-leaf door (or one leaf of a double-leaf door) should be in accordance with Table 2.10 and as indicated in Figure 2.9.

Table 2.10 Minimum effective clear widths of doors

Direction and width of approach	New buildings (mm)	Existing buildings (mm)
Straight on (without a turn or oblique approach)	800	750
At right angles to an access route at least 1500 mm wide	800	750
At right angles to an access route ate least 1200 mm wide	825	775
External doors to buildings used by the general public	1000	775

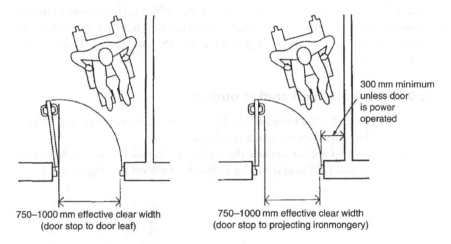

Figure 2.9 Effective clear width and visibility requirements of doors.

Power-operated entrance doors should have a sliding, swinging or folding action controlled manually (by a push pad, card swipe, coded entry, or remote control) or automatically controlled by a motion sensor or proximity sensor such as a contact mat.

Power-operated entrance doors should, when and where necessary for health or safety:

*   be provided with a manual or automatic opening device in the event of a power failure;

- open towards people approaching the doors;
- provide visual and audible warnings that they are operating (or about to operate);
- incorporate automatic sensors to ensure that they open early enough; (and stay open long enough to permit safe entry and exit);
- incorporate a safety stop that is activated if the doors begin to close when a person is passing through;
- have a readily identifiable and accessible stop switch;
- have safety features to prevent injury to people who are struck or trapped (such as a pressure-sensitive door edge which operates the power switch);
- revert to manual control (or fail-safe) in the open position in the event of a power failure;
- when open, not project into any adjacent access route;
- ensure that their manual controls:

  o are located between 750 mm and 1000 mm above floor level;
  o are operable with a closed fist;
  o are set back 1400 mm from the leading edge of the door when fully open;
  o are clearly distinguishable against the background;
  o contrast visually with the background.

 Note: Revolving doors are **not** considered 'accessible', as they create particular difficulties (and possible injury) for people who are visually impaired, people with assistance dogs or mobility problems, and for parents with children and/ or pushchairs.

2.9.17 Switches and socket outlets

- Switches and socket outlets for lighting and other equipment should be located so that they are easily reachable.
- Switches and socket outlets (for lighting) should be installed between 450 mm and 1200 mm from the finished floor level (see Figure 2.10).

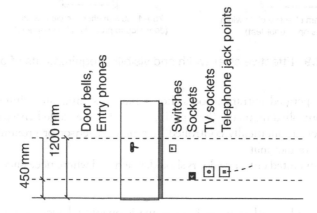

Figure 2.10 Heights of switches and sockets etc.

The aim is to help people with limited reach (e.g. seated in a wheelchair) access a dwelling's wall-mounted switches and socket outlets.

2.9.17.1 Socket outlets

- Non-fused older types of socket outlet plugs must not be connected to a ring circuit.
- Socket outlets that will accept unearthed (2-pin) plugs must **not** be used supply equipment that needs to be earthed.
- Sensitive RCD protection is required for all socket outlets which have a rating of 32 A or less and which may be used to supply portable equipment for use outdoors.
- Socket outlets should comply with the requirements of Part M (*Access to and use of buildings*).

2.9.17.2 Portable equipment for use outdoors

- All socket outlets which have a rating of 32 A or less and which may be used to supply portable equipment for use outdoors must be protected by an RCD.

2.9.17.3 Switched socket outlets

All switched socket outlets should be wall-mounted and should:

- be located no nearer than 350 mm from room corners,
- have front plates which contrast visually with their backgrounds;
- indicate whether they are ON.

Mains and circuit isolator switches should clearly indicate whether they ON or OFF.

The colours red and green should not be used in combination as indicators of ON and OFF for switches and controls.

Individual switches on panels and on multiple-socket outlets should be well separated.

2.9.17.4 Wall sockets

Wall sockets shall meet the following requirements as depicted in Table 2.11.

2.9.18 Telephone points and TV sockets

All telephone points and TV sockets should be located between 400 mm and 1000 mm above the floor.

Table 2.11 Building Regulations requirements for wall sockets

Type of wall	Requirement
Timber-framed	Power points may be set in the linings provided there is a similar thickness of cladding behind the socket box
	Power points should **not** be placed back to back across the wall
Solid masonry	Deep sockets and chases should **not** be used in separating walls
	Stagger the position of sockets on opposite sides of the separating wall
Cavity masonry	Stagger the position of sockets on opposite sides of the separating wall
	Deep sockets and chases should **not** be used in a separating wall
	Deep sockets and chases in a separating wall should **not** be placed back to back
Framed walls with absorbent material	Sockets should: • be positioned on opposite sides of a separating wall • **not** be connected back to back • be staggered a minimum of 150 mm edge to edge

2.9.19 Other considerations

2.9.19.1 Lecture/conference facilities

Artificial lighting should be designed to:

• give good colour rendering of all surfaces;
• be compatible with other electronic and radio-frequency installations.

2.9.19.2 Swimming pools and saunas

Swimming pools and saunas are subject to special requirements specified in Part 7 of BS 7671: 2018.

2.9.20 Cellars or basements

Liquid petroleum gas (LPG) storage vessels and LPG-fired appliances fitted with automatic ignition devices or pilot lights must **not** be installed in cellars or basements.

Author's end note

Now that you are aware of the background to the Building Regulations, Chapter 3 looks at the different types of earthing systems that are available for both domestic and non-domestic buildings.

3

Earthing

Author's start note

This chapter reminds the reader about the different types of earthing systems and earthing arrangements. It then lists the main requirements from the various Approved Documents and Regulations for safety protection (direct and indirect contact), protective conductors and protective equipment, before briefly touching on the test requirements for earthing.

*Similar to other chapters, it should be noted that these lists of requirements are **only** the author's impression of the most important aspects of the Wiring Regulations and electricians should **always** consult the latest edition of BS 7671 to satisfy compliance. More detailed information concerning BS 7671:2018's 'earthing arrangements and protective conductors' is contained in Chapter 54 of the standard.*

From an electrical point of view, the world is effectively a huge conductor at zero potential and is used as a reference point which is called *Earth* (in the UK) or *ground* in the USA. People and animals are normally in contact with the Earth and so if another part, which is open to touch, becomes charged at a different voltage from Earth, a shock hazard will exist.

 One lightning bolt has enough electricity to service 200 000 homes!!!

 Readers are reminded that the previous 2018 edition of BS 7671 included some additional requirements for bonding and earthing in order to maintain technical alignment with CENELEC harmonisation documents. Earthing requirements, therefore, now include the following:

- Protection of low-voltage installations against temporary overvoltages due to Earth faults in high- and low-voltage systems.

Figure 3.1 Typical lightning strike. (Courtesy Stingray.)

New appendices concerning earthing were also inserted into BS 7671:2018, and these include:

- measurement of Earth fault loop impedance (consideration of the increased resistance of conductors with an increased temperature);
- methods for measuring the insulation resistance/impedance of floors and walls to Earth or to the protective conductor system.

 It has been agreed that using a gas, water or other metal service pipe as a means of earthing for an electrical installation is **not** permitted. However, this doesn't rule out equipotential bonding conductors being connected to these pipes.

3.1 What is Earth?

In electrical terms, 'Earth' is defined as:

> *"The conductive mass of the Earth, whose electric potential at any point is conventionally taken as zero."*

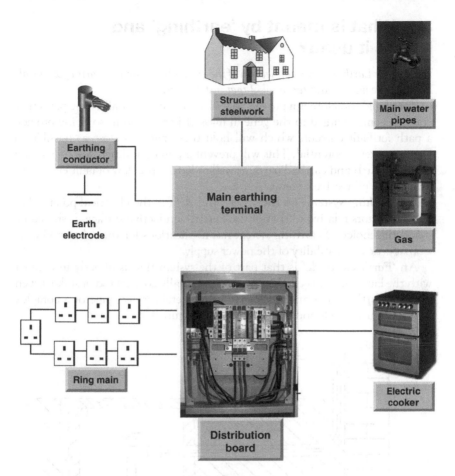

Figure 3.2 Bonding and earthing. (Courtesy Herne.)

From an astronomical and geophysical point of view, on the other hand:

> "Earth (also known as The Earth, Terra, and – mostly in the 19th century – Tellus) is the third planet outward from the Sun. It is the largest of the solar system's terrestrial planets and the only planetary body that modern science [so far!] has confirmed as harbouring life."

The planet formed around 4.57 billion (4.57 × 10⁹) years ago and 'shortly' thereafter (i.e. 4.533 billion years ago, to be precise!) acquired its single natural satellite, the Moon. Its astronomical symbol consists of a circled cross, representing a meridian and the equator.

3.2 What is meant by 'earthing' and how is it used?

Definition: *Earthing is the connection of the exposed-conductive- parts of an installation to the main Earth terminal of that installation.*

'**Earthing**' therefore is a process that is used to connect all of the parts that could become charged to the general mass of Earth, and in so doing provide a path for fault currents which will hold these parts as close as possible to Earth (i.e. zero) potential. This will prevent a potential difference happening between Earth and earthed parts, as well as letting the flow of fault current to operate the protective systems.

An '*earthing system*', on the other hand, defines the electrical potential of the conductors relative to that of the Earth's conductive surface. It should be noted the choice of earthing system has implications for the safety and electromagnetic compatibility of the power supply.

An '*Earth electrode*' is that part of the system that is directly in contact with the Earth and this can be just a metal (usually copper) rod or stake driven into the Earth or a connection to a buried metal service, pipe or a complex system of buried rods and wires as shown in Figure 3.3.

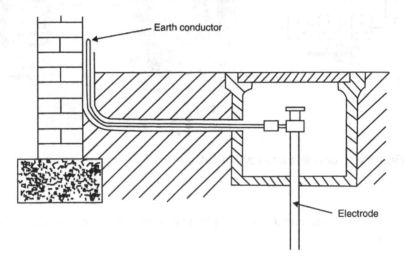

Figure 3.3 Earth conductor and electrode.

The resistance of the electrode-to-Earth connection will determine its quality and this can be improved by:

- increasing the surface area of the electrode that is in contact with Earth;
- increasing the depth to which the electrode is driven;
- using several connected ground rods;
- increasing the moisture of the soil;

- improving the conductive mineral content of the soil; and
- increasing the land area covered by the ground system.

A protective Earth (PE) connection ensures that all exposed conductive surfaces are at the same electrical potential as the surface of the Earth and thus avoids the risk of an electrical shock if a person, or an animal, touches a piece of equipment (or device) in which an insulation fault has occurred. PE also ensures that, if an insulation fault occurs, a high fault current will flow which will trigger an overcurrent protection device (e.g. a fuse) that will disconnect the power supply.

A functional Earth (FE) connection, as well as providing protection against electric shock, can carry a current during the normal operation of a device – a facility that is often required by devices such as surge suppression and electro-magnetic-compatibility filters, some types of antennas as well as a number of measuring instruments.

In a mains (i.e. a.c. power) wiring installation, the 'ground' wire is (directly or indirectly) connected to one or more Earth electrodes and carries currents away under fault conditions. These Earth electrodes may be located nearby or in the supplier's network some distance away, and the ground wire is also usually bonded to pipework so as to keep it at the same potential as the electrical ground during a fault.

3.3 Advantages of earthing

The main advantage to earthing is that the whole electrical system is tied to the potential of the general mass of Earth and cannot *float* at another potential. By connecting Earth to metalwork (that is not intended to carry current), a path is provided for fault current which can be detected by a protective conductor and, if necessary, broken. The path for this fault current is shown in Figure 3.4.

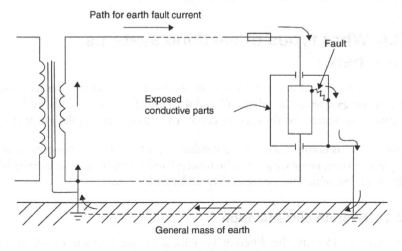

Figure 3.4 Path for Earth fault current (shown by arrows).

On the other hand, the main disadvantage of earthing is primarily the cost of having to provide protective conductors and Earth electrodes, etc.!

Note: Until the mid 1900s, all power outlets generally lacked protective Earth terminals!

Earthing arrangements may be used jointly or separately for protective and functional purposes, according to the requirements of the installation. They should, however, ensure that:

- they are sufficiently robust (or have additional mechanical protection) to external influences;
- the impedance from the consumer's main earthing terminal to the earthed point of the supply, meets the protective and functional requirements of the installation;
- Earth fault currents and protective conductor currents that may occur are carried without danger (particularly from thermal, thermomechanical and electromechanical stresses).

If a number of installations have separate earthing arrangements, then any protective conductor that is common to one of these installations:

- shall either be capable of carrying the maximum fault current likely to flow through them; or
- shall Earth one installation and be insulated from the earthing arrangements of the other installation(s).

Precautions should be taken against possible damage to other metallic parts through electrolysis and, if the protective conductor forms part of a cable, then this can be achieved by earthing the installation containing the associated protective device as shown in Figure 3.5

3.4 What types of earthing systems are there?

BS 7671:2018 defines an electrical system as consisting *"of a single source of electrical energy and an installation"*, and the type of system depends on the link between the source and the exposed conductive parts of the installation, to Earth.

Note: In this context, an *exposed conductive part* means a conductive part of a piece of equipment which can be touched and which is not (i.e. currently) a live part, but which *may* become live under fault conditions.

3.4.1 System classification

In order to identify the different systems, a unique four-letter code is used whereby:

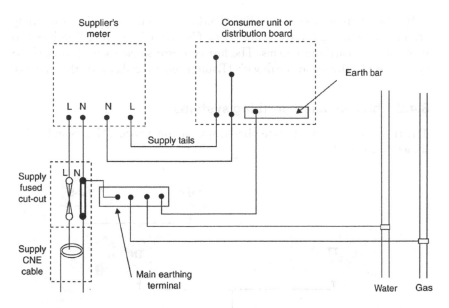

Figure 3.5 Domestic earthing arrangement Earth electrodes.

The **first letter** indicates the type of earthing supply, so that:

- **T** indicates that one or more points of the supply are directly earthed (for example, the earthed neutral at the transformer);
- **I** indicates either that the supply system is not earthed (at all) or that the earthing includes a deliberately inserted impedance, in order to limit fault current.

The **second letter** provides details of the actual earthing arrangements in the installation, so that:

- **T** indicates that all exposed conductive metalwork is connected directly to Earth;
- **N** indicates that all exposed conductive metalwork is connected directly to an earthed supply conductor provided by the electricity supply company.

The **third and fourth letters** show the arrangement of the Earthed supply conductor system so that:

- **S** ensures that neutral and Earth conductor systems are quite separate; and
- **C** ensures that neutral and Earth are combined into a single conductor.

T = **Earth** (from the French word Terre), **N** = **Neutral**, **S** = **Separate**, **C** = **Combined**, **I** = **Isolated** (the source of an IT system is either connected to Earth through a deliberately introduced earthing impedance or is isolated from Earth. All exposed-conductive-parts of an installation are connected to an Earth electrode).

Whilst International standard IEC 60364 identifies three families of earthing arrangements, using the two-letter codes TN and TT, BS 7671:2018 lists two types of main earthing systems. The first concerns single-sourced three-phase systems, whilst the second deals with IT, multiple-source, d.c. and other systems.

3.4.2 Three-phased earthing systems

The three most common three phased systems are TN-S, TN-C-S and TT as shown in Figure 3.6.

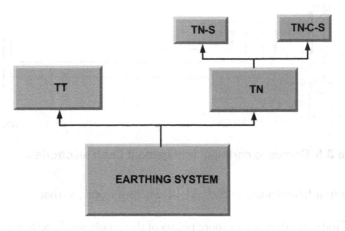

Figure 3.6 Three phased earthing systems.

3.4.2.1 TN-S system

TN-S systems have separate protective Earth (PE) and Neutral (N) conductors that are connected together only near the power source and remain separated throughout the system (see Figure 3.7).

For a TN-S system, the main earthing terminal of the installation is connected to the earthed point of the source of energy and that that part of the connection may be formed by the distributor's lines and/or equipment.

The TN-S is the most common earthing system in the UK and one where the electricity supply company provides an Earth terminal (usually the armour and/or sheath of the underground supply cable) at the incoming mains position.

One of the main advantages of a TN-S system concerns Electromagnetic Compatibility as the consumer has a low-noise connection to Earth and, therefore, does not suffer from the voltage that appears on the neutral conductor because of the return currents and the impedance of that conductor. This is particularly important for some types of telecommunication and measurement equipment and costs will be reduced by TN-S networks having a fairly low-impedance Earth connection near each consumer.

In TN-S systems an RCD can be used as an additional protection.

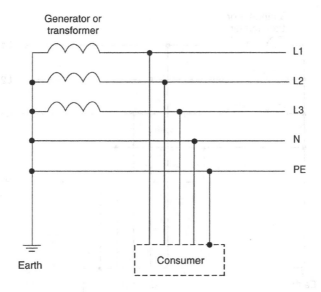

Figure 3.7 TN-S System.

3.4.2.2 TN-C-S system

For a TN-C-S system (where protective multiple earthing is provided), the main earthing terminal of the installation is connected by the distributor to the neutral of the source of energy.

To achieve this requirement, a TN-C-S earthing system uses a combined PEN conductor from transformer to the building's distribution point, but separate PE and N conductors to fixed indoor wiring and flexible power cords. (See Figure 3.8).

 Note: In the UK, this system is also known as *protective multiple earthing* (PME) as it connects the combined neutral and Earth to real Earth at many locations – and thereby reduces the risk of broken neutrals.

 The use of TN-C-S is **not** recommended for locations such as petrol stations etc. where there are lots of buried metalwork and explosive gasses.

 Owing to the possibility of a lost neutral, the use of TN-C-S supplies is **banned** for caravans and boats in the UK and it is often recommended to make outdoor wiring TT with a separate rod.

3.4.2.3 TT system

A TT system has one point of the energy source directly earthed and the exposed-conductive parts of the consumer's installation are provided with a local connection to Earth, independent of any Earth connection at the generator (see Figure 3.9).

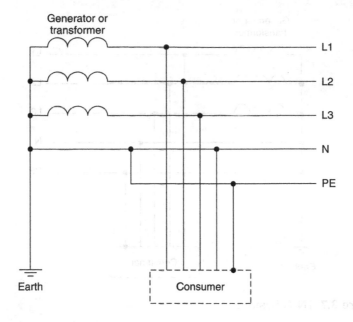

Figure 3.8 TN-C-S system.

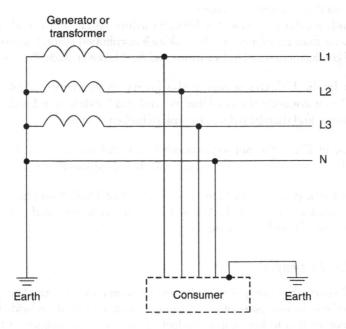

Figure 3.9 TT system.

As each consumer has its own connection to Earth, they will not notice any currents that may be caused by other consumers on a shared PE line.

This type of installation is usually found in rural locations where the system is not provided with an Earth terminal by the electricity supply company and the installation is fed from an overhead supply. Neutral and Earth (protective) conductors must be kept quite separate throughout the installation and the final Earth terminal must be connected to an Earth electrode via an earthing conductor.

TT systems (similar to TN-S systems) have a low-noise connection to Earth, which is particularly important with some types of telecommunication and measurement equipment. (See Figure 3.9).

3.5 IT system

In an IT system, the main earthing terminal is connected via an earthing conductor to an Earth electrode (see Figure 3.10).

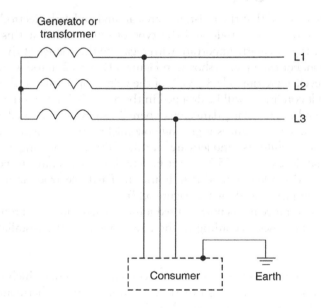

Figure 3.10 IT system.

An IT system is similar to a TT system except that the supply earthing in an IT system can either be from an unearthed supply or one which (although not totally earthed) is connected to Earth through a current-limiting impedance.

In an IT system:

- all live parts shall be insulated from Earth or connected to Earth through a sufficiently high impedance connector, either at the neutral point or midpoint of the system - or at an artificial neutral point;

- precautions shall be taken to avoid the risk of a person being in contact with simultaneously accessible exposed-conductive-parts in the event of two faults occurring at the same time;
- exposed-conductive-parts shall be earthed individually, in groups, or collectively.

This lack of Earth will usually mean that normal protective methods cannot be used, and for this reason IT systems are **not** generally allowed in the UK public supply system – except for hospitals and other medical locations where such systems are recommended for use with circuits supplying medical equipment intended for life-support of patients.

A step-up autotransformer shall **not** be connected to an IT system.

3.6 Earthing points

It has been proved that the resistance area around an Earth electrode depends on the size of the electrode and the type of soil – and that this electrode resistance is particularly important with regard to the voltage at the surface of the ground. For example, as shown in Figure 3.11, for a 2 m rod with its top at ground level, approximately 80–90% of the voltage appearing at the electrode under fault conditions will be dropped in the first 2.5–3 m from the electrode.

This can be particularly dangerous where livestock is concerned. For example, in some circumstances a grazing cow might have its forelegs inside the resistive area, whilst its hind legs are outside of the area. Bearing in mind that a potential difference of 25 V can it be lethal, measures have to be taken to reduce this risk. One method is to house the Earth electrode in a pit that is below ground level (as shown in Figure 3.12).

Earthing arrangements may be used jointly or separately for protective and functional purposes, according to the requirements of the installation, such that:

- earth fault currents and protective conductor currents which may occur are carried without danger (particularly from thermal, thermomechanical and electromechanical stresses); and
- the value of impedance from the consumer's main earthing terminal to the earthed point of the supply is considered to be continuously effective; and
- they are adequately robust or have additional mechanical protection appropriate to the assessed conditions of external influence.

3.7 Earthing terminals

The main earthing terminal acts as the single reference point and can be either a bar, plate or even a copper internal 'ring' conductor. This is usually directly connected to an effective Earth electrode but, because of the risk of

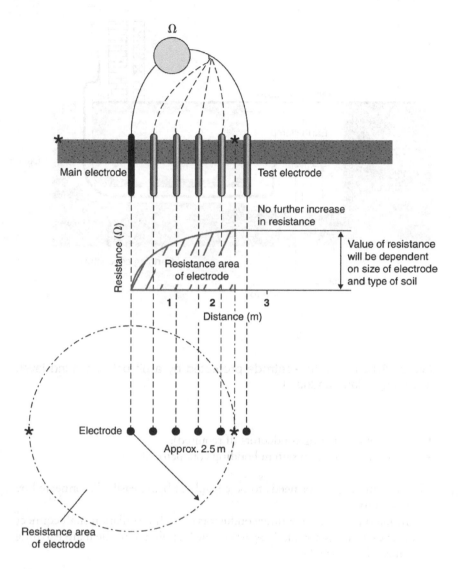

Figure 3.11 The resistance area of an Earth electrode. (Courtesy Brian Scaddan.)

corrosion occurring if aluminium or copper-clad aluminium were used, this connection must be of copper.

The main earthing terminal then connects to the following main earthing protecting conductors:

- circuit protective conductor;
- protective bonding conductor;

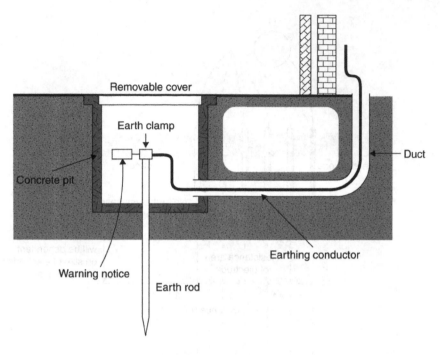

Figure 3.12 An Earth electrode protected by a pit below ground level. (Courtesy Brian Scaddan.)

- functional earthing conductors (if required);
- lightning protection system bonding conductor, if any.

The earthing conductor needs to be capable of being easily disconnected by means of a tool.

The main protective bonding conductors of each installation shall connect all other extraneous-conductive-parts of the installation to the main earthing terminal. These include:

- central heating and air-conditioning systems;
- exposed metallic structural parts of the building;
- gas installation pipes;
- water installation pipes;
- other installation pipework and ducting.

To enable the resistance of the earthing arrangements to be measured, the earthing conductor needs to be capable of being easily disconnected – however:

Joints in an earthing conductor shall **only** be capable of being disconnected by means of a specialist tool!

If an installation serves **more** than one building, then the above requirement shall be applied to **each** building.

The means of connecting the different electrical systems is shown in Table 3.1.

Table 3.1 Connections of main earthing terminals

System	Means of connection
TN	All exposed-conductive-parts of the installation shall be connected by a protective conductor to the main earthing terminal of the installation (which shall, in turn, be connected to the earthed point of the power supply system)
TN-S	The main earthing terminal of the installation shall be connected to the earthed point of the source of energy
TN-C-S	Where protective multiple earthing is provided, the main earthing terminal of the installation shall be connected, by the distributor, to the neutral of the source of energy
TT	Exposed-conductive-parts of the installation shall be protected by a single protective device that is connected (via the main earthing terminal) to a common Earth electrode
IT	The main earthing terminal shall be connected via an earthing conductor to an Earth electrode

The main earthing terminal is connected to Earth as shown in Figure 3.13 and the following text.

The Earth electrode should be positioned as close as possible to the main earthing terminal.

The main earthing terminal connects following to the earthing conductor:

- the circuit protective conductors;
- the protective bonding conductors; and also
- functional earthing conductors (if required);
- lightning-protection system bonding conductor, if any.

3.8 Conductor arrangement and system earthing

In physics and electrical engineering, a conductor is an object or type of material that allows the flow of charge (electrical current) in one or more directions. Materials made of metal are common electrical conductors and the following are the most common current-carrying conductors (as shown in Figure 3.14)

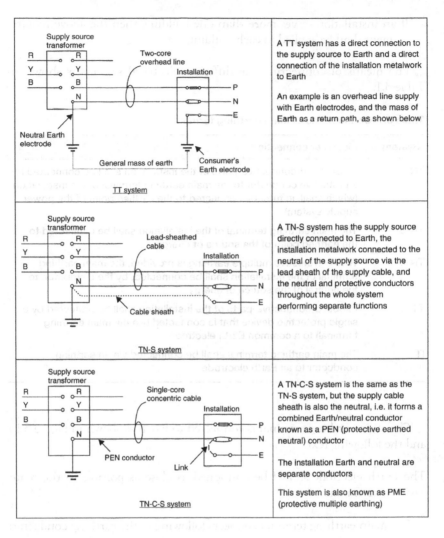

	A TT system has a direct connection to the supply source to Earth and a direct connection of the installation metalwork to Earth

A TT system has a direct connection to the supply source to Earth and a direct connection of the installation metalwork to Earth

An example is an overhead line supply with Earth electrodes, and the mass of Earth as a return path, as shown below

A TN-S system has the supply source directly connected to Earth, the installation metalwork connected to the neutral of the supply source via the lead sheath of the supply cable, and the neutral and protective conductors throughout the whole system performing separate functions

A TN-C-S system is the same as the TN-S system, but the supply cable sheath is also the neutral, i.e. it forms a combined Earth/neutral conductor known as a PEN (protective earthed neutral) conductor

The installation Earth and neutral are separate conductors

This system is also known as PME (protective multiple earthing)

Figure 3.13 Earth provision.

3.9 Current-carrying conductors

The standard method of attaching the electrical supply system to Earth is to make a direct connection between the two at the supply transformer so that the neutral conductor (often the star point of a three-phase supply – see Figure 3.15) is connected to Earth using an Earth electrode or the metal sheath and/or armouring of a buried cable.

Note: Lightning conductor systems must be bonded to the installation's Earth with a conductor that is no larger (i.e. by cross-sectional area) than that of the earthing conductor itself.

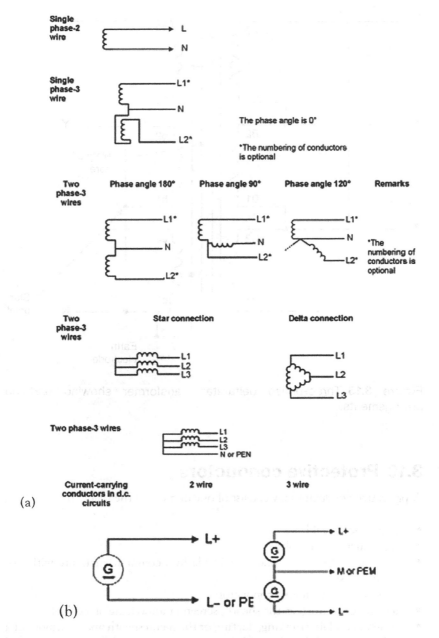

Figure 3.14 Conductor arrangements and system earthing.

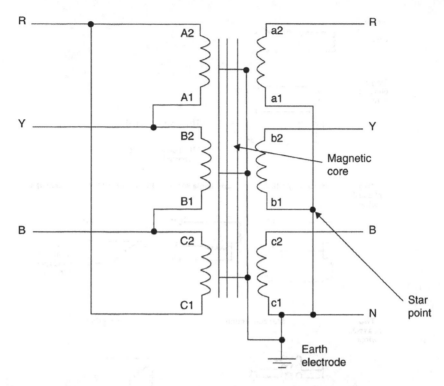

Figure 3.15 Three-phase delta/star transformer showing earthing arrangements.

3.10 Protective conductors

A protective conductor may consist of one or more of the following:

- a single-core cable;
- a conductor in a cable;
- an insulated or bare conductor inside in a common enclosure with live conductors;
- a fixed bar or insulated conductor;
- a metal covering (e.g. the sheath, screen or armouring of a cable);
- a metal conduit, trunking, ducting or the metal sheath and/or armour of a cable (provided that the earthing terminal of each accessory is connected by a separate protective conductor to an earthing terminal incorporated in the associated box or other enclosure.

 A gas pipe, oil pipe, flexible or pliable conduit, support wires or other flexible metallic parts, or constructional parts that are subject to mechanical stress in normal service, **shall NOT** be selected as a protective conductor.

3.11 Installations with separate earthing arrangements

Where a number of installations have separate earthing arrangements, protective conductors common to any of these installations shall either:

- be capable of carrying the maximum fault current likely to flow through them; or
- be earthed within one installation only and insulated from the earthing arrangements of any other installation.

3.11.1 Equipment having a protective conductor current exceeding 10 mA

In these cases, the conductor shall be connected to the supply either:

- permanently via the wiring of the installation; or
- via a flexible cable with a plug and socket outlet; or
- via a protective conductor with an Earth monitoring system.

When this is a permanent connection, it must also be by means of a flexible cable.

3.11.2 Switching devices

A switching device shall **not** be inserted in a protective conductor unless:

- the switch is between the neutral point and the means of earthing; and
- the switch is a linked switch arranged so that it can disconnect and connect the earthing conductor, at substantially the same time as the related live conductors. (Also see BS 7671:2018 Chapter 46 for details of isolation devices.)

3.11.3 Electrical monitoring

Where electrical monitoring of earthing is used, dedicated devices (such as operating sensors, coils etc.) shall **not** be connected in series with the protective conductor.

3.12 Earth electrodes

An Earth electrode is a conductor, or a group of conductors, that connects the main earthing terminal of an installation to an Earth electrode or to other means of earthing.

The type and embedded depth of an Earth electrode must ensure that soil drying and freezing will not increase its resistance above the required level.

if an Earth electrode has parts that need to be connected together, the connection shall be made by either welding or using pressure connectors, clamps or other suitable mechanical connectors.

The following types of Earth electrode may be used for electrical installations:

- Earth rods or pipes;
- Earth tapes or wires;
- Earth plates;
- lead sheaths and other metal cable covers;
- other suitable underground metalwork;
- structural metalwork embedded in foundations that is capable of withstanding corrosion;
- welded metal-reinforced concrete (except pre-stressed concrete) embedded in the Earth.

The use of the lead sheath or other metal covering of a cable, as an Earth electrode, shall be subject to all of the following conditions:

- adequate precautions have been taken to prevent excessive deterioration by corrosion;
- the sheath or covering has an effective contact with Earth;
- the consent of the owner of the cable has been obtained;
- the owner of the electrical installation will be warned of any proposed change to the cable which might affect its suitability as an Earth electrode.

Note: The materials and dimensions of foundation earth electrodes shall be capable of withstanding corrosion.

The following types of Earth electrode may **not** be used for electrical installations:

- metallic pipes used for gases or flammable liquids;
- the metallic pipe from a water facility; or
- metallic objects immersed in water.

Further information on Earth electrodes can be found in BS 7430 and (if a lightning-protection system (LPS) is in place) BS 62305-1.

3.13 Earthing conductors

The earthing conductor is an important part of the Earth fault loop impedance as it is a protective conductor that connects the main earthing terminal of an installation to an Earth electrode, or to some other means of earthing. See Figure 3.16.

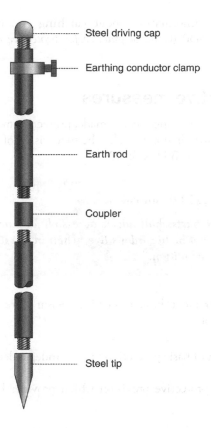

Figure 3.16 An example of an earthing conductor. (Courtesy Brian Scaddan.)

All earthing rods (see Figure 3.16) should be:

- soundly made;
- electrically and mechanically satisfactory;
- suitably labelled;
- suitably protected against corrosion;
- driven into virgin soil (as opposed to backfilled or previously disturbed) so as to make an effective contact with the surrounding material, and should have a cross-sectional area not less than that shown in Table 3.2.

Table 3.2 Minimum-cross sectional area of buried earthing connector

Protected against corrosion	Protected against mechanical damage	Not protected against mechanical damage
Protected by a sheath	2.5 mm^2 copper, 10 mm^2 steel	16 mm^2 copper, 16 mm^2 steel
Not protected	25 mm^2 copper, 50 mm^2 steel	

Note: For further information about earthing conductors, see the latest edition of BS 7430 (*Code of practice for protective earthing of electrical installations*).

3.14 Protective measures

 To avoid unauthorised changes being made, protective measures shall only be applied where the installation is under the supervision of skilled or instructed persons and as stated in BS EN 61140:

> **Note: BS EN 61140 Requirement**
>
> *"Hazardous live parts shall not be accessible and accessible conductive parts shall not be hazardous live, when in use without a fault or in a single fault condition."*

To meet this requirement, basic protective measures are necessary and these will either consist of:

- a combination of basic protection and an independent provision for fault protection; or
- an enhanced protective provision which provides both basic and fault protection;

and one or more of the following protective measures needs to be applied:

3.14.1 Automatic disconnection of supply

Automatic disconnection of supply is a protective measure, in which fault protection is provided by protective earthing, protective equipotential bonding and automatic disconnection in case of a fault.

 Where an alternative system for the automatic disconnection of supply is available, an IT system shall **not** be used.

3.14.2 Electrical separation for the supply to one item of current-using equipment

Electrical separation is a protective measure that provides:

- protection to live parts via enclosures or barriers; and
- fault protection by the separation of one circuit from other circuits and from Earth.

3.14.3 Equipotential bonding

Main protective bonding conductors are required to connect extraneous-conductive-parts to the main earthing terminal in each installation, including the following:

- central heating and air conditioning systems;
- exposed metallic structural parts of the building;
- gas installation pipes;
- water installation pipes;
- other installation pipework and ducting.

 Where an installation serves more than one building, the above requirement shall be applied to each building.

3.14.4 Extra-low-voltage (SELV and PELV)

A SELV (separated extra-low-voltage) circuit is a secondary circuit which has no direct connection to the primary power (a.c. mains) and derives its power via a transformer, converter or equivalent isolation device. It is designed and protected so that under normal and single fault conditions its voltages do not exceed a safe value.

 Typical examples for a SELV circuit are decorative outdoor lighting, a Class III battery charger, fed from a Class II power supply, and modern cordless hand tools.

A PELV (protective extra-low-voltage) circuit is an electrical system in which the voltage cannot exceed the extra-low voltage (ELV) under normal and single-fault conditions, except Earth faults in other circuits.

 A typical example for a PELV circuit is a computer with a Class I power supply.

3.14.5 Reinforced insulation

Reinforced insulation is an improved basic insulation with such mechanical and electrical properties that provide the same degree of protection against electrical shock as double insulation.

The insulation provides a physical separation between the solder tracks or joints, cores, windings, etc., provides protection against electric shocks and (by means of a single layer), and offers sufficient clearance from dangerous voltages and user-touchable circuits.

3.15 Protective devices

Unless backup protection is provided, the rated breaking capacity of a protective device shall **not** be less than the maximum prospective short-circuit or Earth fault current at the point of installation.

3.15.1 Protective earthing

A circuit protective conductor shall be run to (and terminated at) each point in the wiring and at each accessory.

All exposed-conductive-parts:

- **must** be connected to a protective conductor;
- shall be connected to the same earthing system, either individually, in groups or collectively.

3.15.2 Protective multiple earthing

Protective multiple earthing (PME) (see Figure 3.17) is effectively a high-integrity TN-C-S system and is the most common form of earthing provided by electricity supply companies at new installations. Basically, PME utilises a single conductor for the neutral and earthing functions within their network and provides a PME Earth terminal at the customer's installation.

The great virtue of the PME system is that neutral is bonded to Earth so that a phase to Earth fault is automatically a phase to neutral fault. The Earth-fault loop impedance will then be low, resulting in a high value of fault current which will operate the protective device quickly.

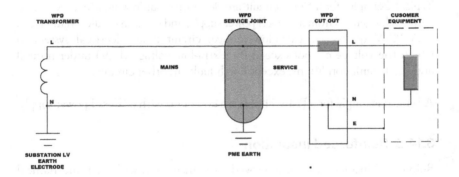

Figure 3.17 Protective multiple earthing. (Courtesy HEC.)

 The Electricity Supply Regulations forbid the use of PME supplies to feed caravans and caravan sites.

3.15.3 Protective and neutral (PEN) conductors

The neutral conductor is also used as a protective conductor, technically referred to as a PEN (protective earth and neutral) conductor, and is connected to a number of Earth electrodes in the installation.

These connections may be to the terminals (or a bar) of the protective earthing conductor **and** the neutral conductor – without any metallic connection (except for the earthing connection) with the distributor's network. PEN connectors may only be used within an installation where the installation is supplied by a privately owned transformer or converter. (See Figure 3.18).

 Note: In Great Britain, Regulation 8(4) of the Electricity Safety, Quality and Continuity Regulations 2002 **prohibits** the use of PEN conductors in consumers' installations.

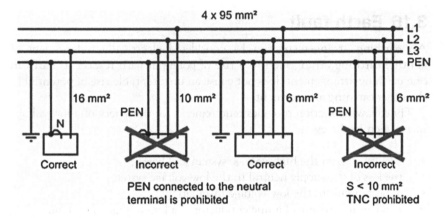

Figure 3.18 PEN conductors in consumers' installations. (Courtesy HEC.)

 PEN conductors shall not be used in medical locations and medical buildings downstream of the main distribution board.

In an IT system, the neutral conductor shall not be distributed unless:

- overcurrent detection is provided for the neutral conductor of every circuit; or
- the neutral conductor is effectively protected against short-circuit by a protective device installed on the supply side; or
- the circuit is protected by an RCD with a rated residual operating current not exceeding 0.2 times the current-carrying capacity of the corresponding neutral conductor.

3.15.4 Protective switches

Switches, circuit-breakers, (except where linked) or fuses shall only be inserted in an earthed neutral conductor.

Any linked switch or linked circuit-breaker that is used with an earthed neutral conductor shall be capable of breaking **all** of the related line conductors.

Where an installation is supplied from more than one source of energy with different earthing means, a switch may be inserted in the connection between the neutral point and the means of earthing, **provided** that the device:

- is a linked switch arranged to disconnect and connect the earthing conductor for the appropriate source, at substantially the same time as the related live conductors; or
- an interlinked switching device is inserted in the related live conductors.

3.16 Earth fault

An earthing arrangement can be considered electrically independent of another earthing arrangement if a rise of potential with respect to Earth in one earthing arrangement does not cause an unacceptable rise of potential in the other earthing arrangement.

The following section provides requirements for the safety of a low-voltage installation in the event of:

- a fault between the high-voltage system and Earth;
- the loss of the supply neutral in the low-voltage system;
- a short-circuit in the low-voltage installation;
- accidental earthing of a line conductor of a low-voltage IT system.

The change of power frequency stress voltages on the low-voltage equipment due to an Earth fault in a high-voltage system should not exceed the requirements given in Table 3.3.

Table 3.3 Permissible power frequency stress voltage

Duration of an Earth fault in the high-voltage system	Accepted power frequency stress voltage on low-voltage equipment
>5 s	Uo + 250 V
<5 s	Uo + 1200 V

3.16.1 Earth fault protection

Earth fault protection may be omitted for:

- unearthed street furniture supplied from an overhead line provided that it is out of arm's reach;
- exposed-conductive-parts which, owing to their size, cannot be gripped or come into significant contact with the human body.

In all other cases:

- fault protection shall be achieved by automatic disconnection of the power supply by means of an overcurrent protective device in each line conductor or by an RCD;

 Note: if an RCD is used, the product of the residual operating current (in amperes) and the Earth fault loop resistance (in ohms) shall not exceed 50 V.

- all exposed-conductive-parts of the reduced low-voltage system shall be connected to Earth;
- the Earth fault loop impedance at every point of utilisation, including socket outlets, shall be such that the disconnection time does not exceed 5 s;
- live parts of the separated circuit shall not be connected at any point to another circuit or to Earth or to a protective conductor;
- flexible cables shall be visible throughout their entire length where they are liable to mechanical damage;
- exposed-conductive-parts of a separated circuit shall be connected to the protective conductor, to exposed-conductive-parts of other circuits, or to Earth.

3.16.2 Earth fault loop impendence

Earth fault loop impedance (Zs) is the impedance of the intended path of an Earth fault current (i.e. the Earth fault loop) starting and ending at the point of the fault to Earth.

As shown in Figure 3.19, the Earth fault loop starts at the point of the fault and comprises:

- the circuit protective conductor (CPC);
- the consumer's main earthing terminal (MET) and earthing conductor;
- (for TT - see Figure 3.19 - and IT systems) the Earth return path, or (for TN systems) the metallic return path;
- the path through the earthed neutral point of the transformer;
- the transformer winding;
- the line (phase) conductor from the transformer to the point of fault.

 It is recommended that radial wiring patterns are used to avoid 'Earth loops' that may cause electromagnetic interference – particularly in medical locations.

3.17 Insulation monitoring devices for IT systems

An IMD (insulation monitoring device) is designed to indicate when an Earth fault is detected and to allow that fault to be located and eliminated, as soon as possible, in order to restore normal operating conditions.

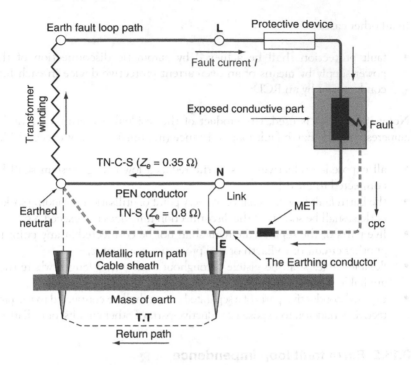

Figure 3.19 Example of a TT Earth fault loop impendence path. (Based on a diagram by Pathos Electrician.)

The IMD shall:

- be permanently connected to an IT system in order to continuously monitor the insulation resistance of the complete system;
- be connected between Earth and a live conductor of the monitored equipment;
- be a switching device that disconnects all live poles and which is only energised in the event of an emergency;
- have its Earth or functional Earth terminal connected to the main Earth terminal of the installation.

 In some d.c. IT two-conductor installations, a passive IMD that does not inject current into the system may be used, **provided** that:

- the insulation of all live distributed conductors is monitored; and
- all exposed-conductive-parts of the installation are interconnected; and
- circuit conductors are selected and installed so as to reduce the risk of an earth fault to a minimum.

 An IMD is **not** intended to provide protection against electric shock.

3.18 RCDs

An RCD (residual current device) is a life-saving device which is designed to prevent you from getting a fatal electric shock if you touch something live, such as a bare wire, as a current of around 30 mA through the human body is potentially sufficient to cause a cardiac arrest or serious harm if it persists for more than a small fraction of a second!

Any difference between the currents in these conductors indicates leakage current, which presents a shock hazard, and the design of the RCD quickly and automatically disconnects a circuit when it detects that the electric current is not balanced between the circuit's supply and return conductors.

RCDs can also provide some protection against electrical fires.

The RCD can be opened and closed manually to switch normal load currents, and it opens automatically when an earth fault current flows which is more than 50 per cent of the rated tripping current.

When a fault current flows there is a difference between the load and return currents which generates a resultant flux in the toroid which induces a current in the detecting winding – which opens the main contacts of the RCD.

3.18.1 The use of RCDs in different types of supplies

- In a TN system, it is the characteristics of a protective device (for a particular piece of equipment in a certain part of the installation) do not satisfy the requirements, then that part may be protected by an RCD. (See Figure 3.20).
- For a TN-S system where the neutral is not isolated, RCDs shall be positioned so as to avoid incorrect operation owing to the existence of any parallel neutral–earth path.
- In a TT system, one or more of the following types of protective device shall be used:

  o   an RCD (the preferred option);
  o   an overcurrent protective device.

- In an IT system (where protection is provided by means of one or more RCDs (with a residual operating current rated not more than 30 mA) and disconnection following a first fault is not envisaged) the non-operating residual current of the device shall be at least equal to the current which circulates on the first fault to Earth.

 In an IT system, an RCD may **not** operate unless one of the Earth faults is on a part of the system that is on the supply side of the device. If a medical IT system is used, additional protection by means of an RCD need not be used.

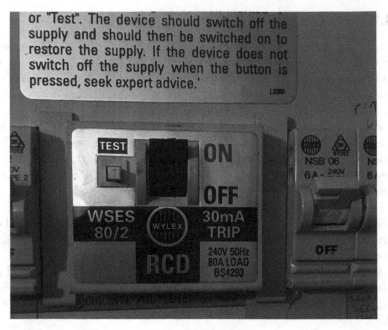

or "Test". The device should switch off the supply and should then be switched on to restore the supply. If the device does not switch off the supply when the button is pressed, seek expert advice.'

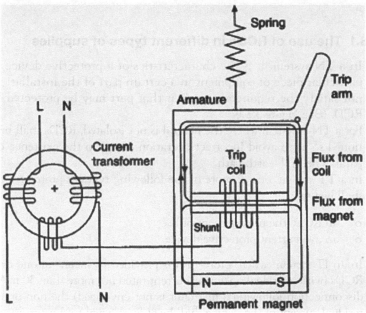

Figure 3.20 Residual current device (RCD). (Courtesy Riddiford.)

Note: Where RCDs are also used for protection against fire, the conditions for protection by automatic disconnection of the supply shall be verified by:

- measuring of the resistance of the Earth electrode for exposed-conductive-parts of the installation;
- verifying of the characteristics and/or effectiveness of the associated protective device.

 An RCD shall **not** be used in a TN-C system.

3.18.2 Power supply

An RCD which is powered from an independent auxiliary source and which does not operate automatically in the case of failure of the auxiliary source shall **only** be used if:

- fault protection is maintained even in the case of failure of the auxiliary source; or
- the device is incorporated in an installation intended to be supervised by an instructed person or a skilled person, and inspected and tested by a competent person.

The generating set shall be connected so that any part of the installation that is protected by an RCD remains effective for every intended combination of sources of supply.

In a TN, TT or IT system, one or more RCDs with a rated residual operating current of not more than 30 mA shall be installed to protect every circuit.

Where RCDs are also used for protection against fire, the conditions for protection by automatic disconnection of the supply shall be verified.

Where RCDs are required for additional protection, the effectiveness of automatic disconnection of supply by RCDs shall be verified using suitable test equipment according to BS EN 61557-6 to confirm that the relevant requirements are met.

3.18.2.1 TN systems

In a TN system:

- an RCD may be used as a protective device for fault protection provided that it incorporates an overcurrent protective device;
- if a protective device in part of an installation does not completely satisfy the requirements, that part may be protected by an RCD;
- the neutral conductor shall be protected against short-circuit current;
- the integrity of the earthing of the installation depends on the reliable and effective connection of the PEN or PE conductors to Earth;
- wiring systems (other than mineral insulated cables, busbar trunking or powertrack systems) shall be protected against insulation faults by an RCD.

When an RCD is used, the maximum values of Earth fault loop impedance (see Table 3.4) may be applied for non-delayed RCDs in accordance with BS EN 61008-1 and BS EN 61009-1.

Table 3.4 Maximum Earth fault loop impedance (Zs) for non-delayed and time-delayed RCDs

Rated residual operating current (mA)	Maximum Earth fault loop impedance 4 (ohms)			
	50 V < U0. 120 V	120 V < U0 230 V	230 V < U05.400 V	Uo > 400 V
30	1667	1667	1533	1667
100	500	500	460	500
300	167	167	153	167
500	100	100	92	100

 If an RCD is used in a TN-C-S system, a PEN conductor shall **not** be used on the load side.

3.18.2.2 TT system

 Where an RCD is used for earth fault protection, the circuit should also incorporate an overcurrent protective device.

Except for mineral insulated cables, busbar trunking or powertrack systems, a wiring system shall be protected against insulation faults in a TT system by an RCD.

If an installation which is part of a TT system is protected by a single RCD it is placed at the origin of the installation.

3.18.2.3 IT system

In an IT system, the neutral conductor shall **not** be distributed unless:

- the circuit is protected by an RCD with a rated residual operating current not exceeding 0.2 times the current-carrying capacity of the corresponding neutral conductor; or
- overcurrent detection is provided for the neutral conductor of every circuit; or
- the neutral conductor is effectively protected against short-circuit by a protective device installed on the supply side.

 In an IT system without a neutral conductor, it is permitted to omit the overload protective device in one of the line conductors **if** an RCD is installed in each circuit.

3.18.2.4 Warning notices: periodic inspection and testing

Where an installation incorporates an RCD, a notice shall be fixed in a prominent position at or near the origin of the installation and shall read as shown in Figure 3.21.

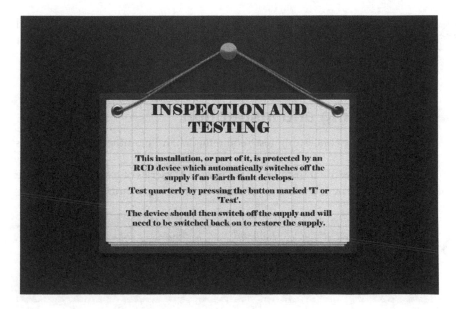

Figure 3.21 Warning notice – RCD protection.

3.19 Residual current monitors

 A Residual Current Monitor (RCM) is **not** intended to provide protection against electric shock.

An RCM permanently monitors any leakage current in the downstream installation or part of it and is intended to provide protection from electric shock.

 Note: Although RCM protection reduces the risk of death or injury from electric shock, it does not reduce the need to protect yourself, your family and your property by having your electrical wiring checked at least every ten years.

RCMs are used to monitor earthed TN and TT systems for fault currents or residual currents. They reliably detect deteriorations of the insulation level at an early stage and are (as long as the fault persists) used to send an audible and/or visual signal to the user before reaching the shutdown threshold of the RCD.

If an RCM is used in an IT system, it is normal to use a directionally discriminating RCM in order to avoid unnecessary signalling of leakage current when high leakage capacitances exist downstream from the point of installation of the RCM.

Any insulation or insulating arrangement of extraneous-conductive-parts shall not pass a leakage current exceeding 1 mA in normal conditions of use.

After the occurrence of a first fault and in the event of a second fault occurring on a different live conductor, automatic disconnection of supply shall occur.

Note: Where an RCD is installed upstream of the RCM, it is recommended that the RCM has a rated residual operating current not exceeding a third of that of the RCD.

For further information about RCMs, see BS EN 62020.

3.20 Testing and inspection

For more details concerning test requirements and maintenance inspections see Chapter 10.

Before any electrical installation is energised, a series of tests will need to be completed and the following are among the most important:

No addition or alteration, temporary or permanent, shall be made to an existing installation unless it has been ascertained that the earthing and bonding arrangements, used as a protective measure for the safety of the addition or alteration, are adequate.

3.20.1 Cables

A cable passing through a joist within a floor or ceiling construction or through a ceiling support (e.g. under floorboards), shall:

- include in an earthed metallic covering; or
- be enclosed in an earthed conduit; or
- be enclosed in earthed trunking or ducting; or
- be mechanically protected against damage sufficient to prevent penetration of the cable by nails, screws etc.; or
- be at least 50 mm measured vertically from the top, or bottom as appropriate, of the joist or batten.

A cable concealed in a wall or partition at a depth of less than 50 mm from a surface of the wall or partition shall (in addition to the above requirements):

- be installed in an area within 150 mm from the top of the wall or partition or within 150 mm of an angle formed by two adjoining walls or partitions.

If the cables of an installation that is not intended to be under the supervision of a skilled or instructed person are concealed in a wall or partition (the internal construction of which includes metallic parts, other than metallic fixings such as nails, screws and the like) then it shall:

- incorporate an earthed metallic covering; or
- be enclosed in earthed conduit: or
- be enclosed in earthed trunking or ducting; or
- be mechanically protected sufficiently to avoid damage to the cable during construction of the wall or partition and during installation of the cable; or
- be provided with additional protection by means of an RCD.

 Note:

1. If a cable is not put in in a conduit or duct but is directly buried in the ground, then it **must** include a suitable protective conductor such as earthed armour or metal sheath or both, together with a warning tape.

3.20.2 Conductors

The continuity of conductors and connections to unprotected or superfluous conductive parts must be confirmed by measuring the resistance of:

- protective conductors;
- protective bonding conductors;
- live conductors (in the case of ring circuits).

3.20.3 Earth electrodes

Where the earthing system incorporates an earth electrode as part of the installation, the electrode resistance to Earth shall be measured.

3.20.4 Earth fault loop impedance

If protective measures are used which require knowledge of Earth fault loop impedance, the relevant impedances shall be measured.

 Note: Further information on the measurement of Earth fault loop impedance can be found in Appendix 14 to BS 7671:2018.

3.20.5 Electrical separation

Electrical separation of individual circuits is designed to prevent shock currents energising exposed conductive parts in the basic circuit insulation and is a protective measure in which:

- basic protection is provided by simple insulation of live parts or by barriers or enclosures; and

- fault protection is provided by ensuring that the electrical circuit is separated from other circuits and from Earth.

Electrical separation may only supply one item of current-using equipment from one unearthed source with simple separation. If there are two or more items from the same electrical source, then a warning notice (see example at Figure 3.22) must be fixed – in a prominent position – beside all points of access.

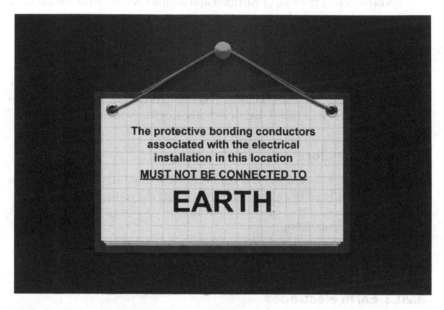

Figure 3.22 Warning notice – protective bonding conductors.

In accordance with BS 951, a durable label using words similar to that shown in Figure 3.23:
   shall be permanently fixed, in a visible position, at or near:

- the point of connection of every earthing conductor to an earth electrode; and
- the point of connection of every bonding conductor to an extraneous-conductive-part; and
- the main earth terminal, if it is separated from the main switchgear.

3.20.6 Electrical services

A voltage Band I circuit shall **not** be contained in the same wiring system as a Band II circuit unless (for a multi-core cable where they are separated by an earthed metal screen) the cores of the Band I circuit are detached from the cores of the Band II circuit by an earthed metal screen of equivalent current-carrying capacity to that of the largest core of the Band II circuit.

Figure 3.23 Warning notice – earthing and bonding.

The neutral (star) point of the secondary windings of three-phase transformers and generators (or the midpoint of the secondary windings of single-phase transformers and generators) shall be connected to Earth.

  The connection of live parts of the generator with Earth may affect the fault protection that is provided by the equipment.

3.20.7 Insulation resistance

The insulation resistance shall be measured:

* between live conductors; and
* between live conductors and the protective conductor connected to the earthing arrangement.

The insulation resistance measured with the test voltages shown in Table 3.5 shall be considered satisfactory if the main switchboard and each distribution circuit tested separately (with all of its final circuits connected but with current-using equipment disconnected) has an insulation resistance not less than the appropriate value given in Table 3.5.

  More stringent requirements are applicable for the wiring of fire alarm systems in buildings (see BS 5839-1).

Table 3.5 Minimum values of insulation resistance

Circuit nominal voltage (V)	Test voltage d.c. (V)	Minimum insulation Resistance (MΩ)
SELV and PELV	250	0.5
Up to and including 500 V, with the exception of the above systems	500	1.0
Above 500 V	1000	1.0

Note: Where the circuit includes electronic devices, which are likely to influence the results or be damaged, only a measurement between the live conductors connected together and the earthing arrangement shall be made.

3.20.8 Isolation

For reasons of safety, it is important that **every** circuit is capable of being isolated from every source of electric energy and from each live supply conductor.

Note: In a TN-S or TN-C-S system, it is **not** necessary to isolate or switch the neutral conductor where it is regarded as being reliably connected to Earth by suitably low impedance, unless an interlocking arrangement is provided to ensure that all the circuits concerned are isolated and:

- if an installation, item of equipment or enclosure contains live parts that are connected to more than one supply, a warning notice must be placed to prevent any person inadvertently gaining access to these live parts; and
- where necessary, suitable means shall be provided to discharge stored electrical energy;
- if an isolating device for a particular circuit is remote from the equipment to be isolated, the means of isolation should always be secured in the open position.

Notes:

1. If the 'isolation' is by means of a lock or removable handle, the key or handle shall be non-interchangeable with any other used for a similar purpose within the premises.
2. If a switch is provided for this purpose:

   o   it shall be capable of cutting off the full load current of the relevant part of the installation; but

   o   if used as a device for switching off for mechanical maintenance, the switch need not necessarily interrupt the neutral conductor.

3.20.9 Insulation resistance/impedance of floors and walls

In a non-conducting location, at least three measurements shall be made in the same location. One must be within 1 m of any accessible extraneous-conductive-part in the location, whilst the other two measurements can be made further away.

 Note: Further information on measurement of the insulation resistance and impedance of floors and walls can be found in Appendix 13 of BS 7671:2018.

3.20.10 Low voltage generating sets

When the generator is being used as a switched alternative to a TN system, the earthed point of the public electricity distribution system cannot be relied on for protection of the system and an alternative suitable means of earthing **shall** be provided.

Potential short-circuit and earth fault currents need to be assessed for each source of supply which can operate independently of other sources.

3.20.11 Polarity

Where necessary, a test of polarity of the supply shall be made at the origin of the installation to verify that:

- every fuse, single pole control and protective device is connected to the line conductor – only; and (except for E 14 and E27 lamp holders according to BS EN 60238)
- wiring has been correctly connected to the socket outlets and similar accessories;
- all wiring has been correctly terminated throughout the installation.

3.20.12 Prospective fault current

The prospective short-circuit current and prospective Earth fault current shall be capable of being measured at the origin, as well as at other relevant points in the installation.

3.20.13 Protection by electrical separation

The separation of the live parts from those of other circuits and from Earth shall be confirmed by measuring the insulation resistance. The values obtained shall be in accordance with Table 3.5 Section 3.20.7).

3.21 Special locations and installations

3.21.1 Monitoring devices in an IT system

In IT systems, continuous insulation monitoring devices shall be provided which give an auditable and visual indication in the event of a first fault which shall continue as long as the fault persists.

The following monitoring devices and protective devices may be used:

- residual current devices (RCDs);
- residual current monitoring devices (RCMs);
- insulation monitoring devices (IMDs);
- insulation fault location systems;
- overcurrent protective devices.

Consideration should be given to the possibility that, if a line conductor of an IT system is earthed accidentally, the insulation (or components) that are normally rated for the voltage between line and neutral conductors can be temporarily stressed with the line-to-line voltage.

The object of this particular regulation is to provide requirements for the safety of a low-voltage installation in the event of:

- a fault between the high-voltage system and Earth in the transformer sub-station that supplies the low-voltage installation;
- loss of the supply neutral in the low-voltage system;
- short-circuit between a line conductor and neutral in the low-voltage installation;
- accidental earthing of a line conductor of a low-voltage IT system.

In locations where there is a strong risk of fire due to the nature of processed or stored materials, the wiring system of an IT system (except for mineral insulated cables, busbar trunking systems or powertrack systems) shall be protected against insulation faults, by an IMD with audible and visual signals.

3.21.2 Agricultural and horticultural premises

 In agricultural and horticultural premises, a TN-C system shall **not** be used.

In locations intended for livestock, normal protective measures used in other locations (like placing obstacles out of reach) shall not be used. Instead, supplementary bonding shall connect all exposed-conductive-parts and extraneous conductive-parts that can be touched by livestock.

In circuits (whatever the type of earthing system is used) the following shall be provided:

- in final circuits supplying socket outlets with rated current not exceeding 32 A, an RCD with an operating time not exceeding 40 ms;

- in final circuits supplying socket outlets with rated current more than 32 A, an RCD with a rated residual operating current not exceeding 100 mA;
- in all other circuits, RCDs with a rated residual operating current not exceeding 300 mA.

Precautions shall also be taken to ensure that:

- persons or livestock cannot unintentionally touch live parts;
- all live parts are inside barriers or enclosures and suitable warning signs will be made available;
- barriers and/or enclosures are secured by a bolt or a key.

3.21.3 Amusement parks and circuses etc.

A PME earthing facility shall **not** be used.

3.21.4 Electrical installations in caravans, motor caravans

 The use of a PME earthing facility for earthing a caravan is **prohibited** by the Electricity Safety, Quality and Continuity Regulations (ESQCR).

The nominal supply voltages for caravans shall (in accordance with BS EN 60038) be as follows:

- a.c. supplies: 230 V single-phase or 400 V three-phase;
- d.c. supplies: 48 V.

BS 7671:2018 also states that:

- structural metal parts within the caravan shall be connected via bonding conductors to the main earthing terminal within the caravan;
- electrical separation shall not be used except for the shaver socket.

3.21.5 Electrical installations at construction and demolition sites

A PME earthing facility shall **not** be used for earthing an installation at a construction or demolition site unless all extraneous-conductive-parts are reliably connected to the main earthing terminal. If a functional earth is required for certain equipment (e.g. measuring and control equipment):

- supplementary equipotential bonding shall be provided between all exposed-conductive- and extraneous-conductive-parts inside a conducting location that has restricted movement and the functional earth;
- the unearthed source shall have simple separation and shall be situated outside any conducting location with restricted movement, unless the source is part of the fixed installation within that particular conducting location.

3.21.6 Electrode water heaters and boilers

When an electrode water heater or boiler is directly connected:

- to a supply exceeding low voltage, the installation shall include an RCD;
- to a three-phase low-voltage supply, it shall be connected to the neutral of the supply as well as to the earthing conductor; and
- is not piped to a water supply or is in physical contact with any earthed metal, a fuse in the line conductor may be substituted for the circuit-breaker and the shell of the electrode water heater or electrode boiler need not be connected to the neutral of the supply.

If the supply to an electrode water heater or electrode boiler is single-phase and one electrode is connected to a neutral conductor earthed by the distributor, then the shell of the electrode water heater or electrode boiler shall be connected to the neutral of the supply as well as to the earthing conductor.

3.21.7 Exhibitions, shows and stands

The following requirements are intended for temporary electrical installations at exhibitions, shows, stands and mobile (or portable) displays and equipment:

- Any cable supplying these temporary structures shall be protected by an RCD.
- All accessible structural metallic parts shall be connected through the main protective bonding conductors to the main earthing terminal within the unit.
- All accessible socket-outlet circuits (other than those for emergency lighting) shall also be protected by and RCD.
- All live parts shall be covered by insulation, and located inside barriers and shelters.
- PME earthing shall not be used unless the installation is continuously monitored and a suitable means of Earth has been confirmed before the connection.
- The protective measures of non-conducting location and Earth-free local equipotential bonding are not permitted.

3.21.8 Medical locations

HTM 06-01(a Government guidance document entitled *Electrical services supply and distribution*) provides the legal requirements, design applications, and operation and maintenance guidelines for the electrical infrastructure within healthcare premises.

Although the requirements of BS 7671:2018, Section 710 mainly refer to patient healthcare facilities (such as hospitals, private clinics, medical and dental practices, healthcare centres and dedicated medical rooms in the work-place), they equally apply to electrical installations in locations designed for medical research and (where applicable) to veterinary clinics. They do not, however, apply to the medical electrical equipment itself, as this is fully cov-ered in ISO 13485!

In patient healthcare facilities, the risk to patients through reduced body resistance is enhanced and it is extremely important that the following rules and regulations are strictly adhered to.

3.21.8.1 Electromagnetic disturbances

In the following medical locations special considerations have to be made with respect to the possibility of electromagnetic interference (EMI) and electro-magnetic compatibility (EMC).

- Group 1 medical locations (where discontinuity of the electrical supply does not represent a threat to the safety of the patient) where current-us-ing equipment or applied parts of it are intended to be used either exter-nally or invasively to any part of the body;
- Group 2 medical locations where applied parts are intended to be used, and where discontinuity (failure) of the supply can cause danger to life applied;
- Group 2 medical locations using SELV and/or PELV circuits shall not exceed 25 V a.c. rms or 60 V ripple-free d.c., and protection by basic insu-lation of live parts or by barriers or enclosures shall be provided;
- Group 2 medical locations where PELV is used, exposed-conductive-parts of equipment (e.g. operating theatre luminaires) shall be connected to the circuit protective conductor.

3.21.8.2 Inspection and testing

3.21.8.2.1   Initial verification

In addition to the requirements of Chapter 64 of the Wiring Regulations and HTM 06-01 (Part A) the following tests **shall** be carried out, prior to commis-sioning, after alteration or repairs and before re-commissioning:

- the functioning of the residual current monitor (RCM) and the insulation monitoring device (IMD) shall be verified;
- measurements of leakage current from the IT transformers of the output circuit and enclosure in no-load condition must be taken;
- measurements must be taken to verify that the resistance of the supple-mentary equipotential bonding is within stipulated limits.

3.21.8.2.2 Periodic inspection and testing

In addition to the requirements of BS 7671:2018 (Chapter 62) periodic inspection and testing should be carried out in accordance with Health Technical Memorandum (HTM) 06-01 (Part B) and local health authority requirements as follows, and at the given intervals:

- **Annually** – complete functional tests of all IMDs associated with the medical IT system, including for insulation failure, transformer high temperature, overload, discontinuity and the acoustic/visual alarms linked to them.
- **Annually** – complete measurements to verify that the resistance of the supplementary equipotential bonding is within the stipulated limits.
- **Every 3 years** – complete measurements of leakage current of the output circuit and of the enclosure of the medical IT transformers in the no-load condition.

The dates and results of each verification **shall** be recorded on an electrical installation report.

In an IT system, the omission of devices for protection against overload is permitted for circuits supplying current-using equipment where unexpected disconnection of the circuit could cause danger or damage (e.g. a circuit supplying medical equipment used for life support in specific medical locations).

3.21.8.3 PEN conductors

PEN conductors **shall not** be used in medical locations and medical buildings downstream of the main distribution board.

3.21.8.4 Medical facility Earth faults

In the event of a first fault to Earth, a total loss of supply in medical group 2 locations shall be prevented.

3.21.8.5 RCDs

Chapter 710 of BS 7671:2018 lists the following requirements and recommendations for the use of RCDs in medical locations:

- Care shall be taken to ensure that simultaneous use of large numbers of equipment connected to the same circuit cannot cause unwanted tripping of the RCD.
- Only type A (according to BS EN 61008 and BS EN 61009) or type B (according to IEC 62423) RCDs shall be used in Group 1 and Group 2 medical locations.
- Type a.c. RCDs shall **not** be used.

- In TN systems, additional protection by RCDs in Group 1 and Group 2 final circuits shall be according to their rated current.
- RCDs shall be used and in TN-S systems and the insulation level of all live conductors shall be monitored.
- In Group 1 and Group 2 medical locations that include a TT system, RCDs shall be used.

 Where a medical IT subsystem is used in a Group 1 location, additional protection via an RCD is **not** required.

- in Group 2 medical locations (except for a medical IT system), protection by automatic disconnection of supply by means of RCDs shall only be used on circuits for:

  o the supply of movements of fixed operating tables; or
  o X-ray units; or
  o large equipment with a rated power greater than 5 kVA.

They should not be used for:

- final circuits supplying medical electrical equipment and systems intended for life support;
- surgical applications; and
- 'other' electrical equipment located in, or that may be moved into, the 'patient environment'.

 Note: For each circuit that is protected by an RCD, the possibility of the RCD's unwanted tripping due to excessive protective conductor currents produced by equipment in normal operation shall be considered.

- in Group 2 medical locations, where PELV is used, exposed-conductive-parts of equipment items (e.g. operating theatre luminaires) shall be connected to the circuit protective conductor.

3.21.8.6 Socket outlets

It is a mandatory requirement that socket-outlet circuits in the medical IT system Group 2 for medical locations:

- intended to supply medical electrical equipment shall be unswitched;
- shall be coloured **blue** and clearly and permanently marked 'Medical equipment only'.

In addition, at each patient's place of treatment (e.g. bedheads):

- each socket outlet shall be supplied by an individually protected circuit; or
- several socket outlets shall be separately supplied by a minimum of two circuits.

3.21.8.7 Equipotential bonding busbar

The equipotential bonding busbar shall be located in or near the medical location using a protective conductor.

3.21.8.8 Supplementary equipotential bonding

In Group 1 and Group 2 medical locations, supplementary equipotential bonding shall be installed for the parts which are located in the patient environment:

* protective conductors;
* connection to conductive floor grids, if installed;
* extraneous-conductive-parts;
* metal screens of isolating transformers;
* screening against electrical interference fields, if installed.

 Note: Supplementary equipotential bonding connection points for the connection of medical electrical equipment shall be provided in each medical location, as follows:

* Group 1: one per patient location.
* Group 2: a minimum of 25% of the total number of individual medical IT socket outlets provided per patient location.

Unless they are intended to be isolated from Earth, fixed conductive non-electrical patient supports such as operating theatre tables, physiotherapy couches and dental chairs should be connected to the equipotential bonding conductor.

The equipotential bonding busbar shall be located in or near the medical location and all connections shall be accessible, labelled, clearly visible, and capable of being easily disconnected individually.

3.21.8.9 Supplies

In medical locations at least two different sources of supply shall be provided; one for the electrical supply system and one for safety services.

Automatic changeover facilities shall comply with BS EN 60947-6-1 and the distribution system shall be designed and installed so that in the case of a mains failure it facilitates the automatic changeover from the main distribution network to an electric safety source within:

* 0.5 s for luminaires in an operating theatre, light sources for essential medical electrical (ME) (e.g. endoscopes, monitors etc.) and life-support ME equipment;

- 15 s for safety lighting and other services (e.g. firefighters' lifts, vigilance systems for smoke extraction, paging communication systems and fire detection, etc.);
- more than 15 s for the maintenance of healthcare installation (e.g. sterilising equipment etc.).

3.21.8.10 Transformers

Transformers **shall** be installed in close proximity to a medical location and with the following additional requirements:

- they shall comply with the requirements of BS EN 61439;
- the leakage current of the output winding to Earth and the leakage current of the enclosure shall not exceed 0.5 mA;
- at least one single-phase transformer per room (or functional group of rooms) shall be used to form the IT systems for mobile and fixed equipment and the rated output shall be not less than 0.5 kVA and shall not exceed 10 kVA;
- if several transformers are required to supply equipment in one room, they shall not be connected in parallel;
- if the supply of three-phase loads via an IT system is also required, a separate three-phase transformer shall be provided for this purpose.

 Capacitors **shall not** be used in transformers for medical IT systems.

3.21.9 Mobile or transportable units

3.21.9.1 Protective measures

Author's hint

For the purpose of this particular requirement, the term 'unit' can mean a vehicle and/or mobile transportable structure in which all or part of an electric structure is contained.

The protective measures of:

- obstacles and placing out of reach, are not permitted;
- non-conducting location, is not permitted;
- Earth-free local equipotential bonding, is not recommended.

 Where an alternative system for the automatic disconnection of supply is available, an IT system **shall not** be used.

3.21.9.2 Protective equipotential bonding

Accessible parts of the unit will be connected through the main protective bonding conductors to the main Earth terminal including the following:

- central heating and air-conditioning systems;
- exposed metallic structural parts of the building;
- gas installation pipes;
- water installation pipes;
- other installation pipework and ducting.

3.21.9.3 TN system

A PME system shall **not** be used as a means of earthing, except:

- where the installation is continuously under the supervision of a skilled or instructed person; and
- the suitability and effectiveness of the means of earthing has been confirmed before the connection is made.

3.21.9.4 IT system

An IT system can be provided by either:

- an isolating transformer or a low voltage generating set; or
- an installation fault-location system; or
- a transformer providing simple separation via an RCD; or
- an earth electrode that has been installed so that it provides automatic disconnection of the supply in case of failure in the transformer. In mobile (or transportable units) additional protection by an RCD shall be provided for every socket outlet intended to supply current-using equipment outside the unit, with the exception of socket outlets which are supplied from circuits with protection by:
    o   SELV; or
    o   PELV; or
    o   electrical separation.

3.21.10 Outdoor lighting installations

BS 7671:2018 Section 714 concerns all outdoor lighting installations comprising one or more luminaires, a wiring system and accessories, and the relevant highway power supplies and street furniture.

3.21.10.1 Protective measure: Automatic disconnection of supply

Where automatic disconnection of supply is used, all live parts of electrical equipment shall incorporate basic protection by either insulation, barriers or enclosures.

The earthing conductor of a street electrical fixture shall have a minimum copper equivalent cross-sectional area not less than that of the supply neutral conductor at that point or not less than 6 mm², whichever is the smaller.

3.21.10.2 Provisions for basic protection

- Enclosures for live parts shall only be accessible with a key or a tool.
- A door giving access to electrical equipment and located less than 2.50 m above ground level shall be locked with a key or shall require the use of a tool for access.
- Access to the light source of a luminaire, which is positioned less than 2.80 m above ground level, shall only be possible after removing a barrier or an enclosure requiring the use of a tool.

3.21.10.3 Additional protection

Lighting in places such as telephone kiosks, bus shelters and advertising panels etc. shall be provided with an RCD for additional protection.

3.21.11 Solar photovoltaic (PV) power supply systems

Section 712 of BS 7671:2018 concerns PV power supply systems (including those with a.c. modules) and the following rules apply:

- Earthing of one of the live conductors of the d.c. side of a solar photovoltaic is permitted, if there is at least simple separation between the a.c. side and the d.c. side.
- Any connections with Earth on the d.c. side should be electrically connected so as to avoid corrosion (see BS EN 13636 and BS EN 15112).

 Earth-free local equipotential bonding **shall not** be used on the d.c. side.

- PV string cables, array cables and d.c. main cables that are selected and erected must minimise the risk of Earth faults and short-circuits.

The protective measures of non-conducting location and Earth-free local equipotential bonding are **not** permitted on the d.c. side.

3.21.12 Rooms and cabins containing sauna heaters

 The protective measures of non-conducting location Earth-free local equipotential bonding are **not** permitted.

3.21.13 Swimming pools and other basins

 Note: The following requirements apply to the basins of swimming pools, the basins of fountains and the basins of paddling pools, as well as to the surrounding zones of these basins. In these areas, in normal use, the risk of electric shock is increased by a reduction in body resistance and contact of the body with Earth potential.

3.21.13.1 Extra-low voltage provided by SELV or PELV

Where SELV is used, whatever the nominal voltage, basic protection shall be provided by insulation, barriers or enclosures.

3.21.13.2 Supplementary protective equipotential bonding

Extraneous-conductive-parts in zones 0, 1 and 2 shall be connected by supplementary protective bonding conductors to the protective conductors of the exposed-conductive-parts of equipment situated in these zones.

It is permitted to install an electric heating unit embedded in the floor, provided that:

- it is protected by SELV; or
- it includes an earthed metallic sheath connected to the supplementary protective equipotential bonding or its supply circuit is additionally protected by an RCD; or
- it is covered by an embedded earthed metallic grid connected to the supplementary protective equipotential bonding and its supply circuit is additionally protected by an RCD.

 The protective measures of non-conducting location and Earth-free local equipotential bonding are **not** permitted.

3.21.13.3 Underwater luminaires for swimming pools

Underwater luminaires or luminaires in contact with the water shall be fixed and shall comply with BS EN 60598-2-18.

 Special requirements may be necessary for swimming pools for medical purposes.

3.21.14 Temporary electrical installations for structures, amusement devices and booths at fairgrounds, amusement parks and circuses

3.21.14.1 Restrictions

- Protective measures such as non-conducting location and Earth-free local equipotential bonding are **not** permitted for installations of this type.
- A PME earthing facility shall not be used for any electrical installation covered by this section. (In addition, the Electricity Safety, Quality and Continuity Regulations (ESQCR) prohibit the use of PME earthing facilities for the supply to caravans or similar constructions.)
- If a TN system is used, a PEN conductor shall not be used downstream of the origin of the temporary electrical installation.

3.21.14.2 Generators

Where a generator supplies a temporary installation, that is part of a TN, TT or IT system, then an Earth electrode shall be capable of withstanding damage and take into account the possibility of corrosion.

3.21.14.3 Automatic disconnection of supply

If an RCD is used as part of the supplies to a.c. motors, then it should be in accordance with BS EN 60947-2 (i.e. of the time-delayed type).

As additional protection, all final circuits for:

- lighting;
- mobile equipment connected by means of a flexible cable with a current-carrying capacity up to 32 A; and
- socket outlets rated up to 32 A,

shall be protected by RCDs.

3.21.14.4 Supplementary protective equipotential bonding

In livestock locations, supplementary bonding shall connect all exposed conductive parts and extraneous-conductive-parts that can be touched by livestock.

 Note: If the extraneous conductive part is in or on the floor, it will need to be connected to supplementary protective equipotential bonding.

3.21.14.5 *Water heaters having immersed and uninsulated heating elements*

All metal parts of the heater or boiler which are in contact with the water (other than current-carrying parts) shall be connected to the metal water pipe which supplies water to the heater or boiler – provided that that water pipe is connected to the main earthing terminal independently of the circuit protective conductor.

Author's end note:

Having looked at how important earthing is to electrical installations, the next thing to consider is the safety precautions that need to be taken into consideration.

With this in mind, Chapter 4 provides a complete resume of the mandatory requirements of not only from the Wiring Regulations but also the Building Regulations. It shows methods for protecting against electric shock, precautions and protections, safety service circuits and disconnecting devices.

4

Safety protection

Author's start note

This chapter lists the mandatory and fundamental requirements for safety protection contained in both the Wiring Regulations and the Building Regulations 2010. It also includes information concerning basic protection against electric shock, fault protection, protection against direct and indirect contact, protective conductors and protective equipment, and lists the test requirements for safety protection.

Figure 4.1 Safety protection.

Over a thousand electrical accidents at work are reported to the UK's Health and Safety Executive (HSE) every year and about 30 people die of their injuries. Indeed, electrocution is one of the top five causes of workplace deaths!

Many of these deaths and injuries arise from:

- use of poorly maintained electrical equipment;
- working near overhead power lines;

- contact with underground power cables during excavation work;
- work on or near 230 V domestic electricity supplies;
- use of unsuitable electrical equipment in explosive areas such as car paint spraying booths etc.

In addition, fires started by poor electrical installations and faulty electrical appliances cause many deaths and injuries. For this reason, protection against electric shock and safety protection methods are an essential part of the Regulations.

Note: *'installation'* in this context is taken to mean either as a whole or in its several parts.

In electrical installations, risk of injury may result from:

- arcing or burning (likely to cause blinding effects, excessive pressure and/or toxic gases);
- excessive temperatures (likely to cause burns, fires and other injurious effects);
- ignition of a potentially explosive atmosphere;
- mechanical movement of electrically actuated equipment;
- power-supply interruptions and/or interruption of safety services;
- shock currents;
- undervoltages, overvoltages and electromagnetic influences likely to cause or result in injury or damage.

4.1 Basic safety requirements

The fundamental safety requirements (as detailed in BS 7671:2018) are as follows:

4.1.1 Mandatory requirements

The following are amongst the most important mandatory notices in the Standard:

- Protective safety measures shall be applied in every installation, part installation and/or equipment.
- Installations shall comply with the requirements for safety protection in respect of:

  o electric shock;
  o fault current (i.e. overcurrent, thermal effects, undervoltage, isolation and switching).

- There shall be no detrimental influence between various protective measures used in the same installation, part installation or equipment.

4.1.2 Fundamental safety requirements

The following are summarised details of the most important elements from BS 7671:2018 that meet these fundamental design requirements.

4.1.2.1 Design

Electrical installations shall be designed for:

- the protection of persons, livestock and property;
- the proper functioning of the electrical installation;
- protection against mechanical and thermal damage; and
- protection of people from an electric shock or fire hazard.

4.1.2.2 Characteristics of available supply or supplies

Detailed design characteristics shall be available for all supplies. These shall include:

- nature of current (a.c. and/or d.c.);
- purpose and number of conductors (as indicated in Table 4.1).

Table 4.1 A.c. and d.c. supplies

For a.c.	For d.c.
<ul><li>Phase conductor(s)</li><li>Neutral conductor</li><li>Protective conductor</li><li>PEN conductor</li></ul>	<ul><li>Outer conductor</li><li>Middle conductor</li><li>Earthed conductor</li><li>Live conductor</li><li>Protective conductor</li><li>PEN conductor</li></ul>

- values and tolerances of:
  - Earth fault loop impedance;
  - nominal voltage and voltage tolerances;
  - nominal frequency and frequency tolerances;
  - maximum current allowable;
  - particular requirements of the distributor;
  - prospective short-circuit current;
  - protective measures inherent in the supply (e.g. Earth, neutral or mid-wire).

4.1.2.3 Electricity distributor's responsibilities

The electricity distributor is responsible for:

- evaluating and agreeing proposals for new installations or significant alterations to existing ones;

- ensuring that their equipment on consumers' premises:

  o   is suitable for its purpose;
  o   is safe in its particular environment;
  o   clearly shows the polarity of the conductors;

- installing the cut-out and meter in a safe location;
- ensuring that the cut-out and meter is mechanically protected and can be safely maintained;
- providing an earthing facility for all new connections;
- maintaining the supply within defined tolerance limits;
- providing certain technical and safety information to the consumer to enable them to design their installations.

4.1.2.4 Installation and erection

All electrical joints and connections shall meet stipulated requirements concerning conductance, insulation, mechanical strength and protection.

- Conductors shall be identified by colour, lettering and/or numbering.
- Connections and joints shall be accessible for inspection, testing and maintenance, unless:

  o   they are in a compound-filled or an encapsulated joint;
  o   the connection is between a cold tail and a heating element;
  o   the joint is made by welding, soldering, brazing or compression tool.

- Design temperatures shall not be exceeded by the installation of electrical equipment.
- Electrical equipment shall be arranged so that it is fully accessible (i.e. for operation, inspection, testing, maintenance and repair) and so that there is sufficient space for later replacement.
- Equipment used for the supply of safety services shall be arranged to allow easy access for periodic inspection, testing and maintenance.
- Exposed parts of electrical equipment shall be located (or guarded) so as to prevent accidental contact and/or injury to persons or livestock.
- Good workmanship and proper materials shall be used.
- Installed electrical equipment shall minimise the risk of igniting flammable materials.
- Installed equipment must be accessible for operational, inspection and maintenance purposes.
- Installations shall be divided into circuits in order to:

  o   avoid danger and minimise inconvenience in the event of a fault;
  o   facilitate safe operation, inspection, testing and maintenance;

- The process of erection shall not impair the characteristics of electrical equipment.

4.1.2.5 Identification and notices

Wiring shall be marked and/or arranged so that it can be quickly identified for inspection, testing, repair or alteration of the installation.

4.1.2.6 Inspection and testing

 Precautions shall be taken to avoid danger to persons and livestock, and to avoid damage to property and installed equipment, during inspection and testing. Figure 4.2 shows the sort of basic protective clothing that an electrician needs to wear when dealing with the main electricity supply.

Figure 4.2 Inspection and testing of electrical installations.

- Every electrical installation must be inspected and tested during, erection and on completion before being put into service.
- Details of the general design characteristics of the electrical installation must be made available. These shall include the result of the assessment of general characteristics.
- Information (e.g. diagrams, charts, tables and/or schedules) must be made available to the person carrying out the inspection and testing and these (as a minimum) shall indicate:

  o the type and composition of each circuit (points of utilisation served, number and size of conductors, type of wiring);
  o the method used;
  o the identification (and location) of all protection, isolation and switching devices;
  o circuits or equipment that are susceptible to a particular test;

- If the inspection and tests are satisfactory, a signed Electrical Installation Certificate together with a Schedule of Inspections and a Schedule of Test

Results (see Chapter 10 in this book) are to be given to the person responsible for ordering the work.

- Precautions shall be taken to avoid danger to persons and to avoid damage to property and installed equipment during inspection and testing.

4.1.2.7 Maintenance

An assessment shall be made of the frequency and type of maintenance (e.g. periodic inspection, testing, maintenance and repair etc.) that an installation can reasonably be expected to receive during its intended life.

4.2 Building Regulations requirements

The following are summarised details of the most important elements of the Building Regulations' Approved Documents and Standards concerning safety protection. As can be seen in Figure 4.3, these include:

- design, installation, inspection and testing of electrical installations;
- conservation of fuel and power;
- access and facilities for disabled people;
- extensions, material alterations and material changes of use.

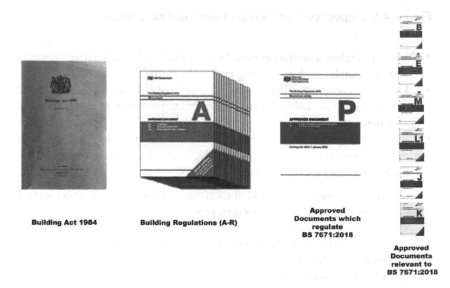

Building Act 1984   Building Regulations (A-R)   Approved Documents which regulate BS 7671:2018

Approved Documents relevant to BS 7671:2018

Figure 4.3 The interoperability of the Building Regulations with BS 7671:2018.

4.2.1 Design, installation, inspection and testing of electrical installations

All proposals to carry out electrical installation work must be notified to the local authority's Building Control Body before work begins, unless the proposed installation work:

- is undertaken by a person who is a competent person registered with an electrical self-certification scheme; and
- does not include the provision of a new circuit.

Any work that involves adding a new circuit to a dwelling needs to be either notified to the Building Control Body (who will then inspect the work) or carried out by a competent person who is registered under a government approved Part P Self Certification Scheme.

 Note: Where a person who is **not** registered to self-certify intends to carry out the electrical installation, then a Building Regulation (i.e. a Building Notice or Full Plans) application will need to be submitted together with the appropriate fee, based on the estimated cost of the electrical installation, to the local council. The council's Building Control Body will then arrange to have the electrical installation inspected at first-fix stage and tested upon completion.

- Reasonable provision shall be made in the design, installation, inspection and testing of electrical installations in order to protect persons from fire or injury.
- Sufficient information shall be provided so that persons wishing to operate, maintain or alter an electrical installation can do so with reasonable safety.

Work involving any of the following will also have to be notified to the Building Control Body:

- electric floor or ceiling heating systems;
- extra-low-voltage lighting installations (other than pre-assembled, CE-marked lighting sets);
- garden lighting or power installations;
- hot-air saunas;
- locations containing a bathtub or shower basin;
- small-scale generators such as microchip units;
- solar photovoltaic (PV) power supply systems;
- swimming pools or paddling pools.

 Note: Whilst Part P of the Building Regulations makes requirements for the safety of fixed electrical installations, this does not cover system functionality (e.g. electrically powered fire alarm systems, fans and pumps) which is covered in other Parts of the Building Regulations and government legislation.

4.2.2 Classifications of electrical equipment

There are three basic different classifications of electrical appliance. Class I and Class II appliances are all powered by mains voltages (and both are required to provide at least two levels of protection to the end user) whilst a Class III appliance is designed to be supplied from a separated/safety extra-low-voltage (SELV) power source.

4.2.2.1 Class I appliances

With Class I appliances, the user is protected by a combination of basic insulation and the provision of an Earth connection, thus providing two levels of protection. If the basic insulation fails, the Earth connection (which is connected to the metal case) will divert the current into the ground. The fuse should then blow, either in the plug or the fuse box, or there should be a power trip.

The required PAT (portable appliance testing) tests for Class I appliances are the Earth Continuity and Insulation Resistance tests which will check the basic insulation and earth connection.

Typically refrigerators, microwaves, kettles, irons, and toasters, are all Class I.

4.2.2.2 Class II appliances

In Class II appliances, the user is protected by at least two layers. The plastic connector supplies the basic insulation, and an added layer of insulation (provided by the plastic casing) provides backup protection. For this reason, Class II appliances are also known as double-insulated appliances.

The only PAT test required is the insulation resistance test. Typical examples of Class II appliances are hairdryers, DVD players, televisions, computers and photocopiers. Most plastic power tools would also be Class II.

4.2.2.3 Class III appliances

Class III appliances use an isolating transformer which has two separate coil windings called (not surprisingly!) the *primary winding*, which is connected to the power source, and the *secondary winding*, which is connected to the appliance itself. Each winding is wrapped around opposite sides of a common closed magnetic circuit called the *core*. Due to the lack of an earth connection, if there is an electromagnetic problem, the current is cut off and cannot continue to flow and a person can safely come into contact with it without risk of electric shock.

PAT testing is not required unless the charging leads fall under Class II.

 If the appliances are for medical use, they are **not** considered to be sufficiently safe for mass consumer usage and must, therefore, meet additional requirements.

4.2.3 Conservation of fuel and power

Energy efficiency measures shall be provided which:

* provide lighting systems that utilise energy-efficient lamps with manual switching controls (in the case of external lighting fixed to the building) or automatic switching, or both manual and automatic switching controls, as appropriate, such that the lighting systems can be operated effectively with regard to the conservation of fuel and power;
* provide information, in a suitably concise and understandable form (including results of performance tests carried out during the works), which shows building occupiers how the heating and hot water services can be operated and maintained.

The person responsible for achieving compliance should either themselves provide a certificate or obtain a certificate from the sub-contractor that commissioning has been successfully carried out. The certificate should be made available to the client and the Building Control Body.

 Responsibility for achieving compliance with these requirements rests with the person carrying out the work. That *person* may be, for example, a developer, a main (or sub-) contractor, or a specialist firm directly engaged by a private client.

4.2.4 Extensions, material alterations and material changes of use

Where electrical installation work is classified as an extension, a material alteration or a material change of use, the work must consider and include:

* confirmation that the mains supply equipment is suitable and can carry the additional loads envisaged;
* the amount of additions and alterations that will be required to the existing fixed electrical installation in the building;
* the earthing and bonding systems being satisfactory and meeting the requirements;
* the necessary additions and alterations to the circuits which feed them;
* the protective measures required to meet the requirements;
* that the rating and the condition of existing equipment (belonging to both the consumer and the electricity distributor) is sufficient.

 Appendix C to Part P of the Building Regulations offers guidance on some of the older types of installations that might be encountered during alteration work, and Appendix D provides guidance on the application of the new harmonised European cable-identification system.

4.2.5 Access and facilities for disabled people

In addition to the requirements of the Disability and Equality Act 2010 [which makes it unlawful to discriminate against employees (including workers) because of a mental or physical disability is that precautions need to be taken to ensure that:

- new, non-domestic buildings and/or dwellings (e.g. houses and flats used for student living accommodation etc.); and
- extensions to existing non-domestic buildings; and
- non-domestic buildings that have been subject to a material change of use (e.g. so that they become a hotel, boarding house, institution, public building or shop)

are capable of allowing people, **regardless** of their disability, age or gender, to be able to safely use the facilities of the buildings (both as visitors and as people who live or work in them).

4.3 Protection from electric shock

Safety requirements for protection against electric shock need to be provided:

- against both direct contact (i.e. basic protection) and indirect contact (i.e. fault protection);
- for persons and livestock against dangers that may arise from contact with:

  o exposed-conductive-parts during a fault;
  o live parts of the installation.

Live parts should be completely covered with insulation which:

- can only be removed by destruction;
- is capable of durably withstanding electrical, mechanical, thermal and chemical stresses normally encountered during service.

Live parts should also be inside enclosures (or behind barriers) protected to at least IP2X (protection against approach by hands) or IPXXB (the degree of protection provided once the cover is off a distribution board or consumer unit).

Bare (or insulated) overhead lines being used for distribution between buildings and structures shall be installed in accordance with the Electricity Safety, Quality and Continuity Regulations (ESQCR) 2002.

Bare live parts (other than overhead lines) shall:

- not be within arm's reach;
- not be within 2.5 m of:

  o an exposed conductive part;

o   an extraneous-conductive-part;
o   a bare live part of any other circuit.

Simultaneously accessible exposed-conductive-parts shall be connected to the same earthing system either individually, in groups or collectively.

Exposed parts of electrical equipment shall be located (or guarded) so as to prevent accidental contact and/or injury to persons or livestock.

The following methods are used for protection against direct contact (basic protection) and against indirect contact (fault protection).

4.3.1 Basic protection against electric shock

Basic protection against electric shock is taken as meaning that:

* the nominal voltage cannot exceed the upper limit of voltage Band I;
* all electrical equipment is protected by some form of basic insulation; barriers, enclosures; obstacles or placing out of reach;
* the SELV and/or PELV supply is from a recognised source such as an a safety isolating transformer, battery, diesel-driven generator, insulation testing equipment, monitoring device; motor-generator; and
* exposed-conductive-parts of a SELV circuit have not been connected:

    o   to Earth; or
    o   to protective conductors; or
    o   to exposed-conductive-parts of another circuit.

In normally dry conditions, basic protection is generally unnecessary for SELV if the nominal voltage does not exceed 25 V a.c. or 60 V d.c. It is a similar situation for PELV circuits, provided any exposed-conductive-part is connected to the main earthing terminal by a protective conductor.

A person may perform work involving direct contact with electrical parts **only** if the electrical part:

* is isolated from all sources of electricity;
* is tested to ensure its isolation from all sources of electricity; and
* is earthed if it is of high voltage.

In all cases, live parts must be completely covered with insulation which can **only** be removed by destruction.

To meet these requirements - and as shown in Figure 4.5 - the Regulations state that one of the following basic measures shall be used for protection against indirect contact:

* insulating live parts;
* using a barrier or an enclosure;
* using obstacles;
* placing equipment out of reach;
* use of a residual current device (RCD).

1–2 mA	Barely perceptible, no harmful effects
5–10 mA	Throw off, painful sensation
10–15 mA	Muscular contraction, can not let go!
20–30 mA	Impaired breathing
50 mA and above	Ventricular fibrillation and death

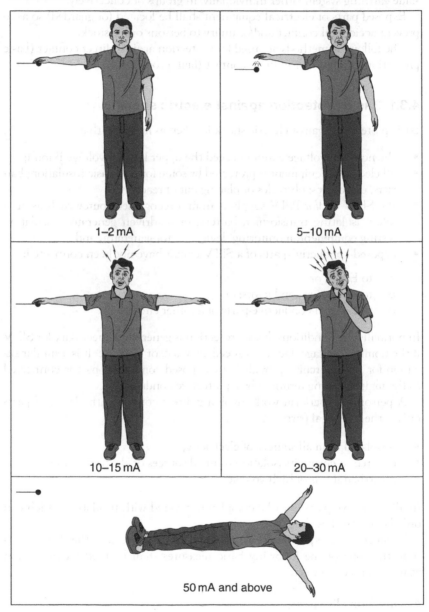

Figure 4.4 The effects of electric shock. (Courtesy Brian Scaddan.)

Figure 4.5 Basic protection against electric shock.

4.3.2 Insulation of live parts

As the heading suggests, this is a basic form of insulation protection and is intended to prevent contact with a live parts of an electrical installation. Paint, lacquers and varnishes do **not** provide adequate protection.

4.3.3 Using a barrier or enclosure

Live parts of any circuit or electromagnetic equipment should always be inside an enclosure or behind a barrier that provides a degree of protection at least that of IPXXB or IP2X and the barrier (or enclosure) must:

- be firmly secured in place;
- have sufficient stability and durability to maintain the required degree of protection;
- be appropriately separated from live parts;
- restrict the removal of a barrier, opening an enclosure or removal of parts of enclosures, which is permitted only:

  o   after disconnection of the supply to live parts; or

- o   where an intermediate barrier (with a degree of protection of at least IPXXB or IP2X) prevents contact with live parts; or
- o   by the use of a key or tool (provided all conductive parts are accessible behind an intermediate insulating barrier).

This Regulation does not apply to:

- a ceiling rose complying with BS 67;
- a cord-operated switch complying with BS 3676;
- a bayonet lampholder complying with BS EN 61184;
- an Edison screw lampholder complying with BS EN 60238.

Note: If an item of equipment (e.g. a capacitor) is installed behind a barrier or is in an enclosure (and that equipment could retain a dangerous electrical charge after it has been switched off) a warning label **must** be provided.

Where the protective measure *'automatic disconnection of supply'* is used, all live parts of electrical equipment need to be protected by insulation, barriers or enclosures to at least IPXXB or IP2X provided that the insulating enclosure:

- cannot be crossed by conductive parts likely to transmit a potential;
- contains no screws (or other fixings) which might need to be removed (e.g. during installation and maintenance) which could, possibly, be replaced by metallic screws or some other type of fixing that could affect the enclosure's insulation;
- has no mechanical joints or connections (e.g. an operating handle of built-in equipment) crossing it which are not protected against shock;
- contains no conductive part that is connected to a protective conductor (unless explicit provision for this is made in the specification for the equipment concerned);
- will not affect the operation of the equipment that it is protecting.

4.3.4 Protection by obstacles and placing out of reach

Protection by obstacles and placing out of reach is primarily intended for installations that are controlled or supervised by skilled persons. It will only provide basic protection; the prime intention is to prevent unintentional contact with a live part, but not an intentional contact caused by deliberately circumnavigating the obstacle.

The intention of this form of protection is to prevent or deter any contact with a live part. Whilst, generally speaking, this method is for protection against direct contact, it also provides a degree of protection against indirect contact.

4.3.4.1 Obstacles

Obstacles are intended and shall be designed to:

- prevent unintentional bodily approach;
- prevent unintentional contact with to live parts during the operation of live equipment in normal service;
- be secured to prevent unintentional removal;
- be capable of being removed without using a key or tool.

4.3.4.2 Placing out of reach

The protective measures of placing out of reach and obstacles shall not be used except where the maintenance of equipment is restricted to skilled persons who are specially trained.

Protection by placing out of reach is **only** intended to prevent unintentional contact with live parts. It also ensures that:

- simultaneously accessible parts at different potentials shall not be within arm's reach;
- a bare live part (other than an overhead line) is not within arm's reach or within 2.5 m of:

    o   an exposed-conductive-part;
    o   an extraneous-conductive-part;
    o   a bare live part of any other circuit.

Bare (or insulated) overhead lines used for distribution between buildings and structures shall be installed in accordance with the Electricity Safety, Quality and Continuity Regulations 2002 (as amended).

 Items of street furniture which are within 1.5 m of a low-voltage overhead line **must** be protected by something other than placing out of reach!

It should also be noted that the protective measure of placing out of reach and installing obstacles is not permitted in:

- agricultural or horticultural premises;
- construction and demolition site installations;
- conducting locations with restricted movement;
- electrical installations in caravan/camping parks and similar locations;
- electrical installations in caravans and motor caravans;
- exhibitions, shows and stands;
- floor and ceiling heating systems;
- locations containing a bath or shower;
- medical locations;
- mobile or transportable units;
- rooms and cabins containing sauna heaters;
- swimming pools and other basins.

4.3.5 Protection by RCDs

In electrical installations, an RCD, or an RCCB (residual current operated circuit breaker with integral overcurrent protection), is a circuit breaker that operates to disconnect a particular circuit whenever it detects that current leaking out of that circuit (e.g. current leaking to Earth through a ground fault) exceeds safety limits.

Figure 4.6 illustrates the construction of an RCD and works on the principle that in a normal (i.e. healthy) circuit, the magnetic effects of the phase and neutral currents will cancel out because the same current will pass through the phase coil, the load and then back through the neutral coil. In a faulty circuit where the phase or the neutral are to Earth, the currents will no longer be equal and the out-of-balance current will produce some residual magnetism in the core. As the magnetism will be alternating, it will link with the turns of the search coil and induce an electromotive force (EMF) in it which will drive a current through the trip coil and cause the tripping mechanism to operate.

 Note: The use of RCDs is not recognised as a sole means of protection and does not obviate the need to apply one of the other protective measures (e.g. automatic disconnection of supply, double or reinforced insulation, SELV or PELV).

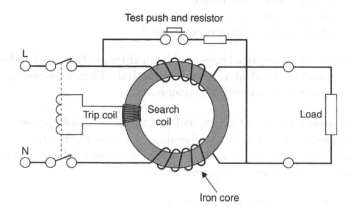

Figure 4.6 The construction of a basic RCD. (Courtesy Brian Scaddan.)

The use of RCDs with a rated residual operating current not exceeding 30 mA and an operating time not exceeding 40 ms is recognised in a.c. systems as providing additional protection in the event of:

- failure of one of the other methods of basic protection against electric shock; or
- failure of the provision for fault protection; or
- carelessness by users.

Although RCDs reduce the risk of electric shock, they should **not** be used as the sole means of protection against direct contact.

The following restrictions and requirements of RCDs apply.

- Every installation shall be divided into circuits to reduce the possibility of unwanted tripping of RCDs due to excessive protective conductor currents produced by equipment.
- In a.c. systems, additional protection by means of an RCD shall be provided for:
- socket outlets with a rated current not exceeding 20 A that are used by ordinary persons, unless:

  o a suitably labelled and/or identified socket outlet is provided for connection of a particular item of equipment;
  o mobile equipment with a current rating not exceeding 32 A is used outdoors; or
  o they are used under the supervision of skilled or instructed persons e.g. in some commercial or industrial locations.

If a generating set is connected, then protection by RCDs shall remain effective for every intended combination of sources of supply.

4.3.6 Fault protection against indirect contact

 All persons and livestock should be protected against dangers that may arise from contact with exposed-conductive-parts during a fault.

Indirect contact (i.e. when part of the body touches or is in dangerous proximity to any object that is in contact with energised electrical equipment or exposed-conductive-parts which might become live under fault conditions) has always been a potential problem to the unwary when installing, maintaining or inspecting electrical installations. It should also be remembered that as voltages increase, the potential for arcing increases, and through arcing, injuries and/or fatalities will often occur – **even** if no actual bodily contact with high-voltage lines and/or equipment is made!

To meet these requirements for protection against indirect contact, (and as shown in Figure 4.7) one of the following basic measures shall be used:

- Earthed Equipotential Bonding and Automatic Disconnection of Supply (EEBADS);
- non-conducting location;
- protection by obstacles and placing out of reach;
- Class II equipment or equivalent insulation.

4.3.6.1 Protection by EEBADS earthed equipotential bonding and automatic disconnection of supplies

EEBADS provides a very good form of protection against indirect contact by joining together (i.e. bonding) all of the metallic parts and then connecting them to Earth.

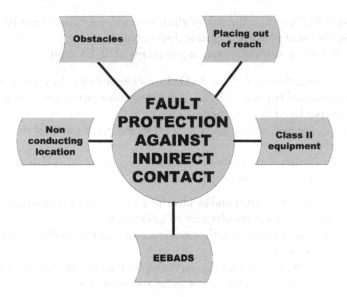

Figure 4.7 Protection against indirect contact.

In each installation, main protective bonding conductors shall connect extraneous-conductive-parts to the main earthing terminal. Such installations include:

- central heating and air-conditioning systems;
- exposed metallic structural parts of the building;
- gas installation pipes;
- water installation pipes;
- other installation pipework and ducting.

This ensures that all metalwork is at (or near) zero volts and so, under fault conditions, all metalwork will rise to a similar potential and simultaneous contact with two metal parts will not result an electric shock as there is no significant potential difference between them.

 Where an installation serves more than one building, it is essential that the above requirement is applied to each building.

For installations and locations with an increased risk of shock (e.g. bathrooms and saunas etc.) additional measures may be required, such as:

- automatic disconnection of supply by means of an RCD with a rated residual operating current not exceeding 30 mA;
- supplementary equipotential bonding;
- reduction of maximum fault clearance time.

 Note: Protection by earthed equipotential bonding (and automatic disconnection of supply) will depend on the requirements of the type of system earthing in use (e.g. TN, TT or IT).

The connection of a lightning protection system to the protective equipotential bonding shall be made in accordance with BS EN 62305.

4.3.6.2 Protection by non-conducting location

A *'non-conducting location'* is a location where there is no earthing or protective system because:

- there is nothing which needs to be earthed;
- exposed-conductive-parts are arranged so that it is impossible to touch two of them (or one exposed-conducting-part and one extraneous-conductive-part) at the same time.

Protection by non-conducting location is intended to prevent simultaneous contact with parts which may be at different potentials (i.e. through the failure of the basic insulation of live parts).

This method of protection is **not** recognised for general application and may not be used in installations such as electrical installations in caravan and camping parks; outdoor or temporary electrical installations at fairgrounds, amusement parks and circuses; construction and demolition sites; marinas; swimming pools or rooms containing a bath or showers; or medical locations that are subject to an increased risk of shock. (For complete details of these restrictions, see BS 7671:2018 Sections 701 to 753.)

4.3.6.3 Protection by Class II equipment or equivalent insulation

Class II equipment is unique in that as well as providing the basic insulation for live parts, it also has a second layer of insulation which can be used to either prevent contact with exposed-conductive-parts or to make sure that there can never be any contact between such exposed-conductive-parts and live parts.

Class II protection is provided by one or more of the following:

- electrical equipment having double or reinforced insulation;
- low-voltage switchgear;
- low-voltage control-gear assemblies;
- supplementary insulation;
- reinforced insulation applied to uninsulated live parts.

4.3.7 Fault protection against both direct and indirect contact

The Regulations state that one of the following (see Figure 4.8) basic, measures shall be used for protection against both direct contact and indirect contact:

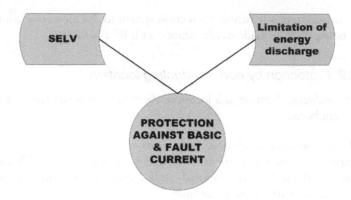

Figure 4.8 Protection against basic and fault contact.

- SELV; or
- limitation of discharge of energy.

 As shown in the following section and Figure 4.9, there are three methods of protection.

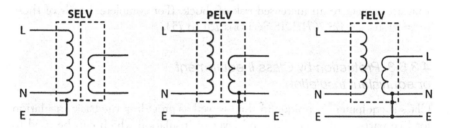

Figure 4.9 SELV, PELV and FELV.

4.3.7.1 Protection by SELV, PELV or FELV

SELV (separated extra-low voltage) is an extra-low-voltage system that is electrically separated from Earth and from other systems so that a single fault cannot give rise to the risk of electric shock. It is a term used to describe the highest voltage level that can be contacted by a person without causing injury. It is usually defined as 60 V d.c.

This type of protection may also be used for electric fences – provided that they are supplied from electric fence controllers complying with BS EN 61011 or BS EN 6101 1-1.

PELV (protective extra-low voltage), on the other hand, is an extra-low-voltage system which is not electrically separated from Earth, but which otherwise satisfies all the requirements for SELV.

FELV (functional extra-low voltage) describes any other extra-low-voltage circuit that does not fulfil the requirements for a SELV or PELV circuit.

In medical locations, FELV is **not** permitted as a method of protection against electric shock.

The separation of the live parts from those of other circuits and from Earth needs to be confirmed by a measurement of the insulation resistance.

For details of more stringent requirements for wiring fire alarm systems in buildings, see BS 5839-1 and Approved Document P.

4.3.7.2 Medical locations

When using SELV and/or PELV circuits in Group 1 and/or Group 2 medical locations, protection by basic insulation of live parts or by barriers or enclosures shall be provided.

In Group 2 medical locations where PELV is used, operating theatre luminaires (and any other exposed-conductive-parts of equipment) will need to be connected to the circuit protective conductor.

4.4 Additional requirements

The following are additional requirements for installations and locations where the risk of electric shock is increased by a reduction in body resistance and/or by contact with Earth potential.

4.4.1 Protective bonding conductors

Equipotential bonding ensures that protective devices will operate and remove all dangerous potential differences before a hazardous shock can be delivered. This is done by making sure that all of the installation's earthed metalwork (i.e. exposed-conductive-parts) is connected to other metalwork (i.e. extraneous-conductive-parts) via the Earth conductor to provide an Earth fault current path that ensures dangerous potential differences cannot occur As the title, Figure 4.10 pondweeds an indication of the majority of typical fixed installations that are part of new (or upgraded) dwellings.

4.4.2 Main equipotential bonding conductors

Main equipotential bonding conductors (see Figure 4.11) connect together the installation earthing system and the metalwork of other services such as gas, electricity and water as close as possible to their point of entry to the building.

In accordance with the requirements of BS 7671:2018, main equipotential bonding conductors for every electrical installation must be connected to the

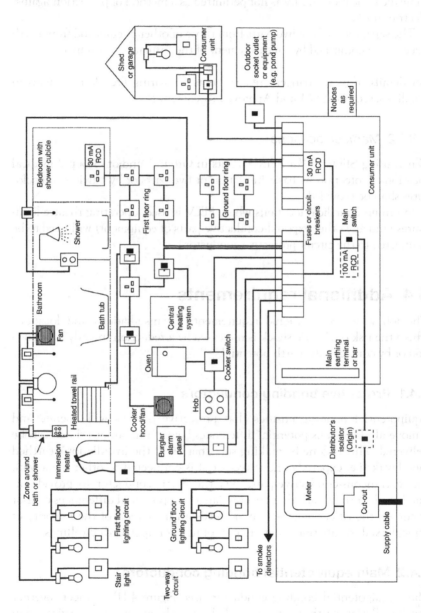

Figure 4.10 Typical fixed installations that might be encountered in new (or upgraded) existing dwellings.

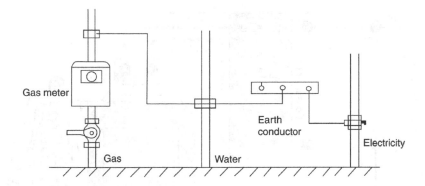

Figure 4.11 Main equipotential bonding.

main earthing terminal of that particular installation, and these shall include the following:

- water service pipes (but see requirements for domestic buildings in Chapter 2);
- gas installation pipes;
- other service pipes and ducting;
- central heating and air-conditioning systems;
- exposed metallic structural parts of the building;
- the lightning protective system.

 Note: Where an installation serves more than one building, the above requirement shall be applied to each building.

4.4.3 Protective earthing

Automatic disconnection of supply is a protective measure in which fault protection is provided by protective earthing, and the following sign shall be included.

4.4.4 Supplementary bonding conductors

Supplementary bonding conductors connect together extraneous-conductive-parts (i.e. metalwork which is not associated with the electrical installation but which may provide a conducting path that could give rise to shock).

For installations and locations where there is an increased risk of shock (e.g. agricultural and horticultural premises, building sites etc.) additional measures may be required, such as reduction of maximum fault clearance time and the use of supplementary equipotential bonding.

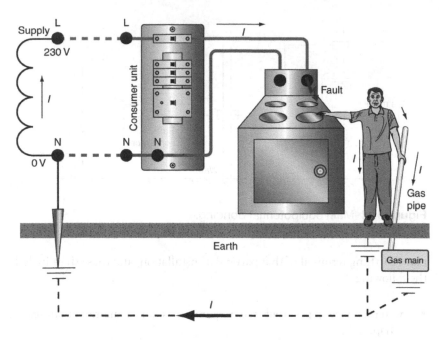

Figure 4.12 Earthed equipotential bonding.

Locations which contain a bath or shower (and where body resistance is lowered as a result of water) are potentially very hazardous environments and it is important to ensure that no dangerous potentials exist between exposed and extraneous conductive parts. For this reason, local supplementary equipotential bonding needs to be provided to connect together the terminals of the

Figure 4.13 Earthing and bonding notice.

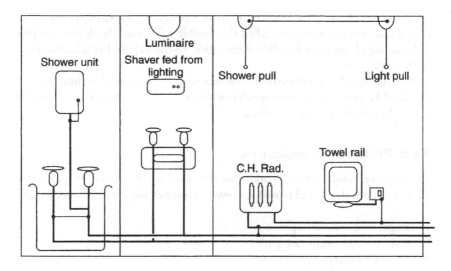

Figure 4.14 Supplementary equipotential bonding. (Courtesy Brian Scaddan.)

protective conductors of each circuit supplying Class I and Class II equipment with extraneous-conductive-parts in those zones, such as:

* metallic pipes supplying services and metallic waste pipes (e.g. water, gas);
* metallic central heating pipes;
* air-conditioning systems;
* accessible metallic structural parts of the building;
* metallic baths and shower basins.

Amongst the requirements for supplementary bonding conductors are that: two exposed conductive-parts that are conjoined shall together have a conductance greater than the smaller protective conductor when they are connected to the exposed-conductive-parts.

If mechanical protection is not provided, the cross-sectional area of the supplementary bonding conductors shall be not less than 4 mm². A supplementary bonding conductor:

* that connects an exposed-conductive-part to an extraneous conductive part shall have a conductance not less than half that of the protective conductor connected to the exposed conductive part;
* connecting two extraneous-conductive-parts shall have a cross-sectional area not less than 2.5 mm² if sheathed or 4 mm² if mechanical protection is not provided.

Supplementary bonding shall be provided by a supplementary conductor, or a conductive part of a permanent (and reliable nature), or by a combination of these.

Where supplementary bonding is to be applied to a fixed appliance, the circuit protective conductor within the flexible cable shall be deemed to provide the supplementary bonding connection to the exposed-conductive-parts of the appliance.

In agricultural and horticultural premises, supplementary bonding conductors shall be protected against mechanical damage and corrosion, and shall be selected to avoid electrolytic effects.

4.4.5 Protective conductors

A protective conductor is a conductor that provides a measure of protection against electric shock and is used to connect together any of the following parts:

- exposed-conductive-parts;
- extraneous-conductive-parts;
- the main earthing terminal;
- Earth electrode(s);
- the earthed point of the source.

A circuit protective conductor, on the other hand, is an arrangement of conductors that join all of the exposed-conductive-parts together and connect them to the main earthing terminal. There are many types of circuit protective conductor, such as:

- a separate conductor;
- a conductor included in a sheathed cable with other conductors;
- the metal sheath and/or armouring of a cable;
- a conducting cable enclosure (e.g. conduit or trunking);
- exposed-conductive-parts (e.g. the conducting cases of equipment).

 A gas pipe, an oil pipe, flexible or pliable conduit, support wires or other flexible metallic parts, or constructional parts that are subject to mechanical stress in normal service, shall **NOT** be selected as a protective conductor.

4.4.6 Protective equipment (devices and switches)

The type of protective equipment chosen will depend on the type of protection that is required (e.g. whether overcurrent, Earth fault current, overvoltage or undervoltage).

4.4.7 Protection against overvoltage

Overvoltage is the hazardous condition that occurs when the voltage in a circuit (or part of a circuit) is suddenly raised over its upper limit. An overvoltage incident can be permanent or transient and it is often referred to as a 'voltage spike'.

A typical example of a naturally occurring transient over voltages is by lightning. Man-made sources are usually due to electromagnetic induction when switching on or off inductive loads (e.g. electric motors or electromagnets).

Transient overvoltage might last microseconds and reach hundreds – sometimes thousands – of volts in amplitude.

In accordance with the requirements of BS 7671:2018, additional protection against overvoltages of atmospheric origin is **not** necessary for:

- installations that are supplied by low-voltage systems which do not contain overhead lines;
- installations that are supplied by low-voltage networks which contain overhead lines and their location is subject to less than 25 thunderstorm days per year;
- installations that contain overhead lines and their location is subject to less than 25 thunderstorm days per year;

provided that they meet the required minimum equipment impulse to withstand the voltages shown in Table 4.2.

Suspended cables with insulated conductors that have earthed metallic coverings are considered to be *underground cables*.

Table 4.2 Required minimum impulse to withstand voltage (kV) (data from BS 7671:2008)

Required minimum impulse to withstand voltage				
Nominal voltage of the installation (V)	Category IV (equipment with very high impulse voltage) (e.g. energy meter, or telecontrol systems)	Category III (equipment with high impulse voltage) (e.g. boards, switches and socket outlets)	Category II (equipment with normal impulse voltage) (e.g. domestic appliances and tools)	Category I (equipment with reduced impulse voltage) (e.g. sensitive electronic equipment such as alarm panels, computers and home electronics)
120/208	4	2.5		0.8
230/240	6	4	2.5	1.5
277/480	6	4	2.5	1.5
400/690	8	6	4	2.5
1000	12	8	6	4

4.4.8 Accessibility of electrical equipment

Electrical equipment should always be arranged so that:

- there is sufficient space for the initial installation and (if required) later replacement of some individual items of electrical equipment;
- the equipment is completely accessibility for operation, inspection, testing, fault detection, repair and maintenance.

4.4.9 Additions and alterations to an installation

No addition or alteration (temporary or permanent) shall be made to an existing installation, unless:

- it has been ascertained that the rating and the condition of any existing equipment (including that of the distributor) will be adequate for the altered circumstances;
- the earthing and bonding arrangements used as a protective measure for the safety of the addition or alteration are adequate.
- It has been verified that alterations to an existing installation fully comply with the Regulations and do not weaken the safety of the existing installation.

4.4.10 Automatic supply

Automatic safety services may be required to operate at all certain times where people or livestock are at risk – including during mains and local supply failure and through fire conditions. To meet this requirement, specific sources, equipment, circuits and wiring are necessary. The automatic safety service must:

- be capable of maintaining a supply of adequate duration;
- have equipment, either by construction or by erection, that is capable of fire-resistance of adequate duration.

 Note: The safety source is generally additional to the normal source. The normal source is, for example, the public supply network.

4.5 Circuits for safety services

4.5.1 Safety service supply

An electrical safety supply service is either:

- a non-automatic supply, initiated by an operator; or
- an automatic supply, opened independent of an operator.

It is, however, extremely important that all circuits to a safety service:

- are independent of other circuits;
- do not pass through zones exposed to an explosion risk;
- do not pass through locations exposed to fire risk – unless they are fire-resistant.

In all cases:

- the equipment used for the supply of safety services must be arranged to allow easy access for periodic inspection, testing and maintenance;
- if equipment is supplied by two different circuits, then a fault occurring in one circuit does not affect the protection against electric shock – or the correct operation of – the other circuit;
- overcurrent protective devices must avoid an overcurrent in one circuit weakening the correct operation of other circuits of safety services;
- switchgear and controlgear shall be clearly identified and grouped in locations accessible only to skilled or instructed persons;
- safety circuit cables (other than metallic-screened, fire-resistant cables) shall be separated from other safety circuit cables.

4.5.1.1 Lifts

Safety service supplies should not be installed in lift shafts (or other flue-like openings) unless they have been specifically designed for lifts supplying cables for the fire and rescue service lift or the wiring is just for lifts with special requirements.

Whilst fire-resistant cables will survive most fires, if they are located in an unstopped vertical shaft the upward air draught of a fire can generate excessive temperatures which can damage the (otherwise) fire-resistant cable.

4.5.2 Safety sources

Protection against fault current and against electric shock in case of a fault shall be ensured whether the installation is supplied separately by either of the two sources or by both in parallel.

Examples of safety services include:

- carbon monoxide (CO) detection and alarm systems;
- emergency lighting;
- essential medical systems;
- fire detection and alarm systems;
- fire evacuation systems;
- fire pumps;
- fire rescue service lifts;
- fire services communication systems;

- industrial safety systems;
- smoke ventilation systems.

A safety source shall:

- be placed in a suitable location;
- be accessible only to skilled persons or instructed persons;
- be installed as fixed equipment, so that they cannot be adversely affected by failure of the normal source;
- be properly and adequately ventilated so that exhaust gases, smoke or fumes from the safety source cannot penetrate areas occupied by persons;
- have sufficient capability to supply its related services.

Separated independent feeders from a supply network shall not serve as electrical safety sources unless assurance can be obtained that the two supplies are unlikely to fail concurrently.

A safety source may, in addition, be used for purposes other than safety services, **provided** that a fault occurring in a circuit for purposes other than safety services does not cause the interruption of any circuit for safety services.

The following sources for safety services are recognised:

- storage batteries;
- primary cells;
- generator sets independent of the normal supply;
- a separate feeder of the supply network effectively independent of the normal feeder.

4.5.3 Equipment details

A list (indicating the nominal electrical power, rated nominal voltage, current and starting current, together with its duration) of all current-using equipment that is permanently connected to the safety power supply shall be available. (This information may, of course, be included in the circuit diagrams.)

4.5.4 Diagrams and other information

As well as a general schematic diagram, details of **all** electrical safety sources must be provided adjacent to the distribution board. In addition, drawing(s) of the electrical safety installations need to be available showing the exact location of all:

- electrical equipment and their distribution boards (together with their equipment designations);
- safety equipment (together with their final circuit designation and purpose);
- special switching and monitoring equipment for the safety power supply (e.g. area switches, visual or acoustic warning equipment).

Operating instructions for all safety equipment and electrical safety services shall be available.

4.5.5 Cross-sectional area of conductors

The cross-sectional area of conductors shall be determined for both normal operating conditions and, where appropriate, for fault conditions, according to:

- the admissible maximum temperature;
- the voltage drop limit;
- the electromechanical stresses likely to occur due to short-circuit and Earth fault currents;
- other mechanical stresses to which the conductors are likely to be exposed;
- the maximum impedance for operation of short-circuit and Earth fault protection;
- the method of installation;
- harmonics;
- thermal insulation.

4.5.6 The design of an electrical installation

The electrical installation shall be designed to provide for:

- the protection of persons, livestock and property;
- the proper functioning of the electrical installation for the intended use.

4.6 Disconnecting devices

Disconnecting devices shall be provided so as to allow electrical installations, circuits or individual items of equipment to be switched off or isolated for the purposes of operation, inspection, fault detection, testing, maintenance and repair.

4.6.1 Earthing arrangements and protective conductors

Where protective bonding conductors are installed (especially in photovoltaic (PV) power supply systems) they shall be parallel to and in as close contact as possible with d.c. cables and a.c. cables and accessories.

Where overcurrent protective devices are used for fault protection, the protective conductor shall be incorporated in the same wiring system as the live conductors or in their immediate proximity.

4.6.2 Earthing requirements for the installation of equipment having high protective conductor currents

Equipment having a protective conductor current greater than 3.5 mA but less than 10 mA shall be connected to an installation either permanently or by means of a plug and socket outlet complying with BS EN 60309-2.

Equipment having a protective conductor current exceeding 10 mA shall be connected to the supply:

- permanently via the wiring of the installation; or
- via a flexible cable with a plug and socket outlet; or
- via a protective conductor with an Earth monitoring system.

The wiring for final distribution circuits supplying one or more items of equipment with a total protective conductor current exceeding 10 mA shall have one or more of the following:

- a single copper protective conductor with a cross-sectional area greater than 4 mm²;
- a single protective conductor with a cross-sectional area greater than 10 mm²;
- two individual protective conductors;
- an Earth monitoring system that will automatically disconnect the supply to the equipment in the event of a continuity fault occurring in the protective conductor;
- connection to the supply by means of a double-wound transformer or equivalent unit.

Where two protective conductors are used, the ends of the protective conductors shall be terminated independently of each other at all connection points throughout the circuit (e.g. the distribution board, junction boxes and socket outlets etc.).

At the distribution board, information shall be provided indicating those circuits having a high protective conductor current.

In agriculture and horticultural premises, protective bonding conductors shall be protected against mechanical damage and corrosion and shall be selected to avoid electrolytic effects.

The socket outlet protective conductors used for electrical installations in caravan/camping parks (or similar locations) shall **not** be connected to any PEN conductor of the electricity supply.

4.6.3 Emergency control

An interrupting device shall be installed in such a way that it can be easily recognised and effectively (and rapidly) operated in the case of danger.

4.6.4 Environmental conditions

The design of an electrical installation shall take into account the environmental conditions to which it will be subjected – particularly where the equipment's surroundings are susceptible to risk of fire or explosion.

4.6.5 Erection of electrical installations

Electrical equipment shall be installed in accordance with the instructions provided by the manufacturer of the equipment to ensure that during the process of erection:

- design temperatures are not exceeded;
- electrical joints and connections are properly constructed with regard to conductance, insulation, mechanical strength and protection;
- exposed parts of electrical equipment that could cause injury to persons or livestock are suitably protected;
- the electrical equipment (and its characteristics) are not damaged;
- the positioning of electrical equipment that is likely to cause high temperatures or electric arcs minimises the risk of igniting flammable materials.

 Note: Where necessary, suitable safety warning signs and/or notices shall be provided.

4.6.6 Initial verification

The person or persons responsible for the design, construction, inspection and testing of the installation shall provide the person ordering the work with a certificate which takes account of their respective responsibilities for the safety of that installation.

4.7 Precautions and protections

4.7.1 Omission of devices for protection against overload for safety reasons

Omission protection devices against overload is permitted for circuits supplying current-using equipment if unexpected disconnection of the circuit could cause danger or damage.

Examples of such circuits are:

- the exciter circuit of a rotating machine;
- the supply circuit of a lifting magnet;
- the secondary circuit of a current transformer;
- a circuit supplying a fire-extinguishing device;

- a circuit supplying a safety service, such as a fire alarm or a gas alarm;
- a circuit supplying medical equipment used for life support in specific medical locations where an IT system is incorporated.

In such situations consideration should be given to installing an overload alarm.

4.7.2 Precautions within a fire-segregated compartment

The spread of fire can be restricted by subdividing buildings into a number of discrete compartments. These fire compartments are separated from one another by compartment walls and compartment floors made of a fire-resisting construction which hinders the spread of fire.

The risk of spread of fire needs to be minimised by:

- the selection and erection of appropriate materials and erection;
- wiring systems being installed so that the general building structural performance and fire safety are not reduced;
- ensuring that where safety depends on the direction of rotation of a motor, precautions have been taken to prevent reverse operation due to, for example, a phase reversal.

4.7.3 Preservation of electrical continuity of protective conductors

Protective conductors should:

- be suitably protected against mechanical and chemical deterioration and electrodynamic effects;
- ensure that every joint in metallic conduit is mechanically and electrically continuous;
- ensure that every connection and joint is accessible for inspection, testing and maintenance;

See Regulation 526.3 of BS 7671:2018 for exceptions to this rule.

- be protected by insulating sleeving complying with BS EN 60684;
- not have any dedicated devices such as operating sensors or coils etc. connected in series with the protective conductor – particularly where electrical Earth monitoring is used;
- ensure that any exposed, or conductive, part of an equipment is not used to form a protective conductor for another equipment item.

See Regulation 543.2 of BS 7671:2018 for exceptions to this rule

4.7.4 Prevention of harmful effects

Electrical equipment shall not cause any harmful effects to other equipment or interfere with the supply during normal service – particularly during switching operations.

4.7.5 Protection against fault current

All electrical equipment, including conductors, needs some kind of mechanical protection against electromechanical stresses caused by fault currents in order to prevent injury or damage to persons, livestock or property.

4.7.6 Protection against overcurrent

All persons and livestock shall be protected against injury due to excessive temperatures or electromechanical stresses caused by any overcurrents that might develop in live conductors.

4.7.7 Protection against power supply interruption

Where danger or damage is expected to arise due to an interruption of supply, suitable provisions shall be made in the installation or installed equipment.

4.7.8 Protection against thermal effects

Electrical installations shall be so arranged that:

- the risk of ignition of flammable materials due to high temperature or electric arc is minimised;
- during normal operation, there shall be minimal risk of burns to persons or livestock;
- persons, livestock, fixed equipment and fixed materials that are adjacent to electrical equipment are protected against any harmful effects of heat or thermal radiation emitted by the equipment.

4.7.9 Protection against voltage disturbances, and measures against electromagnetic influences

Persons and livestock shall be protected against injury, and property shall be protected against any harmful effects, as a consequence of:

- a fault between live parts of circuits supplied at different voltages;
- overvoltages such as those originating from atmospheric events or from switching;
- undervoltage and any subsequent voltage recovery.

The design of installations shall:

- ensure that they have an adequate level of immunity against electro-magnetic disturbances.
- have taken into consideration the anticipated electromagnetic emissions generated by the installation or the installed equipment.

4.7.10 Seismic effects

Wiring systems shall be selected and erected with due regard to the seismic hazards of the physical location of the installation.

Where the seismic hazards experienced are of low severity (i.e. AP2) or higher, particular attention shall be paid to:

- the fixing of wiring systems to the building structure;
- the connections between the fixed wiring and all items of essential equip-ment (e.g. safety services) being selected for their flexible quality.

4.8 Testing and inspection

4.8.1 Inspection

Note: Inspection shall precede testing and shall normally be done with that part of the installation under inspection being disconnected from the supply.

The inspection shall be made to verify that the installed electrical equipment is:

- in compliance; and
- correctly selected and erected; and
- not visibly damaged or defective so as to impair safety.

4.8.2 Periodic inspection and testing

Periodic inspection should be carried to confirm that the requirements for the disconnection times of protective devices are sufficient in order to ensure:

- safety of persons and livestock against the effects of electric shock and burns;
- protection against damage to property by fire and heat arising from an installation defect;
- confirmation that the installation is not damaged or weakened so as to impair safety;
- the identification of installation defects that may give rise to danger.

Appropriate safety precautions should be used when testing in a potentially explosive atmosphere (see BS EN 60079-17 and BS EN 61241-17 for further details).

4.8.3 Isolation and switching

It is essential that all voltage can be cut off (when required) from every installation, circuit and equipment, so as to prevent or remove danger. In order to
achieve this, fixed electric motors should be provided with an accessible and
efficient means of switching off that is instantly accessible, easily operated and
located so as to prevent danger.

4.8.4 Luminaires

A luminaire is the construction around the light source such as the mounting,
lampholder, reflector, shade or glass cover.

Any light source that could eject flammable materials in case of failure
should be equipped with a safety protective shield and any flexible cabling
between the fixing means and the luminaire shall be installed so that any
expected stresses in the conductors, terminals and terminations do not interfere with the safety of the installation.

4.8.5 Transformers and converters

Safety isolating transformers for extra-low-voltage lighting installations shall
comply with BS EN 61558-2-6 and:

- either the transformer shall be protected on the primary side by a protective device; or
- the transformer shall be short-circuit proof (both inherently and
  non-inherently).

4.8.6 Type of wiring and method of installation

The type of wiring system used and installation method should be
based on:

- the accessibility of the wiring system to persons and livestock;
- electromagnetic interference;
- the electromechanical stresses likely to occur due to short-circuit and
  Earth fault currents;
- the nature of the location;
- the nature of the structure supporting the wiring;
- voltage;
- other external influences (e.g. mechanical, thermal or those associated
  with fire) to which the wiring is likely to be exposed during the installation or whilst in service.

4.8.7 Uninterruptible power supply sources

A static-type uninterruptible power supply (UPS) source shall be capable of:

- operating distribution circuit-protective devices; and
- starting safety devices when operating in the emergency condition.

Author's end note

Now that we are aware of the mandatory and fundamental require-ments for safety protection contained in the Wiring Regulations, the Building Regulations and in their associated British, European and International Standards, in the next chapter we will look at the dif-ferent types of equipment, components, accessories and supplies for electrical installations that are currently available to satisfy these requirements.

5

Electrical equipment, components, accessories and supplies

Author's start note

The amount of different types of equipment, components, accessories and supplies for electrical installations currently available is enormous and any attempt to cover every type, model and/or manufacture would prove an impossible task for a book such as this.

The intention of this chapter, therefore, is to provide a catalogue of all the different types identified and referred to in the Wiring Regulations (e.g. luminaires, RCDs, plugs and sockets etc.) and then make a list of the specific requirements that are sprinkled throughout the Regulations.

*Similar to other chapters, please remember that these lists of requirements are **only** the author's impression of the most important aspects of the Wiring Regulations and electricians should **always** consult the latest edition of BS 7671 to satisfy compliance!*

5.1 Connection

Except where specifically designed for direct connection to flexible wiring, electrical equipment (see examples shown in Figure 5.1) should always be fixed so that connections between wiring and equipment will **not** be subject to undue stress or strain resulting from the normal use of the equipment.

Unenclosed equipment shall be mounted in a suitable mounting box or enclosure.

Socket outlets, connection units, plate switches and similar accessories shall be fitted to a mounting box.

Wherever equipment is fixed on or in cable trunking, skirting trunking or in mouldings, it shall not be fixed on covers which can be removed inadvertently.

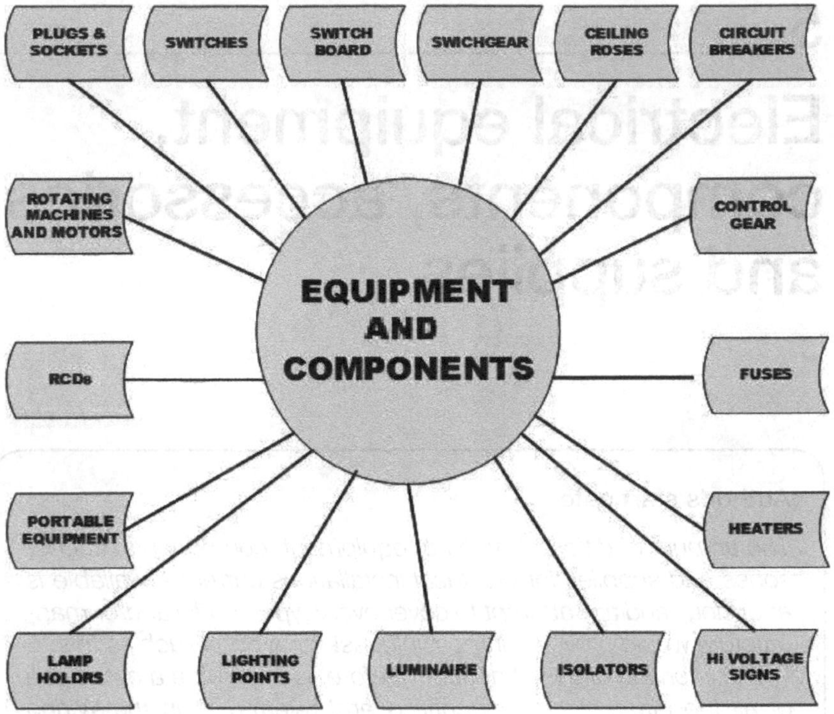

Figure 5.1 Electrical equipment and components.

Any wiring system within Group 2 medical locations shall be exclusively for the use of equipment and accessories within those locations.

5.1.1 Circuit breakers

Every circuit shall be provided with a means of isolation from all live supply conductors by a linked switch or a linked circuit-breaker which shall:

- provide protection against both overload and fault current;
- be capable of 'making' any overcurrent up to and including the maximum prospective fault current at the point where the device is installed;
- be inserted in the line conductor only;
- be inserted in an earthed neutral conductor (except where linked);
- control the supply to an electrode water heater or electrode boiler;
- be designed and installed so that it is not possible to modify the setting or the calibration of its overcurrent release without the use of either a key or a tool which results in a visible indication of its modified setting or calibration.

A main linked switch or linked circuit-breaker shall be provided as near as practicable to the origin of every installation as a means of switching the supply on load and as a means of isolation.

Linked circuit-breakers inserted in an earthed neutral conductor must be capable of breaking all of the related line conductors.

5.1.1.1 Protection against fault current only

A device shall be capable of breaking (and for a circuit-breaker, making) the fault current up to and including the prospective fault current.

 A circuit-breaker with a short-circuit release providing protection against fault current shall **only** be installed where overload protection is achieved by other means.

5.1.2 Isolators

The location of each isolator (frequently described as the '*disconnector*') shall include a suitable notice to avoid any possibility of confusion.

 Off-load isolators shall **not** be used for functional switching.

5.2 Fuses

Single-pole fuses (see descriptions shown in Figure 5.2) shall:

* only be inserted in the line conductor or an earthed neutral conductor;
* provide an indication of their intended rated current.

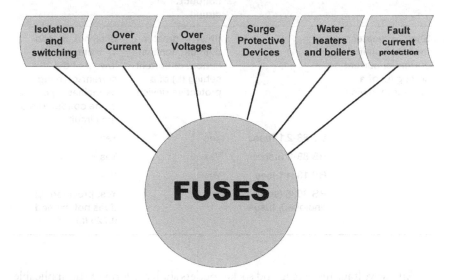

Figure 5.2 Fuses.

5.2.1 Devices for isolation and switching

For TN systems where it is intended that isolation and switching is only carried out by instructed persons, the means of switching the supply on load and the means of isolation can be provided by a suitably rated fuse carrier.

Table 5.1 provides an indication of the maximum values of earth fault loop impedance corresponding to a disconnection time of 5 s for a nominal voltage to Earth (U_o) of 230 V.

Table 5.1 Maximum earth fault loop impedance (Zs) for circuit breakers

General-purpose fuses								
Rating (A)	6	10	16	20	25	32	40	50
Z_s (Ω)	12.8	7.19	4.18	2.95	2.30	1.84	1.35	1/04

If the device is a general-purpose type fuse to BS 88-2.1, BS 88-6, or BS 1361 (or a semi-enclosed fuse to BS 3036) the conditions shown in Table 5.2 shall apply.

Table 5.2 Coordination between conductor and protective device

Design current (I_B) Type of fuse of the circuit		Lowest current-carrying capacity (I_z) of any conductor in a circuit	Operating current of any protective device (I_2)
Shall be greater than the nominal current or current setting (I_n) of a protective device		Shall be greater than the nominal current or current setting (I_n) of a protective device	Shall not exceed 1.45 times the lowest of the current-carrying capacities (I_z) of any of the conductors of the circuit
	BS 88-2.1 fuses	Yes	Yes
	BS 88-6 fuses	Yes	Yes
	BS 1361 fuses	Yes	Yes
	BS 3036 (semi-enclosed) fuses		Yes, provided (I_n) does not exceed 0.725 (I_z)

All low-voltage fused plug and socket outlets shall conform to the applicable British Standard listed in Table 5.3.

Table 5.3 Plugs and socket outlets for low-voltage circuits

Type of plug and socket outlet	Rating (A)	Applicable British Standard
Fused plugs and shuttered socket outlets, two-pole and Earth, for a.c.	13	BS 1363 (fuses to BS 1362)
Plugs, fused or non-fused, and socket-outlets, two-pole and Earth	2, 5, 15, 30	BS 546 (fuses, if any, to BS 646)
Plugs, fused or non-fused, and socket-outlets, protected-type, two-pole and Earth	5, 15, 30	BS 196
Plugs and socket-outlets (industrial type)	16, 32, 63, 125	BS EN 60309-2

5.2.2 Devices for protection against overcurrent

Every fuse shall preferably be of the cartridge type and shall provide an indication of its intended rated current applicable to the circuit it protects.

Fuse bases shall be arranged so as to stop the possibility of the fuse carrier making contact with the conductive parts of two adjacent fuse bases.

A fuse base using screw-in fuses shall be connected so that the centre contact is connected to the conductor from the supply and the shell contact is connected to the conductor to the load.

Fuses that are likely to be removed or replaced shall either:

- be marked with (or have an adjacent indication of) information concerning what type of replacement fuse link should be used; or
- be of a type such that there is **no possibility** of inadvertently replacing the fuse with one that has a higher rated current or a higher fusing factor than intended.

 Fuses (or combination units containing fuse links) should be installed in such a manner that they can **only** be removed or replaced without contact with live parts.

If a semi-enclosed fuse is used, it must either be in accordance with the manufacturer's instructions, or (in the absence of such instructions) fitted with a single element of tinned copper wire whose diameter is shown in Table 5.4.

5.2.3 Devices for protection against overvoltages

Surge protective devices (SPDs) are intended to:

- limit transient overvoltages of atmospheric origin that are transmitted via the supply distribution system and against switching overvoltages to electrical installations of buildings;
- protect against transient overvoltages caused by direct lightning strikes (or lightning strikes in the near vicinity) of buildings protected by a lightning protection system.

Table 5.4 Sizes of tinned copper wire for use in semi-enclosed fuses

Rated current of fuse (A)	Diameter of tinned copper wire (mm)
3	0.15
5	0.2
10	0.35
15	0.5
20	0.6
30	0.85
60	1.53
80	1.8
100	2.0

5.2.3.1 Selection and erection of surge protective devices (SPDs)

Where required or otherwise specified, SPDs shall be installed:

* near the origin of an installation; or
* in the main distribution assembly nearest the origin of an installation.

If additional SPDs are required to protect sensitive and critical equipment (e.g. hospital equipment and fire/security alarm systems etc.), these SPDs shall be coordinated with the SPDs installed upstream and installed as close as practicable to the equipment to be protected.

5.2.3.2 Connection of SPDs

SPDs at or near the origin of the installation shall be connected between specific conductors according to Table 5.5.

 Notes:

1. If more than one SPD is connected on the same conductor, they should be done so in a coordinated manner.
2. The SPD assembly shall be installed as close as possible to the origin of the installation.
3. Where the equipment to be protected is located close to the main distribution board, one SPD assembly may be sufficient.

 Sensitive and critical equipment will require protection in both common and differential modes to ensure further protection against switching transients.

Table 5.5 Types of protection for various LV systems

SPDs connected between:	TN-C-S, TN-S or TT: Installation in accordance with:		IT without distributed neutral
	Connection type 1	Connection type 2	
Each line conductor and neutral conductor	Optional	SPD required	Not applicable
Each line conductor and PE conductor	SPD required	Not applicable	SPD required
Neutral conductor and PE conductor	SPD required	SPD required	Not applicable
Each line conductor and PEN conductor	Not applicable	Not applicable	Not applicable
Line conductors	Optional	Optional	Optional

5.2.3.3 Selection of SPDs

SPDs shall be selected in accordance with the following regulations.

5.2.3.3.1 Selection with regard to voltage protection level (U$_p$)

SPDs shall be selected:

- in accordance with the impulse-withstand voltage of the equipment (or impulse immunity of critical equipment) to be protected, and the nominal voltage of the system shall be considered in selecting the preferred voltage protection level (U$_p$) value of the SPD;
- to provide a voltage protection level lower than the impulse-withstand capability of the equipment or (where the continuous operation of the equipment is critical) lower than the impulse immunity of the equipment;
- which will protect equipment from failure, remain operational during surge activity and withstand most temporary overvoltage conditions.

 Note: If the distance between the SPD and equipment to be protected ('*protective distance*') is greater than 10 m, oscillations could lead to a voltage at the equipment terminals of up to twice the SPD's voltage protection level. Consideration should, therefore, shall be given to providing additional coordinated SPDs, closer to the equipment, or the selecting SPDs with a lower voltage protection level.

Where the voltage protection level required cannot be obtained with a single SPD, additional, coordinated SPDs should be installed to ensure the required voltage protection level.

5.2.3.3.2 Selection with regard to continuous operating voltage

The maximum continuous operating voltage U_c of SPDs shall be equal to (or be higher than) that required by Table 5.6.

Table 5.6 Minimum required continuous operating voltage of the SPD, depending on supply system configuration

SPDs connected between:	TN-C-S, TN-S or TT	IT without distributed neutral
Line conductor and neutral conductor	$1.1\ U_{aspd}$	Not applicable
Each line conductor and PE conductor	$1.1\ U_{aspd}$ (Note 2)	Line to line voltage
Neutral conductor and PE conductor	U_{aspd}	Not applicable
Each line conductor and PEN conductor	Not applicable	Not applicable

Notes:

1. U_{aspd} is the nominal a.c. rms line voltage of the low-voltage system to Earth.
2. These values are related to worst-case fault conditions; therefore, the tolerance of 10% is not taken into account.
3. In extended IT systems, higher values of U_c may be necessary.

5.2.3.3.3 Selection with regard to temporary overvoltages (TOVs)

- SPDs shall be selected and erected in accordance with manufacturers' instructions.
- SPDs are expected to fail safely.

5.2.3.3.4 Selection with regard to nominal discharge current (I_{nspd}) and impulse current (I_{imp})

SPDs shall be selected according to their withstand capability, as classified in BS EN 61643-11 (for power systems) and in BS EN 61643-21 for telecommunication systems.

The nominal discharge current I_{nspd} of the SPD shall be not less than 5 kA, with a waveform characteristic 8/20 for each mode of protection.

Type 1 SPDs shall be installed where a structural lightning protection system is fitted or where the installation is supplied by an overhead line that is at risk of a direct lightning strike.

Where Type 1 SPDs are required, the value of I_{imp} shall be not less than 12.5 kA for each mode of protection, if I_{imp} cannot be calculated.

Where Connection Type 2 (CT 2) installations are required, the lightning impulse current I_{imp} for the SPD connected between the neutral conductor and the protective conductor shall be calculated in accordance with BS EN 62305-4.

 If the current value cannot be established, the value of I_{imp} must be not less than 50 kA for three-phase systems and 25 kA for single-phase systems.

5.2.3.3.5 Selection with regard to the prospective fault current and the follow current interrupt rating

The short-circuit withstand of the combination SPD and overcurrent protective device (OCPD), as stated by the SPD manufacturer shall be equal to, or higher than, the maximum prospective fault current expected at the point of installation.

When a follow current interrupt rating is declared by the manufacturer, it shall be equal to, or higher than, the prospective line to neutral fault current at the point of installation.

SPDs connected between the neutral conductor and the protective conductor in TT or TN systems, which allow a power frequency current follow after operation (e.g. spark gaps), shall have a follow current interrupt rating greater than (or equal to) 100 A.

In IT systems, the current follow interrupt rating for SPDs connected between the neutral connector and the protective conductor shall be the same as for SPDs connected between line and neutral.

 The OCPD may be either internal or external to the SPD.

5.2.3.3.6 Co-ordination of SPDs

SPDs shall be selected and erected such as to ensure coordination in operation.

5.2.3.3.7 Protection against overcurrent and consequences of the end of the life of the SPD

Protection against SPD short-circuits is provided by OCPDs.

5.2.3.4 Fault-protection integrity

 Fault protection shall remain effective in the protected installation, even in case of failures of SPDs.

In TN systems, automatic disconnection of supply shall be obtained by correct operation of the OCPD on the supply side of the SPD.

In TT systems this shall be obtained by the installation of SPDs either on the load side or on the supply side of an RCD.

Notes:

- If SPDs are installed on the load side of an RCD, they should be immune to surge currents of at least 3 kA.
- SPDs shall be provided with a status indicator (e.g. electrical, visual or audible alarm system) to indicate when the SPD no longer provides overvoltage protection.

5.2.3.4.1 Critical length of connecting conductors

To gain maximum protection:

- supply conductors shall be kept as short as possible so as to minimise additive inductive voltage drops across the conductors;
- current loops shall be avoided;
- the total lead length of supply conductors should not exceed 1.0 m (preferably no longer than 0.5 m);
- if an SPD is fitted in-line, the protective conductor shall also not exceed 1.0 m (preferably not longer than 0.5 m).

To ensure that SPD connections are as short and their inductance as low as feasible, SPDs may be connected to the main earthing terminal (or to the protective conductor) via the metallic enclosures of the assembly being connected to the protective conductor.

5.2.3.4.2 Cross-sectional area of connecting conductors

The connecting conductors of SPDs shall either:

- have a cross-sectional area of not less than 4 mm² copper (or equivalent) if the cross-sectional area of the line conductors is greater than or equal to 4 mm²; or
- have a cross-sectional area not less than that of the line conductors, where the line conductors have a cross-sectional area less than 4 mm².

Note: The minimum cross-sectional area for Type 1 SPDs shall be 16 mm² copper, or equivalent, where there is a structural lightning protection system.

5.2.4 Electrode water heaters and boilers

If an electrode water heater or electrode boiler is not piped to a water supply or in physical contact with any earthed metal (and where the electrodes and the water in contact with the electrodes are so shielded in insulating material that they cannot be touched while the electrodes are live), a fuse in the line conductor may be used as a substitute for the circuit-breaker. In this case, the shell of the electrode water heater or electrode boiler need not be connected to the neutral of the supply.

Single-phase water heaters and boilers with an **uninsulated** heating element immersed in the water shall **not** have a fuse fitted in the neutral conductor, in any part of the circuit between the heater or boiler or the origin of the installation.

5.2.5 Functional switching devices

Functional switching devices:

- shall be suitable for the most onerous duty that they are intended to perform;
- may control the current without necessarily opening the corresponding poles.

 Fuses and links shall **not** be used for functional switching.

5.2.6 Protection against fault current only

Fuses that provide protection against fault current shall **only** be installed where overload protection is achieved by other means.

5.3 Heaters

Electrical equipment items (such as heaters – see Figure 5.3) need to be prevented from exceeding the following temperatures:

- 90 °C under normal conditions; and
- 115 °C under fault conditions.

5.3.1 Electrode water heaters and boilers

Electrode water heaters and boilers:

- shall **only** be connected to an a.c. system that is controlled by a linked circuit-breaker;
- shall include an RCD if it is directly connected to a supply whose voltage exceeds low voltage;
- shall have a shell bonded to the metallic sheath and/or armour of the incoming supply cable;
- shall have a shell which, in a three-phase low-voltage supply, shall be connected to the neutral of the supply as well as to the earthing conductor.

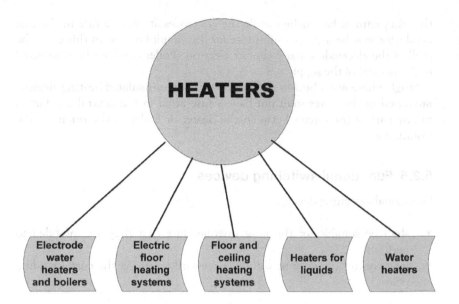

Figure 5.3 Heaters.

If the supply to an electrode water heater or boiler is single-phase and one electrode is connected to a neutral conductor earthed by the distributor, then the shell of the electrode water heater or boiler shall be connected to the neutral of the supply as well as to the earthing conductor.

If the electrode water heater or boiler is not piped to a water supply or in physical contact with any earthed metal, a fuse in the line conductor may be substituted for the circuit-breaker and the shell of the equipment need not be connected to the neutral of the supply.

5.3.2 Floor and ceiling heating systems

This section applies to the installation of electric floor and ceiling heating systems which are erected as either thermal storage heating systems or direct heating systems. It does not apply to the installation of wall heating systems.

5.3.2.1 Compliance with standards

- Flexible sheet heating elements shall comply with the requirements of BS EN 60335-2-96.
- Heating cables shall comply with IEC 60800.

5.3.2.2 Additional protection – RCDs

If a circuit is supplying heating equipment of Class II (or equipment with equivalent insulation) then it shall be provided with additional protection by the use of an RCD with a rated residual operating current not exceeding 30 mA.

 Time-delayed type RCDs are not permitted.

5.3.2.2.1 Electric floor heating systems

The protective measure *'protection by electrical separation'* is not permitted for underfloor installations.

Electric floor heating systems in locations containing a bath or shower shall only utilise heating cables or thin-sheet flexible heating elements provided that they have either a metal sheath or a metal enclosure or a fine-mesh metallic grid – unless the protective-measure SELV is provided for the floor heating system.

5.3.2.3 *External influences*

Heating units for installation in ceilings shall have a degree of protection of not less than IPX1.

Heating units for installation in a floor of concrete or similar material shall have a degree of ingress protection not less than IPX7 and shall have the appropriate mechanical properties.

 Note: For cold tails (circuit wiring) and control leads installed in the zone of heated surfaces, the increase of ambient temperature shall be taken into account.

5.3.2.4 *Heating units*

To avoid the overheating of floor or ceiling heating systems in buildings, one or more of the following measures shall be applied within the zone where heating units are installed, to limit the temperature to a maximum of 80 °C:

*   appropriate design of the heating system;
*   appropriate installation of the heating system in accordance with the manufacturer's instructions;
*   use of protective devices.

Heating units shall be:

*   connected to the electrical installation via cold tails or suitable terminals;
*   inseparably connected to cold tails (e.g. by a crimped connection).

As the heating unit may cause higher temperatures or arcs under fault conditions, special measures should be taken when the heating unit is installed close to easily ignitable building structures, such as placing on a metal sheet, in metal conduit or at a distance of at least 10 mm in air from the ignitable structure.

5.3.2.5 Heating-free areas

For the necessary attachment of room fittings, heating-free areas shall be provided so that the heat emission is not prevented by such fittings.

5.3.2.6 Identification and notices

The designer or installer of the installation/heating system shall provide a plan (fixed to, or adjacent to, the distribution board of the heating system) for each heating system, containing the following details:

- number of heating units installed;
- manufacturer and type of heating units;
- product information and instructions for installation;
- layout of the heating units in the form of a sketch, drawing, or picture;
- heated area;
- length/area of heating units;
- position/depth of heating units;
- position of junction boxes;
- position of cables, earthed conductive, shields etc.;
- rated voltage;
- rated resistance (cold) of heating units;
- rated current of the OCPD;
- rated residual operating current of the RCD;
- surface power density;
- the insulation resistance of the heating installation and the test voltage used;
- the leakage capacitance.

5.3.2.7 Operational conditions

Precautions should be taken not to stress the heating unit mechanically (e.g. the material that is going to be used to protect the finished installation must cover the heating unit as soon as possible).

5.3.2.8 Presence of solid foreign bodies

Where heating units are installed there shall be heating-free areas where drilling and fixing by screws, nails and the like are permitted.

The installer shall inform other contractors that no penetrating means (such as screws for door stoppers) shall be used in the area where floor or ceiling heating units are installed.

5.3.2.9 Prevention of mutual detrimental influences

Heating units shall not cross expansion joints of the building or structure.

5.3.2.10 Protection against burns

In floor areas where contact with skin or footwear is possible, the surface temperature of the floor shall be limited (e.g. to 35 °C).

5.3.2.11 Protection against electric shock

The following protective measures are **not** permitted:

- placing out of reach;
- non-conducting location and Earth-free local equipotential bonding;
- electrical separation.

5.3.3 Heaters for liquids or other substances having immersed heating elements

Heaters for liquid and/or other substances shall have an automatic device to prevent a dangerous rise in temperature.

5.3.4 Water heaters having immersed and uninsulated heating elements

All metal parts of the heater or boiler which are in contact with the water (other than current-carrying parts) shall be solidly and metallically connected to a metal water pipe through which the water supply to the heater or boiler is provided, and that water pipe shall be connected to the main earthing terminal by means independent of the circuit protective conductor.

Water heaters and boilers shall be permanently connected to the electricity supply via a double-pole linked switch which is either:

- separate from and within easy reach of the heater/boiler; or
- part of the boiler/heater.

 If the heater or boiler is installed in a room containing a fixed bath, it is essential that the switch complies with Section 701 of BS 7671:2018.

Single-phase water heaters and boilers with an uninsulated heating element immersed in the water shall **not** have a single-pole switch, non-linked circuit-breaker or fuse fitted in the neutral conductor, in any part of the circuit between the heater or boiler or the origin of the installation.

5.4 Luminaires

A *'Luminaire'* is an equipment which distributes, filters or transforms the light transmitted from one or more lamps. It includes all the parts necessary for supporting, fixing and protecting the lamps (but not the actual lamps themselves),

Figure 5.4 Luminaires.

and where necessary circuit auxiliaries, together with the means of connecting them to the supply. (See Figure 5.4 for some types of luminaires).

Luminaires shall be installed so that dust or fibres cannot accumulate in dangerous amounts, particularly in locations where dust and fibres are likely to present a fire hazard.

5.4.1 General requirements for installations

Every luminaire shall comply with the relevant standard for the manufacture and test of that luminaire and shall also be selected and erected in accordance with the manufacturer's instructions.

 Note: Luminaires without transformers or converters (but which are fitted with extra-low-voltage lamps connected in series) are normally considered as low-voltage equipment as opposed to extra-low-voltage equipment.

5.4.2 Installing luminaires

Electrical installations shall be so arranged that they are appropriately controlled, and:

- the risk of ignition of flammable materials due to high temperature or electric arc is minimised;
- during normal operation of the electrical equipment, there shall be minimal risk of burns to persons or livestock.

Adequate means to fix the luminaire shall be provided, such as mechanical accessories (e.g. hooks or screws), boxes or enclosures which are able to support luminaires or associated supporting devices for connecting a luminaire.

5.4.2.1 Ceiling roses

A ceiling rose or lampholder for a filament lamp shall **not** be installed in any circuit operating at a voltage normally exceeding 250 volts.

A ceiling rose shall **not** be used for the attachment of more than one outgoing flexible cable unless it is specially designed for multiple pendants.

5.4.2.2 Electronic switching devices

Electronic switching devices should, where possible, include a neutral conductor.

BS EN 61184 Requirement

"Bayonet lampholders B15 and B22 shall comply with BS EN 61184 and shall have the temperature rating T2 described in that standard."

5.4.2.3 Fixed lighting points

At each fixed lighting point one of the following Standards shall be used:

- a ceiling rose to BS 67;
- a luminaire-supporting coupler to BS 6972 or BS 7001;
- a batten lampholder or a pendant set to BS EN 60598;
- a luminaire to BS EN 60598;
- a suitable socket outlet to BS 1363-2, BS 546 or BS EN 60309-2;
- a plug-in lighting distribution unit to BS 5733;
- a connection unit to BS 1363-4;
- appropriate terminals enclosed in a box complying with the relevant part of BS EN 60670 series or BS 4662;
- a device for connecting a luminaire outlet according to BS IEC 61995-1.

Note: In suspended ceilings one plug-in lighting distribution unit may be used for a number of luminaires.

5.4.2.4 Lampholders

A lampholder for a filament lamp shall **not** be installed in any circuit operating at a voltage normally exceeding 250 volts.

Bayonet lampholders B15 and B22 shall comply with BS EN 61184 and shall have the temperature rating T2 described in that standard.

In circuits of a TN or TT system (except for E14 and E27 lampholders complying with BS EN 60238) the outer contact of every Edison screw or single-centre bayonet-cap type lampholder shall be connected to the neutral conductor. This regulation also applies to track-mounted systems.

Insulation-piercing lampholders shall **not** be used unless the cables and lampholders are compatible, and providing the lampholders are non-removable once fitted to the cable.

In exhibitions, shows and stands, lighting circuits incorporating B15, B22, E14, E27 or E40 lampholders shall be protected by an OCPD of maximum rating 16 A.

Lampholders with ignitability characteristic 'P' as specified in BS 476 Part 5 (or where separate overcurrent protection is provided) shall not be connected to any circuit where the rated current of the overcurrent protective device exceeds the appropriate value stated in Table 5.7, unless the wiring is enclosed in earthed metal or insulating material.

Table 5.7 Overcurrent protection of lampholders

Type of lampholder			Maximum rating of the overcurrent device protecting the circuit (A)
Bayonet lampholder (BS EN 61184)	B15	SBC	6
	B22	BC	16
Edison screw lampholder (BS EN 60238)	E14	SES	6
	E27	ES	16
	E40	GES	16

5.4.2.5 Maximum permissible weight

In places where the fixing means is intended to support a pendant luminaire, the fixing (unless enhanced by the installer) shall be capable of carrying a mass of more than 5 kg.

The weight of luminaires and their eventual accessories shall be compatible with the mechanical capability of the ceiling or suspended ceiling, or supporting structure where installed.

5.4.2.6 Protection against fire

When selecting and erecting a luminaire, the thermal effects of radiant and converted energy on the surroundings shall be taken into account, including:

- the maximum permissible power dissipated by the lamps;
- the fire resistance of adjacent material:
  - o at the point of installation; and
  - o in the thermally affected areas;

- the minimum distance to combustible materials, including material in the path of a spotlight beam.

Electrical installations shall be so arranged that:

- the risk of ignition of flammable materials due to high temperature or electric arc is minimised;
- during normal operation of the electrical equipment, there shall be minimal risk of burns to persons or livestock.

5.4.2.7 Safe installation of luminaires

Luminaires shall be:

- kept at an adequate distance from combustible materials;
- protected from foreseeable mechanical stresses.

A luminaire with a lamp that could eject flammable materials in case of failure should be equipped with a safety protective shield.

Circuits supplying luminaires at a voltage exceeding low voltage shall be provided with:

- an interlock on a self-contained luminaire;
- an effective local means for the isolation of the circuit from the supply;
- a switch with a lock or removable handle, or a distribution board which can be locked.

Every luminaire shall:

- be appropriate for the location; and
- be provided with an enclosure providing a degree of protection of at least IP5X; and
- have a limited surface temperature in accordance with BS EN 60598-2-24; and
- be of a type that prevents lamp components from falling from the luminaire.

If a luminaire is installed in a pelmet, care should be taken to ensure that there are no adverse effects due to the presence or operation of curtains or blinds.

Luminaires should only be installed and used at a reasonable (i.e. sufficient) distance from combustible materials.

 All luminaire components (e.g. lamps and other components) shall be protected against all foreseeable mechanical stresses. Such protective means should not be fixed to lampholders unless they are actually an integral part of the luminaire or are fitted in accordance with the manufacturer's instructions.

The construction of a luminaire (e.g. parts of a cable or flexible cord mounted inside a luminaire) shall be suitable for the temperatures likely to be encountered, or shall be provided with additional insulation suitable for those temperatures.

The connection of fixed, suspended, current-using equipment (such as a luminaire for a fixed installation) shall be made by cable with flexible cores.

Luminaire-supporting couplers are designed specifically for the mechanical support and electrical connection of luminaires and must not be used for the connection of any other equipment.

5.4.2.8 Small spotlights and projectors

Unless otherwise recommended by the manufacturer, small spotlights or projectors should be installed at the following minimum distance from combustible materials:

- Rating up to 100 W: 0.5 m;
- over 100 and up to 300 W: 0.8 m;
- over 300 and up to 500 W: 1.0 m.

5.4.2.9 Stroboscopic effect

Lighting for premises where machines with moving parts are in operation should consider the stroboscopic effects which can give a misleading impression of moving parts being stationary. Such effects may be avoided by selecting luminaires with a suitable lamp controlgear (such as high-frequency controlgear) or by distributing lighting loads across all phases of a three-phase supply.

5.4.2.10 Through wiring

The installation of through wiring in a luminaire is **only** permitted if the luminaire is designed for such wiring.

A cable for through wiring shall be selected in accordance with the temperature information on the luminaire or on the manufacturer's instruction sheet.

Groups of luminaires divided between the three line conductors of a three-phase system with only one common neutral conductor shall be provided with at least one device that simultaneously disconnects all three line conductors.

5.4.3 Underwater luminaires for swimming pools

A luminaire for use in the water or in contact with the water shall be fixed and shall comply with BS EN 60598-2-18.

Underwater lighting located **behind** watertight portholes, and serviced from behind, shall comply with the appropriate part of BS EN 60598 and be installed in such a way that no intentional or unintentional conductive connection between any exposed-conductive-part of the underwater luminaires and any conductive parts of the portholes can occur.

 For more information about the requirements for underwater luminaires in swimming pools, please see BS 7671:2018 Chapter 702.

5.4.4 Luminaires in caravans and motor caravans

Each luminaire in a caravan shall preferably be fixed directly to the structure or lining of the caravan.

Where a pendant luminaire is installed in a caravan, it shall be made secure so as to prevent damage when the caravan is in motion.

Accessories for the suspension of pendant luminaires shall be chosen with respect to the mass suspended and the forces associated with vehicle movement.

 For more information about the requirements for caravans, please see BS 7671:2018 Chapter 721.

5.4.5 Luminaires and lighting installations in exhibitions shows and stands

Stand installations containing a concentration of electrical equipment, luminaires or lamps which are liable to generate excessive heat shall not be installed unless adequate ventilation is available (e.g. a well-ventilated ceiling constructed of incombustible material).

Extra-low-voltage (ELV) lighting systems for filament lamps shall comply with BS EN 60598-2-23.

Insulation-piercing lampholders shall **not** be used unless the cables and lampholders are compatible, and providing the lampholders are non-removable once they have been fitted to the cable.

Luminaires mounted below 2.5 m (arm's reach) from floor level, or otherwise accessible to accidental contact, shall be firmly and adequately fixed, and sited or guarded so as to prevent possible risk of injuring persons or igniting materials.

 For more information about the requirements for installations in exhibitions shows and stands, please see BS 7671:2018 Chapter 711.

5.4.6 Luminaires in fountains

A luminaire installed in zones 0 or 1 shall be fixed and shall comply with BS EN 60598-2-18.

Electrical equipment in zones 0 or 1 shall be provided with mechanical protection to medium severity (AG2) – for example by use of mesh glass or by grids which can only be removed by the use of a tool.

 For more information about the requirements for luminaires in fountains, please see BS 7671:2018 Chapter 702.

5.4.6.1 Electric discharge lamp installations

Any luminous tube, sign or lamp acting as an illuminated unit on a stand, or as an exhibit, which has a nominal power supply voltage higher than 230/400 V a.c., shall meet the following:

- location – the sign or lamp shall be installed out of arm's reach or shall be adequately protected to reduce the risk of injury to persons;
- installation – the facia or stand fitting material behind luminous tubes, signs or lamps shall be non-ignitable;
- emergency switching devices – a separate circuit shall be used to supply signs, lamps or exhibits, which shall be controlled by an emergency switch. This switch must be easily visible, accessible and clearly marked.

5.4.7 Luminaires in temporary installations

Every luminaire and decorative lighting chain shall:

- have a suitable IP rating;
- be installed so as not to impair its ingress protection; and
- be securely attached to the structure or support intended to carry it.

Its weight shall not be carried by the supply cable, unless it has been selected and erected for this purpose.

Luminaires and decorative lighting chains mounted less than 2.5 m (arm's reach) above floor level, or otherwise accessible to accidental contact, shall be firmly fixed and so sited or guarded as to prevent risk of injury to persons or ignition of materials.

Where transportable floodlights are used, they shall be mounted so that the luminaire is inaccessible.

Supply cables shall be flexible and have adequate protection against mechanical damage.

Luminaires and floodlights shall be fixed and protected so that any focusing or concentration of heat is unlikely to cause ignition of any material.

 For more information about the requirements for temporary installations, please see BS 7671:2018 Chapter 740.

5.4.8 Medical locations

In Group 2 medical locations, where PELV is used, exposed-conductive-parts of equipment (e.g. operating theatre luminaires) shall be connected to the circuit protective conductor.

In the event of mains power failure, the changeover period to the safety services source shall not exceed 15 s.

Escape route luminaires shall be arranged on alternate circuits.

For more information about the requirements for medical locations, please see BS 7671:2018 Chapter 701.

5.4.9 Outdoor lighting installation

An outdoor lighting installation comprises one or more luminaires, a wiring system and accessories. Such lighting installations include those for:

- roads, parks, car parks, gardens, places open to the public, sporting areas, illumination of monuments and floodlighting;
- places such as telephone kiosks, bus shelters, advertising panels and town plans;
- road signs;
- temporary festoon lighting.

For more information about the requirements for outdoor lighting, please see BS 7671:2018 Chapter 714.

5.4.9.1 Highway power supplies and street furniture

The protective measures of placing out of reach and obstacles shall **not** be used, and nor shall Earth-free local equipotential bonding, except where:

- the maintenance of equipment is restricted to skilled persons who are specially trained;
- items of street furniture are within 1.5 m of a low-voltage overhead line.

Access to the light source of a luminaire which is at a height of less than 2.80 m above ground level shall only be possible after removing a barrier or an enclosure requiring the use of a tool.

The earthing conductor of a street electrical fixture shall have a minimum copper equivalent cross-sectional area not less than that of the supply neutral conductor at that point or not less than 6 mm², whichever is the smaller.

For more information about the requirements for all outdoor lighting, please see BS 7671:2018 Chapter 714.

5.4.9.2 Double or reinforced insulation

For an outdoor lighting installation where the protective measure for the whole installation is by double or reinforced insulation:

- no protective conductor is necessary; and
- the conductive parts of the lighting column shall not be intentionally connected to the earthing system.

Suspension devices for extra-low-voltage luminaires, including supporting conductors, shall be capable of carrying five times the mass of the luminaires (including their lamps) intended to be supported. In all cases, the amount of support provided shall **not** be less than 5 kg.

5.4.9.3 Electric discharge lamp installations

The installed location of a luminous tube, sign or lamp shall be out of arm's reach or shall be adequately protected to reduce the risk of injury to persons.

5.5 Plug and socket outlets

As shown in Figure 5.5, there are numerous requirements for outdoor sockets and the following are amongst the most important requirements as stated in BS 7671:2018 and other BS, EN and ISO associated Standards.

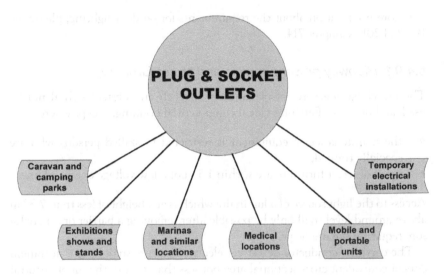

Figure 5.5 Plug and socket outlets.

 A plug and socket outlet shall **not** be selected as a device for emergency switching.

A plug and socket outlet may be inserted in the main supply circuit to enable it to be switched off for mechanical maintenance.

Every plug and socket outlet (except for SELV) shall be of the non-reversible type, with facilities for the connection of a protective conductor.

If the rating of the plug and socket outlet does not exceed 16 A, it shall be capable of cutting off the full load current of the relevant part of the installation.

A socket outlet on a wall or similar structure shall be mounted at a height above the floor or a working surface that minimises the risk of mechanical damage to the socket outlet (or to an associated plug and its flexible cord) which might be caused during insertion, use or withdrawal of the plug.

Equipment which has a protective conductor current exceeding 3.5 mA but not exceeding 10 mA shall either be permanently connected to the fixed wiring of the installation without the use of a plug and socket outlet or be connected by means of a plug and socket complying with BS EN 60309-2.

Equipment which has a protective conductor current exceeding 10 mA shall be connected to the supply:

- via a flexible cable with a plug and socket outlet; or
- permanently via the wiring of the installation; or
- via a flexible cable with a plug and socket outlet complying with BSEN 60309-2; or
- via a protective conductor with an Earth monitoring system.

It shall **not** be possible for any pin of a plug to make contact with any live contact of any socket outlet within the same installation other than the type of socket outlet for which the plug is designed.

Socket outlets for household and similar use shall be of the shuttered type and, for an a.c. installation, shall preferably be of a type complying with BS 1363.

A plug and socket outlet not complying with BS 1363, BS 546, BS 196 or BS EN 60309-2 may be used in single-phase a.c. or two-wire d.c. circuits operating at a nominal voltage not exceeding 250 volts for:

- the connection of an electric clock;
- the connection of an electric shaver;
- a circuit having special characteristics (example where it is necessary to distinguish the function of the circuit).

Examples of the rating for low voltage plugs and sockets is shown in Table 5.8

Table 5.8 Plugs and socket outlets for low-voltage circuits

Type of plug and socket outlet	Rating (A)	Applicable British Standard
Fused plugs and shuttered socket outlets	13	BS 1363 (fuses to BS 1362)
Plugs, fused and non-fused, and socket outlets, two-pole and Earth	2, 5, 15, 30	BS 546 (fuses to BS 646)
Plugs and sockets of an industrial type	16, 32, 63, 125	BS EN 60309-2

Except for the plug of a plug and socket outlet identified in Table 55.1 of BS 7671:2018, equipment of overvoltage categories I and II should **not** be used for isolation.

A plug and socket outlet shall **not** be used as a device for connecting a water heater and/or boiler to the supply.

If mobile equipment is likely to be used, then it should be fed from an adjacent and conveniently accessible socket outlet.

For final circuits containing a number of socket outlets (where it is known that the total protective conductor current in normal service will exceed 10 mA) the circuit shall be provided with a high-integrity protective conductor connection.

The following arrangements of final circuit are acceptable:

- a radial final circuit with a ring protective conductor;
- a radial final circuit with a single protective conductor.

Plugs and socket outlets in a SELV system shall **not** have a protective conductor contact.

For more information about the requirements for plug and socket outlets, please see the relevant sections contained in BS 7671:2018 Chapters 701–753.

5.5.1 Caravan and camping parks

It is recommended that at least one socket outlet is provided for each caravan pitch. Each socket outlet and its enclosure forming part of the caravan pitch electrical supply equipment shall:

- comply with BS EN 60309-2 (i.e. the main standard for plugs, socket outlets and couplers for industrial purposes);
- meet the degree of protection of at least IP44 in accordance with BS EN 60529;
- be provided with individual overcurrent protection;
- be protected individually by an RCD;
- have a current rating of not less than 16 A.

The socket outlets shall be placed at a height of 0.5 m to 1.5 m from the ground to the lowest part of the socket outlet. In special cases (due to environmental conditions such as risk of flooding or heavy snowfall) the maximum height is permitted to exceed 1.5 m.

Socket outlet protective conductors shall **not** be connected to any PME earthing service.

For more information about the requirements for caravan and camping parks, please see BS 7671:2018 Chapters 708, 721 and 722.

5.5.2 Exhibitions, shows and stands

An adequate number of socket outlets shall be installed to allow user requirements to be met safely.

Floor-mounted socket outlets shall be sufficiently protected from accidental ingress of water and have sufficient strength to be able to withstand the
expected traffic load.

 For more information about the requirements for exhibitions, shows and
stands, please see BS 7671:2018 Chapter 711.

5.5.3 Marinas and similar locations

With regard to the current regulations concerning marinas and similar locations, there is a requirement that:

 Socket outlet protective conductors shall **not** be connected to a PME earthing
facility.

In general:

- 200–250 V, 16 single-phase socket outlets shall be provided;
- a maximum of four socket outlets shall be grouped together in one
  enclosure;
- one socket outlet shall supply only one pleasure craft or houseboat.

 BS 7671:2018 Fig 709.3 indicates the recommended instruction notice that
should be placed in marinas adjacent to each group of socket outlets.

Every socket outlet shall:

- meet the degree of protection of at least IP44 or be protected by an
  enclosure;
- be located as close as practicable to the berth to be supplied;
- be installed in the distribution board or in a separate enclosure;
- comply with BS EN 60309-1 above 63 A and BS EN 60309-2 up to 63 A;
- be placed at a height of not less than 1 m above the highest water level.

 Note: In the case of floating pontoons or walkways only, this height may be
reduced to 300 mm above the highest water level.

 For more information about the requirements for marinas and similar locations, please see BS 7671:2018 Chapter 709.

5.5.4 Medical locations

 In medical locations, **all** socket outlets intended to supply medical electrical
equipment shall be unswitched.

In Group 1 and 2 medical locations the resistance between the Earth terminals of a protective conductor's socket outlets and any extraneous-conducive-part (e.g. the equipotential bonding busbar) shall not exceed 0.2 Ω.

Socket outlets and switches installed below any medical-gas outlets (such as those used for oxidising or flammable gases) shall be at least 0.2 m from the outlet (centre to centre), in order to minimise the risk of ignition of the flammable gases.

5.5.4.1 Protection by RCDs

Socket outlets in a medical location that are protected by an RCD should reduce the possibility of unwanted tripping of the RCD due to excessive protective conductor currents produced by equipment in normal operation.

5.5.4.2 Socket outlets in the medical IT system for Group 2 medical locations

Each socket outlet at each patient's place of treatment in a medical location (e.g. bedheads), the configuration of socket outlets shall be as follows:

- each socket outlet supplied by an individually protected circuit; or
- several socket outlets separately supplied by a minimum of two circuits.

Socket outlets used on medical IT systems **must** be coloured blue and clearly and permanently marked 'Medical equipment only'.

For more information about the requirements for medical locations, please see BS 7671:2018 Chapter 710.

5.5.5 Mobile and transportable units

Where mobile equipment is likely to be used, the equipment can be fed from an adjacent and conveniently accessible socket outlet, taking into account the length of flexible cord normally fitted to portable appliances and luminaires.

The socket outlet, plugs and connecting devices used to connect the unit to the supply shall comply with BS EN 60309-2 and meet the following requirements:

- plugs shall be within an enclosure of insulating material;
- connecting devices located outside the unit shall afford a degree of protection of not less than 1P44 when in use;
- enclosures containing the connector shall provide a degree of protection of at least 1P55 when not connected.

5.5.5.1 Additional RCD protection

In a.c. systems, additional protection by means of an RCD shall be provided for mobile equipment with a current rating not exceeding 32 A when used outdoors.

For more information about the requirements for mobile and transportable units, please see BS 7671:2018 Chapter 717.

5.5.6 Temporary electrical installations

An adequate number of socket outlets shall be installed to allow the user's requirements to be met safely.

 Note: In booths, stands and for fixed installations, one socket outlet for each square metre or linear metre of wall is generally considered adequate.

The design and position of temporary electrical installations should always take into account that:

- socket outlets dedicated to lighting circuits that are placed out of arm's reach shall be encoded or marked according to their purpose;
- when used outdoor, plugs, socket outlets and couplers must comply with BS EN 60309;
- all plug and socket outlets (except for SELV) must be of the non-reversible type and capable of being connected to a protective conductor;
- all socket outlets deigned for household and similar use shall be of the shuttered type and, for an a.c. installation, preferably of a type complying with BS 1363.

 In certain instances, additional requirements may be necessary for isolation and switching, and BS 7671:2018 Table 537.4 summarises the functions provided by the devices for isolation and switching, together with indication of the relevant product standards.

 A plug and socket outlet shall **not** be used as a device for connecting a water heater and/or boiler to the supply.

If a final circuit has a number of socket outlets (or connection units intended to supply two or more items of equipment and where it is known that the total protective conductor current in normal service will exceed 10 mA), the circuit must be provided with:

- a high-integrity protective conductor connection; or
- a radial final circuit with a ring protective conductor; or
- a radial final circuit with a single protective conductor.

 Plugs and socket outlets in a SELV system shall **not** have a protective conductor contact.

 For more information about the requirements for temporary installations, please see BS 7671:2018 Chapter 740.

5.6 Rectifiers

A device for protection against fault current need **not** be provided for a
conductor connecting a rectifier where the protective device is placed
in the panel, provided that the following conditions are simultaneously
fulfilled:

- the wiring is carried out in such a way as to reduce the risk of fault to a
  minimum; and
- the wiring is installed in such a manner as to reduce to a minimum the
  risk of fire or danger to persons; and
- all booths (e.g. in fairgrounds etc.) containing rectifiers are adequately
  ventilated and the vents are not obstructed when in use.

5.7 Rotating machines and motors

All equipment, cables and circuits of rotating machines that carry the starting, accelerating and/or load currents of a motor shall be capable of sustain-
ing a current at least equal to the full-load current rating of the rotating
machine.

If the electric motor is intended to work intermittently, the cumulative
effects of the starting or braking currents upon the temperature rise of the
equipment of the circuit shall be taken into account.

Every electric motor having a rating exceeding 0.37 kW shall include a system for protection against overload of the motor.

Every motor shall be provided with means to prevent automatic restarting
after a stoppage due to a drop in voltage (or failure) of supply as unexpected
restarting of the motor might cause danger, **except** where failure to restart is
likely to cause greater danger.

Note: These requirements do not exclude arrangements for starting a motor at
intervals by an automatic control device, provided that other precautions are
taken against danger from unexpected restarting.

Lighting for premises which have machines with moving parts in operation
should consider the stroboscopic effects, as these can give a misleading impression of moving parts being stationary.

These sort of side effects can be avoided by using luminaires which have a suitable lamp with a high frequency controlgear, or by distributing lighting loads
across all the phases of a three-phase supply.

5.8 Supplies

As shown in Figure 5.6 there are numerous types of electrical supply depending on their intended individual usage.

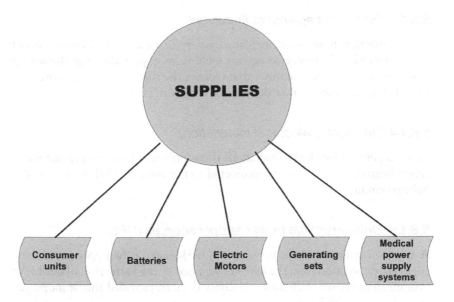

Figure 5.6 Types of supplies.

5.8.1 Batteries

In 1957 a *'battery'* was discovered in Baghdad. It consisted of a ceramic pot, a tube of copper, and a rod of iron, and was made by the Parthians, who ruled Baghdad from 247 to 224 BC. It is assumed to have functioned as a galvanic cell used to electroplate silver!

For safety, all batteries should have some form of basic protection against misuse (e.g. insulation or enclosure), and large industrial batteries be installed in a secure location.

A battery may be used as a source for SELV and PELV systems.

5.8.1.1 Central power supply sources

Batteries which are used as a central power source shall be of a maintenance-free, heavy-duty and industrial design that complies with BS EN 60623. They should have a minimum life span of ten years and be either vented or valve-regulated.

5.8.1.2 Low-power supply sources

The power output of a low-power supply systems is limited to 1500 W for 1-hour duration. The batteries shall be of a heavy-duty industrial design, comply with the requirements stated in BS EN 60623, and have a minimum declared life of 5 years.

5.8.1.3 Protection against fault current

The connection from an accumulator battery to its associated control panel does not need to be protected against fault current, **provided** that the wiring has been completed in a manner that reduces the risk of a fault occurring and installed so that there is no risk of fire or danger to persons.

5.8.1.4 Temporary electrical installations

In a temporary electrical installation, the supply to a battery-operated emergency lighting circuit shall be connected to the same RCD that protects the lighting circuit.

5.8.1.5 Uninterruptible power supply sources (UPS)

A static, uninterruptible power supply (UPS) source (when operating in the emergency condition from the inverter supplied by the battery) shall be able to operate all distribution circuits and safety devices, provided that it meets the requirements of BS EN 62040.

5.8.2 Consumer units

A consumer unit (sometimes known as a 'consumer control unit' or 'electricity control unit') is a particular type of distribution board for the control and distribution of electrical energy, primarily in domestic premises. It includes a manually operated method for isolation (both poles of the incoming circuit(s)) and assemblies of one or more fuses, circuit-breakers, RCDs, signalling and other devices that have been purposely manufactured for this purpose.

One of the requirements of the Regulations is that all consumer units shall comply with the requirements of BS EN 61439-3 and specifically, **all** circuits and final circuits **shall** be provided with a means of switching for interrupting the supply on load

 Note: This regulation particularly applies to circuits and parts of an installation that (for safety reasons) need to be switched independently of other circuits and/or installations. It does not apply to short connections between the origin of the installation and the consumer's main switchgear.

5.8.3 Electric motors

Electric motors which are automatically or remotely controlled and unsupervised shall be safeguarded against excessive temperature by a protective device with manual reset.

 A motor with star-delta starting should be protected against excessive temperature in both the star and delta configurations.

All equipment, and its associated cabling, of all circuits carrying the motor's starting, accelerating and load currents shall be cable of accepting a current at least equal to the full-load current rating of the motor.

Electric motors with a rating in excess of 0.37 kW shall be provided with control equipment to protect against overloading the motor.

Equipment shall be capable of preventing an automatic restart following a stoppage due to a drop in voltage or a failure of the supply.

Fixed electric motors shall be provided with an efficient means of switching off that is readily accessible, easily operated and so placed as to prevent danger.

Where reverse-current braking of a motor is provided, provision shall be made for the avoidance of reversal of the direction of rotation at the end of braking – if such reversal may cause danger.

If the electric motor is only intended for intermittent duty – and, therefore, subject to frequent starting and stopping – the probability of this causing the equipment to rise in temperature must be taken into account.

If safety depends on the direction of rotation of a motor, the possibility of reverse operation (e.g. due to a reversal of phases) must be prevented.

5.8.4 Generating sets

Generator sets (independent of the normal supply) are a recognised supply source for safety services such as emergency escape lighting, fire-detection and alarm systems, installations for fire pumps, fire rescue service lifts, and smoke- and heat-extraction equipment.

5.8.4.1 A generator as a power source

The source of supply to a reduced low-voltage circuit shall be either:

- a motor-generator with windings that provide isolation equivalent to that provided by the windings of an isolating transformer; or
- a source independent of other supplies (such as an engine-driven generator).

 The neutral (star) point of the secondary windings of three-phase generators (or the midpoint of the secondary windings of single-phase transformers and generators) must always be connected to Earth.

5.8.4.1.1 Basic fault protection

Basic fault protection is deemed to be provided where the nominal voltage cannot exceed the upper limit of voltage Band I and the supply is from a recognised source such as a diesel-driven generator or motor-generator.

The following sources may be used for SELV and PELV systems:

- a source independent of a higher-voltage circuit (such as a diesel-driven generator);
- a motor-generator with windings providing equivalent isolation to a safety isolating transformer.

A conductor connecting a generator to its associated control panel does not require a device for protection against fault current, **provided** that the protective device is placed in the panel, and **provided** that the wiring is installed in such a manner that the risk of a fault or of fire or danger to persons is reduced to a minimum.

The generating set shall be connected so that any RCD used in the installation remains effective for every intended combination of sources of supply.

5.8.4.2 Low-voltage generating sets

Low-voltage generating sets can be:

- combustion engines;
- electric motors;
- electrochemical accumulators;
- photovoltaic cells;
- turbines;
- other suitable sources.

Where a generating set with an output not exceeding 16 A is to be connected in parallel with a system already distributing electricity to the public, procedures for informing the electricity distributor are given in the Electricity Safety, Quality and Continuity (Amendment) Regulations (ESQCR) 2009, which are made by the Secretary of State for Trade and Industry under powers contained in the Electricity Act 1989 as amended.

 ESQCR Regulation 4 relates to the avoidance of supply interruption caused by trees.

As noted in BS EN 50438, in addition to the ESQCR requirements, where a generating set with an output exceeding 16 A is to be connected in parallel with a public distribution system:

- The requirements of the electricity distributor should be ascertained before the generating set is connected.
- The safety and proper functioning of other sources of supply shall not be impaired by the generating set.
- The prospective short-circuit current and Earth fault current shall be assessed for each source of supply (or combination of sources) which can operate independently of other sources or combinations.

- The short-circuit rating of protective devices within the installation, and, where appropriate, connected to a public distribution system, shall not be exceeded for any of the intended methods of operation of the sources.
- If the generating set is intended to provide a supply (or switched alternative) to an installation which is not connected to a public distribution system, the generating set shall not endanger or damage equipment after the connection or disconnection of any intended load as a result of any voltage or frequency deviation.
- It shall be possible to automatically disconnect such parts of the installation as may be necessary if the capacity of the generating set is exceeded.

5.8.4.3 Protection against overcurrent

If the generating set requires overcurrent protection, it shall be located as near as practicable to the generator terminals.

Where a generating set is intended to operate in parallel with a public distribution system (or where two or more generating sets may operate in parallel), circulating harmonic currents must be limited so that the thermal rating of conductors is not exceeded.

The effects of circulating harmonic currents may be limited by one or more of the following:

- ensuring that the generating set has compensated windings;
- ensuring that there is a capable impedance in the connection to the generator star points;
- confirming that any switches which interrupt the circulatory circuit are all interlocked so that at all times fault protection is not impaired;
- providing filtering equipment.

5.8.4.4 Standby systems and switched alternatives to public supplies

Precautions for isolation shall be taken so that the generator cannot operate in parallel with a public distribution system.

5.8.5 Medical power supply systems

In medical locations a power supply for safety purposes is required, which will energise the installations required for continuous operation in case of failure of the general power system, for a defined period, within a pre-set changeover time.

The safety power supply system shall automatically take over if the voltage of one or more incoming live conductors of the main distribution board of the building with the main power supply has dropped for more than 0.5 s and by more than 10% in regard to the nominal voltage.

A list of examples with suggested reinstatement times is given in Annex A710 of BS 7671:2018).

A few other things to remember are that:

- primary cells are **not** allowed as safety power sources;
- an additional main incoming power supply, from the general power supply, is not regarded as a source of the safety power supply;
- the availability (i.e. readiness for service) of safety power sources shall be monitored and indicated at a suitable location;
- the circuit which connects the power supply source for safety services to the main distribution board shall be considered a safety circuit.

5.8.5.1 General

The distribution system within medical locations shall be designed and installed so that there is an automatic changeover from the main distribution network to the electrical safety source feeding essential loads to operate at all relevant times, including during mains and local supply failure and through fire conditions.

Automatic changeover devices shall be arranged so that safe separation between supply lines is maintained. (See BS EN 60947-6-1for further information.)

In Group 1 and Group 2 medical locations, at least two different sources of supply shall be provided, one of which shall be connected to the electrical supply system for safety services.

5.8.5.2 Failure of the general power supply source

If there is a failure of the general power supply source, the power supply for safety services shall be energised to feed the equipment with electrical energy for a defined period of time, and within a predetermined changeover period.

If there is a voltage failure on one of the distribution board's line conductors, a safety power supply source shall be used that is capable of providing power for at least 3 h for:

- operating theatre table luminaires;
- medical electrical equipment containing light sources essential for the application of the equipment (e.g. endoscopes and monitors etc.).

 The normal power supply for life-supporting medical electrical equipment **shall** be restored within a changeover period **not exceeding 0.5 s**.

Equipment which is required for the maintenance of healthcare installations (e.g. sterilisation equipment, technical building installations (air conditioning, heating and ventilation) and storage battery chargers etc.) shall be connected either automatically or manually to a safety power supply source capable of maintaining it for a minimum period of 24 h.

Other services which may require a safety service supply with a changeover period not exceeding 15 s include the following:

- selected lifts for firefighters;
- ventilation systems for smoke extraction;
- paging systems;
- medical electrical equipment used in Group 2 medical locations that is used for surgical or other procedures of vital importance;
- electrical equipment for medical gas supplies, including compressed air, vacuum supply and narcosis (anaesthetics) exhaustion, as well as their monitoring devices;
- fire detection and fire alarms to BS 5839;
- fire-extinguishing systems.

Where socket outlets are supplied from the safety power supply source they shall be readily identifiable according to their safety services classification.

5.8.5.3 Power supplies for Group 2 medical locations

In the event of fault or supply failure, a total loss of supply to Group 2 locations shall be prevented.

5.9 Switches

The Regulations try to ensure that all persons and livestock are protected against injury (and property against damage) as a consequence of overvoltages such as those originating from switching. There are a number of rules and regulations regarding this goal, including the following:

- electrical equipment shall not cause harmful effects on other equipment or impair the supply during normal service, including switching operations;
- all circuits shall be provided with a means of isolation from all live supply conductors by a linked switch or a linked circuit-breaker;
- switches should be fitted in a suitable mounting box or enclosure;
- every electric discharge lighting installation which has an open circuit voltage exceeding low voltage needs to be include a lockable switch;
- some form of disconnecting device should be provided which will allow electrical installations, circuits or individual items of equipment to be switched off or isolated, to enable their operation, inspection, fault detection, testing, maintenance and repair;
- the purpose and identification of each item of switchgear and controlgear shall be identified by a label or some form of indicator.

5.9.1 Compatibility

All equipment should be selected and erected so that it will neither cause harmful effects to other equipment nor impair the supply during normal service, including switching operations.

As shown in Table 5.9 switchgear, protective devices, accessories and other types of equipment shall **not** be connected to conductors intended to operate at temperatures exceeding 70°C at the equipment in normal service, **unless:**

- the equipment manufacturer has confirmed that the equipment is suitable for such conditions;
- or the current (including any harmonic current) carried by the conductor does not exceed the appropriate temperature limit specified in Table 5.9.

Table 5.9 Maximum operating temperatures for types of cable insulation

Type of insulation	Temperature limit
Silicon	180°C
Thermoplastic insulated cable	70–90°C
XLPE cross-linked polythene	90°C
Polythene	75°C

5.9.2 Devices for isolation and switching

Isolation shall preferably be provided by a multipole switching device which disconnects all applicable poles of the relevant supply.

Where it is intended that isolation and switching is carried out:

- the means of switching the supply on load and the means of isolation shall be provided by a suitably rated fuse carrier;
- isolation shall preferably be provided by a multipole switching device which disconnects all applicable poles of the relevant supply;
- all devices used for isolation shall be clearly identified;
- where the distributor's cut-out is used as the means of isolation of a highway power supply the approval of the distributor shall be obtained.

Every device provided for isolation or switching shall comply with the relevant requirements of this section.

5.9.3 Emergency switching

Emergency switching may be emergency switching **ON** or emergency switching **OFF.**

In an emergency it shall be made possible to switch off any part of an installation in order to control the supply and in order to remove an unexpected danger.

Other than where a risk of electric shock is involved, the emergency switching device shall be an isolating device and shall interrupt **all** live conductors.

Except where the neutral conductor can be regarded as being reliably connected to Earth in a TN-S or TN-C-S system, the neutral conductor need **not** be isolated or switched.

The execution of emergency switching shall ensure that only one single action is required to interrupt the appropriate supply conductors.

The operation of an emergency switching device shall **not** introduce a further danger or interfere with the complete operation necessary to remove the danger.

5.9.3.1 Devices for emergency switching

Emergency switching is basically an operation intended to remove, as quickly as possible, danger, which may have occurred unexpectedly.

A device for emergency switching shall be capable of breaking the full-load current of the relevant part(s) of the installation, taking account of stalled motor currents where appropriate.

Hand-operated switching devices for direct interruption of the main circuit shall be selected where practicable and these shall be clearly identified, preferably by colour.

An emergency switching device may consist of:

- a switch in the main circuit (or pushbuttons in the control (auxiliary) circuit) that is capable of directly cutting off the appropriate supply; or
- a combination of equipment activated by a single action for the purpose of cutting off the appropriate supply.

The means of operating (handle, pushbutton etc.) an emergency switching device shall:

- be easily identifiable (e.g. by red marking) and convenient for their intended use;
- be readily accessible at places where a danger might occur and/or at any additional remote position from which that danger can be removed;
- be capable of latching or being restrained in the OFF or STOP position;
- **not** re-energize the relevant part of the installation;
- have priority over any other utility relative to safety;
- not be erected where they are accessible to livestock or in any position where access may be impeded by livestock.

Note: Circuit-breakers and RCDs are primarily circuit protective devices and, as such, they are not intended for frequent load switching. However, infrequent switching of circuit-breakers on-load is admissible for the purposes of isolation or emergency switching.

5.9.3.2 Functional switching devices

Functional switching, on the other hand, is an operation intended to switch ON or OFF or vary the supply of electrical energy to all or part of an installation.

Functional switching shall be provided for each part of a circuit which may require to be controlled independently of other parts of the installation (in general, all current-using equipment requiring control shall be controlled by an appropriate functional switching device).

 A single functional switching device may control two or more items of equipment intended to operate simultaneously.

A functional switching device that is designed to ensure the changeover of supply from alternative sources:

*   shall affect (but need not necessarily control) all live conductors;
*   shall not be capable of putting the sources in parallel, unless the installation is specifically designed for this condition;
*   may control the current without necessarily opening the corresponding poles;
*   shall be suitable for the most onerous duty it is intended to perform;
*   shall be provided for each part of a circuit which may require to be controlled independently of other parts of the installation.

 Note: Semiconductor switching devices are examples of devices capable of interrupting the current in the circuit but not opening the corresponding poles.

5.9.3.3 Firefighter's switches

All firefighter's switches shall comply with the requirements of BS EN 60669-2-6 or BS EN 60947-3.

A firefighter's switch shall be provided in the low-voltage circuit supplying:

*   exterior electrical installations operating at a voltage exceeding low voltage; and
*   interior discharge lighting installations operating at a voltage exceeding low voltage.

Every firefighter's switch shall comply with the following:

*   for an exterior installation, the switch shall be outside the building and adjacent to the equipment (or alternatively a notice indicating the

position of the switch shall be placed adjacent to the equipment and an appropriate notice shall be fixed near the switch itself);

- for an interior installation, the switch shall be independent of the switch for any exterior installation and in the main entrance to the building;
- the switch shall be placed in a conspicuous position, that is reasonably accessible to firefighters, and at not more than 2.75 m from the ground or from a person standing beneath the switch.
- where more than one switch is installed on any one building, each switch shall be clearly marked to indicate the installation or part of the installation that it controls.

Author's note

I, initially, could not understand why the switch should be 'not more than 2.75 m from the ground', as average firefighters are normally not that tall as far as I am aware! I therefore sought advice from the local fire service and was told that the reason why the switch was positioned so high was to stop an unauthorised person from switching it off. Firefighters use a ceiling hook to operate the switch. I am now no longer confused!!

A firefighter's switch shall:

- be coloured red and have fixed on or near it a permanent nameplate marked with the words **'FIREFIGHTER'S SWITCH'** or **'FIRE SWITCH'**; and
- have its **ON** and **OFF** positions clearly indicated by lettering legible to a person standing on the ground at the intended site, with the OFF position at the top; and
- be provided with a device to prevent the switch being inadvertently returned to the ON position.

Note: An insulation monitoring device (IMD) shall be used on a circuit comprising safety equipment which is normally de-energised by a switch that disconnects all live poles and which is only energised in the event of an emergency (provided that the IMD is automatically deactivated whenever the safety equipment is activated).

5.9.4 Isolation and switching

Non-automatic, local and remote isolation and switching measures for the prevention or removal of dangers associated with electrical installations or electrically powered equipment and machines is an obvious necessity.

BS 7671:2018 covers these aspects for all electrical installations or electrically powered equipment and machines in great detail throughout the Standard by providing advice on isolation or switching for devices and precautions for:

- emergency stopping;
- emergency switching;
- supply distribution cabinets;
- live supply conductors;
- protection against overcurrent;
- live conductors (including the neutral conductor).

 BS 7671:2018 Table 537.4 summarises the functions provided by the devices for isolation and switching, together with indication of the relevant product standards.

5.9.5 Main switches

Where an installation is supplied from more than one source:

- a main switch shall be provided for each source of supply; and
- a warning notice shall be permanently fixed in a prominent position so that any person seeking to operate any of these main switches will be warned of the need to operate all such switches to achieve isolation of the installation; or
- a suitable interlock system shall be provided;
- Where an installation is supplied from more than one source of energy (one of which requires a means of earthing independent of the means of earthing of other sources and it is necessary to ensure that no more than one means of earthing is applied at any time) a switch may be inserted in the connection between the neutral point and the means of earthing. This can be achieved, **provided** that the switch is a linked switch that is capable of disconnecting and connecting the earthing conductor for the appropriate source, at substantially the same time as the related live conductors.

5.9.6 Main linked switches

An electrical installation must be capable of isolating itself from each supply via a main linked switch.

A main switch that is intended to be operated by ordinary persons (e.g. a householder) shall interrupt both live conductors of a single-phase supply.

 Note: In a TN-S or TN-C-S system the neutral conductor need not be isolated or switched where it can be regarded as being reliably connected to Earth by suitably low impedance.

5.9.7 Mechanical maintenance

The capability of switching off for mechanical maintenance shall be provided where mechanical maintenance could involve a risk of physical injury.

 Suitable means need to be provided to prevent electrically powered equipment from becoming unintentionally reactivated during mechanical maintenance.

5.9.7.1 Devices for switching off for mechanical maintenance

Where a switch is provided for switching off for mechanical maintenance, it must be capable of cutting off the full-load current of the relevant part of the installation. A device such as a:

- multipole switch;
- circuit-breaker;
- control and protective switching device (CPS);
- control switch operating a contactor;
- plug and socket outlet;

may be inserted in the main supply circuit for switching off for mechanical maintenance.

 The open position of the contacts of the device should always be visible or be clearly and reliably indicated by the use of the symbols '0' and 'I' to indicate the open and closed positions, respectively.

A device for switching off for mechanical maintenance shall:

- be inserted in the main supply circuit;
- require manual operation;
- be designed and/or installed so as to prevent inadvertent or unintentional switching on;
- be in accordance with the requirements of BS EN 60204;
- be capable of cutting off the full-load current of the relevant part of the installation;
- be so placed as to be readily identifiable and convenient for the intended use.

 Note: A plug and socket outlet or similar device of rating not exceeding 16 A may be used as a device for switching off for mechanical maintenance.

5.9.7.1.1 Heaters and/or boilers

A heater or boiler must be permanently connected to the electricity supply through a double-pole linked switch, which is both separate from and within easy reach of the heater or boiler or is incorporated therein.

The wiring from the heater or boiler **must** be connected directly to that switch without the use of a plug and socket outlet. All metal parts of the equipment which are in contact with the water (other than current-carrying parts) need to be connected to the metal water pipe which supplies the water to the heater or boiler which will be directly connected to the main earthing terminal by means independent of the circuit protective conductor.

If the heater or boiler is installed in a room containing a fixed bath, the switch must comply with Section 701 of BS 7671:2018.

Where a step-up transformer is used, a linked switch shall be provided for disconnecting the transformer from all live conductors of the supply.

5.9.8 Protective devices

Switches shall only be inserted in the line conductor.

Non-linked switches need to be inserted in an earthed neutral conductor.

Any linked switch or linked circuit-breaker inserted in an earthed neutral conductor must be capable of breaking all of the related line conductors.

5.9.9 Single-pole switching devices

Single-pole switching devices shall not be:

- inserted in the neutral conductor of a multiphase circuit;
- alone in the neutral conductor single-phase circuits.

Where single-pole switching devices are not permitted in the neutral conductor, a test should be made to verify that all such devices are connected in the line conductor(s) only.

5.9.10 Switchboards

Passageways and working platforms which have access to an open type switchboard or an equipment item that has dangerous exposed live parts need to allow persons, without hazard, to:

- operate and maintain the equipment;
- pass one another as necessary with ease; and (most important!)
- be able to back away from the equipment when necessary.

If equipment items are carrying different types of current (or at different voltages) and they are all grouped together on a common assembly, such as a switchboard, equipment items belonging to any one type of current or one type of voltage must be kept apart so as to avoid mutual detrimental influence.

Figure 5.7 Certainly NOT the right sort of installation! (Courtesy Stingray.)

 Figure 5.7 is an example of a definite hazardous installation!

In 2006, to avoid any confusion, the UK changed its cable colours so that they were the same as the European cable colours. (see Figure 5.8).

 Note: Before commencing any work on an old UK property it is always wise to check the property or installation to see if the wiring colours are out of date

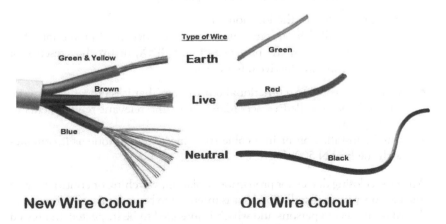

New Wire Colour   **Old Wire Colour**

Figure 5.8 The new versus the old wiring colours.

or have deteriorated over time, and are in compliance with the BS 7671:2018 Wiring Regulations.

 BS 7671:2018 Table 5.51 provides more detailed information regarding the identification of conductors.

5.9.11 Switching devices

A switching device:

- shall be protected against overcurrent;
- without integral overcurrent protection shall be co-ordinated with an appropriate overcurrent protective device;
- shall not be inserted in a protective conductor, unless the switch:

  o   has been inserted in the connection between the neutral point and the means of earthing;
  o   is a linked switch arranged to disconnect and connect the earthing conductor at substantially the same time as the related live conductors;

    –   is a multipole linked switch or plug-in device in which the protective conductor circuit
    –   has not been interrupted before the live conductors and
    –   re-established not later than when the live conductors are reconnected.

5.9.12 Switchgear

All switchgear:

- shall be accessible only to authorised persons;
- when placed in an escape route, shall be enclosed in a cabinet or an enclosure constructed of non-combustible or not readily combustible material;
- shall be installed outside the location, unless:

  o   it is suitable for the location; or
  o   it is installed in an enclosure providing a degree of protection of at least IP4X or, in the presence of dust, IP5X, or (in the presence of electrically conductive dust) IPX6;

- shall be subjected to a functional test to show that it is properly mounted, adjusted and installed in accordance with the relevant requirements of the Regulations;
- may be installed on or in a cable trunking system as long as it complies with the BS EN 50085 series.

 An auto-reclosing device for protection, isolation, switching or control may be installed **only** in an installation that is intended to be under the supervision of skilled or instructed persons and which is intended to be inspected and tested by competent persons.

Means of access to all live parts of switchgear where different nominal voltages exist shall be marked to indicate the voltages present.

Where a metal enclosure or frame of a low-voltage switchgear assembly is used as a protective conductor:

- its electrical continuity should be protected against mechanical, chemical or electrochemical deterioration;
- its cross-sectional area should not be less than 2.5 mm^2;
- it should permit the connection of other protective conductors at every predetermined tap-off point.

5.9.13 Special installations and locations

5.9.13.1 Agricultural or horticultural premises

The electrical installation of each building or part of a building shall be isolated by a single isolation device.

These isolation devices need to be clearly marked to show which part of the installation they belong to.

 Emergency stopping devices or emergency switching shall **not** be erected where they are accessible to livestock or in any position where access may be impeded by livestock.

 See BS 7671:2018 Chapter 705 for more detailed requirements relating to agricultural and horticultural premises.

5.9.13.2 Construction and demolition site installations

Each assembly for construction sites (ACS) shall include a suitable device for switching and isolating the incoming supply. Ideally these devices should be capable of securing in the OFF position (e.g. by padlock or location of the device inside a lockable enclosure).

 See BS 7671:2018 Chapter 704 for more detailed requirements relating to construction and demolition site installations.

5.9.13.3 Electrical installations in caravan/camping parks and similar locations

All switchgear and controlgear assemblies used in caravan/tent pitch supplies shall comply with the requirements of BS EN 61439-7.

If an installation consists of just one final circuit, the isolating switch may be the overcurrent protective device.

 A permanent notice **must** be fixed near the main isolating switch inside the caravan.

See BS 7671:2018 Chapter 721 for more detailed requirements relating to electrical installations in caravan, camping parks and similar locations.

5.9.13.4 Exhibitions, shows and stands

Switchgear and controlgear needs to be placed in closed cabinets which can only be opened by the use of a key or a tool, except for those parts designed and intended to be operated by ordinary persons.

An emergency switch (easily visible, accessible and clearly marked) shall be installed to control a separate circuit supplying signs, lamps or exhibits.

 See BS 7671:2018 Chapter 711 for more detailed requirements relating to exhibitions, shows and stands.

5.9.13.5 Extra-low-voltage lighting installations

The primary circuits of transformers that are operated in parallel should be permanently connected to a common isolating device.

 See BS 7671:2018 Chapter 715 for more detailed requirements relating to extra-low-voltage lighting installations.

5.9.13.6 Locations containing a bath or shower

Unless the switches and their controls are incorporated in fixed current-using equipment suitable for use in a particular zone or to insulate pull cords of cord-operated switches, the following rules apply:

- zone 0: switchgear or accessories shall not be installed;
- zone 1: only switches of SELV circuits supplied at a nominal voltage not exceeding 12 VAC rms or 30 V ripple-free d.c. shall be installed;
- zone 2: switchgear, accessories incorporating switches or socket outlets shall not be installed, with the exception of:

  o switches and socket outlets of SELV circuits; and
  o shaver supply units complying with BS EN 61558-2-5.

 See BS 7671:2018 Chapter 701 for more detailed requirements relating to locations containing a bath or shower.

5.9.13.7 Marinas and similar locations

There shall be one means of separation (i.e. an isolating switching device) for a maximum of four socket outlets, and these:

- shall be installed in each distribution cabinet;
- shall disconnect all live conductors including the neutral conductor.

See BS 7671:2018 Chapter 709 for more detailed requirements relating to marinas and similar locations.

5.9.13.8 Medical locations

Requirements for medical electrical equipment for use in conjunction with flammable gases and vapours are contained in BS EN 60601.

Socket outlets intended to supply ME equipment shall be unswitched.

More detailed requirements relating to medical locations and devices is to be found in BS 7671:2018 Chapter 710.

5.9.13.9 Rooms and cabins containing sauna heaters

Switchgear which forms part of the sauna heater equipment (or of other fixed equipment installed in zone 2) may be installed within the sauna room or cabin.

Lighting and other switchgear will have to be placed outside the cabin.

See BS 7671:2018 Chapter 703 for more detailed requirements relating to rooms and cabins containing sauna heaters.

5.9.13.10 Solar photovoltaic power supply systems

To allow maintenance of the photovoltaic (PV) convertor, means of isolating the PV convertor from the d.c. side and the a.c. side shall be provided.

Switchgear assemblies shall be in compliance with BS EN 60439-1 and appropriate parts of BS EN 61439-1.

In the selection and erection of devices for isolation and switching to be installed between the PV installation and the public supply, the public supply shall be considered the source and the PV installation shall be considered the load.

A switch-disconnector shall be provided on the d.c. side of the PV convertor.

Junction boxes containing a PV generator or PV array must carry a warning label indicating that parts inside the boxes may still be live after isolation from the PV convertor.

See BS 7671:2018 Chapter 712 for more detailed requirements relating to solar photovoltaic power supply systems.

5.9.13.11 Swimming pools and other basins

The following regulations apply to switchgear in swimming pools etc.:

* zones 0 and 1: switchgear, controlgear or a socket outlet shall not be installed;

- zone 2: a socket outlet or a switch is permitted only where the supply circuit is protected by:

    o   SELV (if its supply circuit is protected by an RCD);
    o   automatic disconnection of supply using an RCD.

 See BS 7671:2018 Chapter 702 for more detailed requirements relating to swimming pools and other basins.

5.9.13.12 Temporary electrical installations

Every electrical installation in a booth, stand or amusement device shall have its own means of isolation and switching which shall disconnect all live conductors (i.e. line and neutral conductors).

Switchgear shall be placed in cabinets which can be opened only by the use of a key or a tool, except for those parts designed and intended to be operated by ordinary persons.

Temporary electrical installations for amusement devices (and distribution circuits supplying outdoor installations) shall be provided with their own readily accessible and properly identified means of isolation.

A separate circuit shall be used to supply luminous tubes, signs or lamps, which shall be controlled by an emergency switch.

This switch must be easily visible, accessible and marked in accordance with the requirements of the local authority.

 See BS 7671:2018 Chapter 740 for more detailed requirements relating to temporary electrical installations.

5.10 Transformers

As shown in Figure 5.9, there are numerous types of Transformers in current use today.

In all cases:

- the neutral (star) point of the secondary windings of three-phase transformers and generators (or the midpoint of the secondary windings of single-phase transformers and generators) shall be connected to Earth.
- for SELV and PELV systems, a safety isolating transformer in accordance with BS EN 61558-2-6 or BS EN 61558-2-8 may be used.
- where an auxiliary circuit is supplied by more than one transformer, they shall be connected in parallel on both the primary and secondary sides.

5.10.1 Autotransformers and step-up transformers

Where an autotransformer is connected to a circuit having a neutral conductor, the common terminal of the winding shall be connected to the neutral conductor.

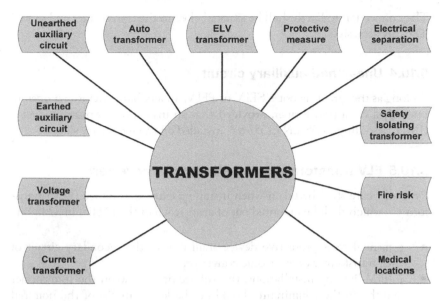

Figure 5.9 Transformers.

Where a step-up transformer is used, a linked switch shall be provided for disconnecting the transformer from all live conductors of the supply.

 A step-up autotransformer shall **not** be connected to an IT system.

5.10.2 Current transformer

Where a measurement device is connected to the main circuit via a current transformer, the following requirements need to be taken into account:

- the secondary side of the transformer in a low-voltage installation is not earthed;
- protective devices interrupting the circuit are not to be used on the secondary side of the transformer;
- conductors on the secondary side of the transformer are either insulated for the highest voltage of any live parts or installed so that their insulation cannot come into contact with other live parts such as busbars;
- terminals for temporary measurements shall be provided.

 Note: The use of self-resetting overcurrent protective devices is permitted only for transformers up to 50 VA.

5.10.3 Earthed auxiliary circuit

An earthed auxiliary circuit supplied via a transformer shall only be connected to Earth at one point on the secondary side of the transformer.

The connection to Earth shall be situated close to the transformer and should be easily accessible.

5.10.4 Unearthed auxiliary circuit

So long as the circuit is not a SELV or PELV, an auxiliary circuit can operate unearthed via a transformer, **provided** that an insulation monitoring device (IMD) according to BS EN 61557-8 is installed on the secondary side.

5.10.5 ELV transformers and electronic converters

Particular care shall be taken when installing extra-low-voltage (ELV) transformers, which shall be mounted out of arm's reach of the public. In addition:

- a manual reset protective device shall protect the secondary circuit of each transformer or electronic convertor;
- in ELV lighting installations, the voltage drop between the transformer and the furthest luminaire should not be less than 5% of the nominal voltage

 Electronic converters **must** always conform to BS EN 61347-1.

5.10.6 Fire risk of transformers

Transformers shall be either be:

- protected on the primary side by a compliant protective device; or
- short-circuit proof (both inherently and non-inherently).

Only the following are permitted to be mounted on a flammable surface:

- a class P thermally protected ballast/transformer;
- a temperature declared thermally protected ballast/transformer.

5.10.7 Medical locations

Transformers shall be in accordance with BS EN 61558-2-15 and should be installed in close proximity to the medical location, with the following additional requirements:

- The leakage current of the output winding to Earth and the leakage current of the enclosure, when measured in no-load condition (and the transformer supplied at rated voltage and rated frequency), shall not exceed 0.5 mA.
- At least one single-phase transformer per room or functional group of rooms shall be used to form the IT systems for mobile and fixed equipment

and the rated output shall be not less than 0.5 kV A and shall not exceed 10 kV A. Where several transformers are needed to supply equipment in one room, they shall not be connected in parallel.

- If the supply of three-phase loads via an IT system is also required, a separate three-phase transformer shall be provided for this purpose.

 Capacitors shall **not** be used in transformers for medical IT systems.

Overload current protection shall not be used in either the primary or secondary circuit of the transformer of a medical IT system.

Overcurrent protection against short-circuit and overload current is required for each final circuit.

5.10.8 Protective measure: electrical separation

An IT system at mobile and/or transportable units can be provided by:

- an isolating transformer or a low voltage generating set, with an insulation monitoring device installed; or
- a transformer offering simple separation (e.g. in accordance with BS EN 61558-1and providing:

  o   automatic disconnection of the supply in case of a first fault between live parts and the frame of the unit; or
  o   the use of an RCD.

5.10.9 Safety isolating transformers

A safety isolating transformer for an extra-low voltage lighting installation shall comply with BS EN 61558-2-6 and:

- either the transformer shall be protected on the primary side by a protective device; or
- the transformer shall be short-circuit proof (both inherently and non-inherently).

5.10.9.1 At booths in fairgrounds etc.

Safety isolating transformers shall comply with BS EN 61558-2-6 or provide an equivalent degree of safety and:

- a manually reset protective device shall protect the secondary circuit of each transformer or electronic convertor;
- the transformers shall be mounted out of arm's reach or be mounted in a location that provides equal protection, and shall have adequate ventilation.

5.10.9.2 Galvanic separation

Where a fixed onshore isolating transformer is used to prevent galvanic currents circulating between the hull of the vessel and metallic parts on the shore side, equipment complying with BS EN 61558-2-4 shall be used.

The protective conductor of the supply to the isolating transformer shall not be connected to the Earth terminal in the socket-outlet supplying the inland navigation vessel.

5.10.10 Voltage transformer

The secondary side of a voltage transformer shall be protected by a short-circuit protective device.

Author's end note

One of the difficulties in working with the 18th edition of the Wiring Regulations is trying to find the full requirements for a particular type of cable or conductor that you might want to use. The main reason for this, of course, is that as there are so many different types of cables and conductors (not to mention conduits, cable ducting and cable trunking) for you to choose from and these are spread throughout the Standard's chapters and appendices. You could, therefore, quite easily miss an essential requirement in your research.

Granted, BS 7671:2018's enormous index covers virtually everything, but unfortunately, as previously mentioned, there are so many cross references between different sections of the Standard that relevant points could easily be overlooked.

The aim of the next chapter, therefore, is to simplify the problem by listing the essential requirements under three main headings, namely 'cables', 'conductors' and 'conduits'.

6

Cables, conductors and conduits

> ### Author's start note
>
> *Within BS 7671:2018 there is frequent reference to different types of cables (e.g. single-core, multi-core, fixed, flexible etc.), conductors (such as live supply, protective, bonding etc.) and conduits, cable ducting, cable trunking and so on. Unfortunately, similar to equipment and components, the requirements for these items are liberally sprinkled throughout the Standard. The aim of this chapter, therefore, is to provide a catalogue of all the different types identified and referred to in the Wiring Regulations under two main headings (i.e. 'cables', and 'conductors and conduits'), and then make a list of their essential requirements.*
>
> *Similar to other chapters, please remember that these lists of requirements are only the author's impression of the most important aspects of the Wiring Regulations, and electricians should always consult BS 7671 to satisfy compliance!*

6.1 Cables

As shown in Figure 6.2, **there are five main types of cables found in electrical installations. These are:**

1. single-core cables;
2. multi-core cables;
3. flexible cables;
4. heating and warm floor cables;
5. fixed wiring.

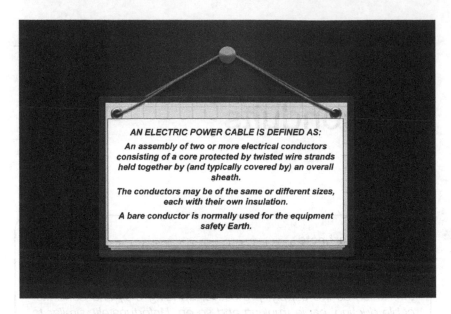

AN ELECTRIC POWER CABLE IS DEFINED AS:

An assembly of two or more electrical conductors consisting of a core protected by twisted wire strands held together by (and typically covered by) an overall sheath.

The conductors may be of the same or different sizes, each with their own insulation.

A bare conductor is normally used for the equipment safety Earth.

Figure 6.1　Definition of an electric power cable.

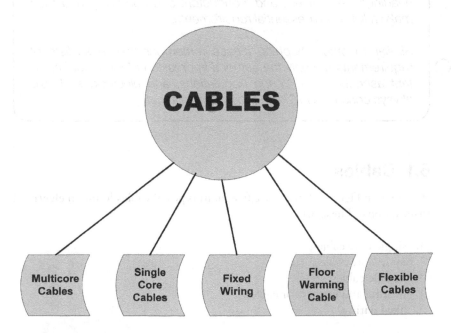

CABLES

Multicore Cables

Single Core Cables

Fixed Wiring

Floor Warming Cable

Flexible Cables

Figure 6.2　Cables.

6.1.1 General

 All cables should comply with the requirements of BS EN 50265-2-1 or 2-2.

6.1.2 Types of cable

6.1.2.1 Single-core cables

The general requirements for single-core cables include the following:

Metallic sheaths and/or the non-magnetic armour of single-core cables that are in the same circuit must either be bonded together:

- at both ends of their run (solid bonding); or
- at one point in their run (single-point bonding);

provided that, at full load, voltages from sheaths and/or armour to Earth:

- do not exceed 25V; and
- do not cause corrosion; and
- do not cause danger or damage to property.

Single-core cables may be used as protective conductors but shall be coloured **green-and-yellow** throughout their length.

The conductance of the outer conductor of a concentric single-core cable shall not be less than the internal conductor.

When non-twisted single-core cables with a cross-sectional area greater than 50 mm² in copper (or 70 mm² in aluminium) are connected in parallel, the load current shall be shared equally between them. (This ruling does not apply to final ring circuits.)

 Owing to possible electromagnetic effects, single-core cables that are armoured with steel wire or tape should **not** be used for a.c. circuits.

 Notes:

- The steel wire or steel tape armour of a single-core cable is regarded as a ferromagnetic enclosure.
- For single-core armoured cables, the use of aluminium armour may be considered.

6.1.2.1.1 Armoured single-core cables

The metallic sheaths and/or non-magnetic armour of single-core cables in the same circuit are normally bonded together at both ends of their run (commonly referred to as 'solid bonding').

6.1.2.2 Multi-core cables

The general requirements for multi-core cables include the requirements that:

- for telecommunication circuits, data transfer circuits and similar, consideration shall also be given to electrical interference, both electromagnetic and electrostatic. (See BS EN 50081 and BS EN 50082 for further details.)

A Band I circuit must **not** be contained in the same wiring system as a Band II voltage circuit unless it is in a multi-core cable and the cores of the Band I circuit are:

- insulated for the highest voltage present in the Band II circuit;
- separated from the cores of the Band II circuit by an earthed metal screen, and the cables:

  o   are insulated for their system voltage;
  o   are installed in a separate compartment of a cable ducting or trunking system;
  o   are installed on a tray or ladder separated by a partition;
  o   use a separate conduit, trunking or ducting system.

Separated extra-low-voltage (SELV) circuit conductors that are contained in a multi-core cable with other circuits should be insulated for the highest voltage present in that cable.

Separated circuits shall, preferably, use a separate wiring system. If this is not feasible, then multi-core cables (without a metallic sheath or insulated conductors) may be used.

The conductance of the outer conductor of a concentric cable for a multi-core cable:

- serving a number of points contained within one final circuit (or where the internal conductors are connected in parallel) should not be less than that of the internal conductors;
- in a multi-phase or multi-pole circuit should not be less than that of one internal conductor.

A voltage Band I circuit shall not be contained in the same wiring system as a Band II circuit, unless:

- each conductor of a multi-core cable is insulated for the highest voltage present in the cable;
- the cores of the Band I circuit are separated from the cores of the Band II circuit by an earthed metal screen.

The circuit shall be arranged so that the conductors are **not** distributed over different multi-core cables, conduits, ducting systems, franking systems or tray or ladder systems.

The line and neutral conductors of each final circuit shall be electrically separate from those of every other final circuit, in order to prevent indirect energising of a final circuit.

Where multi-core cables are installed in parallel, each cable shall contain one conductor of each line.

Note: In caravans and motor caravans, all protective conductors must be incorporated in a multi-core cable or in a conduit together with the live conductors. (For more details see BS 7671:2018 Chapter 721.)

6.1.2.3 Flexible cables

Flexible cables:

* can be used to move normally stationary equipment for the purposes of connecting, cleaning etc.(e.g. a cooker);
* may be used for equipment that is intended to be moved whilst in use unless that particular equipment has been supplied with contact rails;
* shall be visible throughout any part of their length that is liable to mechanical damage;
* shall either be:

    o   a heavy-duty type; or
    o   suitably protected against mechanical damage.

* shall include a protective bonding conductor(unless they supply equipment with double or reinforced insulation);
* shall only be used for fixed wiring where the relevant Regulations permit;
* that are liable to mechanical damage shall be visible throughout its length;
* that are used as an overhead low-voltage line shall comply with the Relevant British or Harmonised Standard.

Non-flexible cables (and flexible cables that are not part of a portable appliance or luminaire) which are sheathed with lead, polyvinyl chloride (PVC) or an elastomeric material may include a catenary wire (or hard-drawn copper conductor) for aerial use or when the cable is suspended.

In caravans and motor caravans, flexible cables shall not be laid in areas accessible to the public, unless they are protected against mechanical damage (for more detailed requirements see BS 7671:2018 Chapter 711.

6.1.2.4 Floor-warming and heating cables

Underfloor heating is a very cost-effective way to warm a room because, unlike radiators, stoves or a traditional solid-fuel fire, it provides and distributes heat evenly and gently. There are no cold spots and very little heat is wasted.

The maximum conductor operating temperature for floor-warming cable is between 70 and 85°C depending on the type of cable used.

The general requirements for heating cables contained in BS 7671:2018 include the following:

- A heating cable laid directly in soil, a road or the structure of a building shall be installed so that it:

    o   is completely embedded in the substance it is intended to heat;
    o   does not suffer damage in the event of movement by the substance in which it is embedded;
    o   is capable of withstanding external mechanical damage;
    o   is resistant to damp and/or corrosion;
    o   complies in all respects with the manufacturer's instructions and recommendations.

- Where a heating cable is required to pass through, or be in close proximity to, material which presents a fire hazard, the cable:

    o   shall be enclosed in material having the ignitability characteristic **'P'** as specified in BS 476-12; and
    o   shall be adequately protected from any mechanical damage reasonably foreseeable during installation and use.

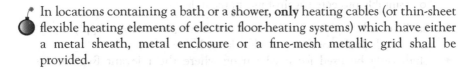 In locations containing a bath or a shower, **only** heating cables (or thin-sheet flexible heating elements of electric floor-heating systems) which have either a metal sheath, metal enclosure or a fine-mesh metallic grid shall be provided.

6.1.2.5 Fixed wiring

Fixed wiring cables are not suitable for mobile/static equipment and are mostly used as power supply cables for sockets, switches and light fittings across residential, commercial and industrial environments. They are designed to be installed in a fixed position, fastened to a support or laid in a specific location, and if a flexible cable is used, it must be a heavy-duty type, unless it is protected against mechanical damage.

If non-sheathed cables are used for fixed wiring, they shall be enclosed in conduit, ducting or trunking.

 A full description of light, ordinary and heavy-duty types of fixed cables is contained in BS EN 50565-1.

Although BS7671 does not actually define 'fixed wiring', it does nevertheless stipulate a number of specific requirements as follows:

- equipment with a protective conductor current exceeding 3.5 mA but not exceeding 10 mA shall be permanently connected to the fixed wiring of an installation or connected via a plug and socket outlet complying with BS EN 60309-2;

- the termination of the wiring system at a fixed lighting point shall use one of the following:

  o a batten lampholder or a pendant set complying with BS EN 60598;
  o a box (enclosing he terminals of the fixed wiring) complying with BS EN 60670 series or BS 4662;
  o a ceiling rose complying with BS 67;
  o a connection unit complying with BS 1363-4;
  o a device for connecting a luminaire (DCL) outlet complying with BS IEC 61995-1;
  o an installation coupler complying with BS EN 61535;
  o a luminaire complying with BS EN 60598;
  o a luminaire-supporting coupler (LSC) complying with BS 6972 or BS 7001;
  o a plug-in lighting distribution unit complying with BS 5733;
  o a suitable socket outlet complying with BS 1363-2, BS 546 or BS EN 60309-2.

 In suspended ceilings one plug-in lighting distribution unit can be used for a number of luminaires.

- the cross-sectional areas of the fixed wiring should be such that the permissible voltage drop is not exceeded;
- in a caravan the connection between its internal fixed wiring should be via a terminal block, with a protective cover. (Complete wiring details for caravans are contained in BS 7671:2018 Chapter 721.)

6.1.3 Cable construction and manufacture

Electrical cables and conduits are integral parts of an electrical system. Whilst cables are used for transmitting power, conduits keep a bunch of cables safely together. Cable trunking on the other hand is an enclosure (normally with a rectangular cross section) with one removable or hinged side that is used to protect cables and provide space for other electrical equipment.

BS 7671:2018 contains the following requirements for electrical conduits, trunking and ducts:

- cable conduits shall comply with the appropriate part of the BS EN 61386 series;
- cable tray and ladder systems shall comply with BS EN 61537;
- cable trunking or ducting shall comply with the appropriate part of the BS EN 50085 series.

6.1.3.1 Cross-sectional areas of conductors of cables

The cross-sectional area of each conductor shall not be less than 1.5 mm.

6.1.3.2 Identification of cables

The cores of cables shall be identified at their terminations (and preferably throughout their length) by:

- colour (see BS 7671:2008 Regulation 514.4); and/or
- lettering and/or numbering (see BS 7671:2008 Regulation 514.5).

 The two-colour combination of green-and-yellow shall **only** be used for identifying a protective conductor.

6.1.4 Cables for special installations and locations

Where an electrical service is located in close proximity to one or more non-electrical services, it shall meet the following conditions:

- the wiring system shall be suitably protected against the hazards that are likely to arise from the presence of the other services in normal use;
- fault protection shall be provided by automatic disconnection of supply.

6.1.4.1 Agricultural and horticultural premises

The particular requirements of this section apply to fixed electrical installations that are indoors and/or outdoors in agricultural and horticultural premises. Some of the requirements are also applicable to other locations that are in common buildings belonging to the agricultural and horticultural premises.

Where vehicles and mobile agricultural machines are operated, the following methods of installation shall be applied:

- cables shall be buried in the ground at a depth of at least 0.6 m, with added mechanical protection;
- cables in arable or cultivated ground shall be buried at a depth of at least 1 m;
- self-supporting suspension cables shall be installed at a height of at least 6 m.

Wiring systems shall be erected so that they are inaccessible to livestock or suitably protected against mechanical damage.

Overhead lines shall be insulated and special attention shall be given to the presence of different kinds of fauna (e.g. rodents).

 See BS 7671:2018 Section 705 for more detailed requirements relating to cabling in agricultural and horticultural premises (note that Section 705 does not cover electric fence installations).

6.1.4.2 Caravan and camping parks

 Note: In order not to mix regulations on different subjects, such as those for electrical installation of caravan parks with those for electrical installation inside caravans, two sections are available in BS 7671:2018:

- Section 708, which concerns electrical installations in caravan parks, camping parks and similar locations; and
- Section 721, which concerns electrical installations in the actual caravans and motor caravans themselves.

The following is a summary of the most important requirements that form BS 7671:2018 for caravans, motor caravans and camping parks.

In a caravan and camping site, underground cables shall be buried at a depth of at least 0.6 m and (unless additional mechanical protection has been made available) be placed outside any caravan pitch or away from any surface where tent pegs or ground anchors are expected to be present.

As the wiring in a caravan or a motor caravan will be subjected to vibration, all wiring shall be protected against mechanical damage either by location or by enhanced mechanical protection.

 No more than four socket outlets should be grouped in one location, in order to avoid the supply cable crossing a pitch other than the one intended to be supplied.

In caravans and motor caravans:

- To meet the requirements of BS EN 61386, the wiring systems shall be installed using one or more of the following:

  - insulated single-core cables, with flexible class 5 conductors, in non-metallic conduit;
  - insulated single-core cables, with stranded class 2 conductors (minimum of seven strands), in non-metallic conduit;
  - sheathed flexible cables.

- All cables (unless enclosed in rigid conduit) and all flexible conduits shall be supported at intervals no greater than 0.4 m for vertical runs and 0.25 m for horizontal runs.
- All cables shall, as a minimum, meet the requirements of BS EN 60332-1-2.
- Low-voltage cable systems shall be run separately from the cables of extra-low-voltage systems, so that there is no possibility of physical contact between the two wiring systems.
- If cables have to run through a compartment, they shall be protected against mechanical damage by installation within a conduit system or within a ducting system.
- Non-metallic conduits shall comply with BS EN 61386-21.

- If a cable crosses (or is in the proximity of) an underground telecommunication cable or underground power cable:

    o   a minimum clearance of 100 mm shall be maintained;
    o   a fire-retardant partition shall be provided between the cables; and
    o   mechanical protection between the cables shall be provided.

 See BS 7671:2018 Sections 708 and 721 for more detailed requirements relating to caravans, motor caravans and camping parks.

6.1.4.3 Construction and demolition site installations

- Cables shall **not** be installed across a road or a walkway unless the cable is sufficiently protected against mechanical damage.
- Surface-run and overhead cables shall be protected against mechanical damage.
- For reduced low-voltage systems, low temperature, thermoplastic cable shall be used.
- For applications exceeding reduced low voltage, a heavy-duty, flexible cable shall be used.

 See BS 7671:2018 Chapter 704 for more detailed requirements relating to construction and demolition site installations.

6.1.4.4 Exhibitions, shows and stands

The particular requirements of this section apply to the temporary electrical installations in exhibitions, shows and stands (including mobile and portable displays and equipment), to protect users.

- A cable intended to supply temporary structures shall be protected at its origin by an RCD whose rated residual operating current does not exceed 300 mA.
- Armoured cables or cables protected against mechanical damage shall be used wherever there is a risk of mechanical damage.
- Flexible cable shall **not** be laid in areas accessible to the public unless they are protected against mechanical damage.
- Insulation-piercing lampholders shall **not** be used unless the lampholders are non-removable once fitted to the cable.
- Joints shall not be made in cables except where necessary as a connection into a circuit. Where joints are made, these shall either use connectors that are in accordance with relevant standards or be in enclosures with a degree of protection of at least IP4X or IPXXD.
- If no fire alarm system has been installed in a building which is going to be used for exhibitions etc., the cable systems shall be either:

    o   flame retardant to BS EN 60332-1-2 and low smoke to BS EN 61034-2; or o   single-core or multi-core unarmoured cables enclosed in metallic or non-metallic conduit or trunking, providing a degree of fire protection of at least IP4X.

- Where strain can be transmitted to terminals, the connection shall incorporate suitable cable anchorage(s).
- Wiring cables shall be copper, have a minimum cross-sectional area of 1.5 mm², and comply with an appropriate British Standard for either thermoplastic or thermosetting insulated electric cables.

See BS 7671:2018 Chapter 711 for more detailed requirements relating to exhibitions, shows and stands.

6.1.4.5 Marinas and similar locations

The particular requirements of this section are applicable only to circuits that are intended to supply pleasure craft or houseboats in marinas and similar locations. They do **not** apply to houseboats if these are directly supplied from the public network or to the internal electrical installations of pleasure craft or houseboats.

The following wiring systems are suitable for marina distribution circuits:

- cables with copper conductors and thermoplastic or elastomeric insulation and sheath;
- cables with armouring and serving of thermoplastic or elastomeric material;
- mineral-insulated cables with a PVC protective covering;
- overhead cables or overhead insulated conductors;
- underground cables.

The following wiring systems shall **not** be used on or above a jetty, wharf, pier or pontoon:

- cables in free air suspended from or incorporating a support wire;
- non-sheathed cables in cable management systems, trunking etc.;
- cables with aluminium conductors;
- mineral-insulated cables.

Underground distribution cables shall, unless provided with additional mechanical protection, be buried at a sufficient depth to avoid being damaged (e.g. by heavy vehicle movement).

Note: A depth of 0.5 m is normally considered as the '*minimum depth*' to fulfil this requirement.

Cable management systems shall be installed to allow the drainage of water through drainage holes (an alternative would be to install the equipment on an incline).

Cables shall be selected and installed so that mechanical damage due to tidal and movement of floating structures is prevented.

All overhead conductors shall be insulated.

Overhead conductors shall be at a height above ground of not less than 6 m in all areas subjected to vehicle movement and 3.5 m in all other areas.

Poles and other supports for overhead wiring shall be located or protected so that they are unlikely to be damaged by any foreseeable vehicle movement.

 See BS 7671:2018 Chapter 709 for more detailed requirements relating to marinas and similar locations.

6.1.4.6 Mobile or transportable units

For the purposes of this section, the term *'unit'* is intended to mean a vehicle and/or mobile (self-propelled or towed) or transportable structure (such as a container or cabin) in which all or part of an electrical installation is contained and which is provided with a temporary supply by means of, for example, a plug and socket outlet.

- Flexible cables that are used to connect the unit to the supply shall have a minimum cross-sectional area of 2.5 mm² copper.
- Flexible cables shall enter the unit by an insulated inlet so as to minimise the possibility of any insulation damage or fault which might energise the exposed-conductive-parts of the unit.
- The wiring system shall be installed using one or more of the following:

  o sheathed flexible cable with thermoplastic or thermosetting insulation;
  o unsheathed flexible cable with thermoplastic or thermosetting insulation installed in either a conduit or trunking or ducting.

- Where cables have to run through a compartment, they shall be installed within a conduit system or within a ducting system in order to protect them against mechanical damage.

 See BS 7671:2018 Chapter 717 for more detailed requirements relating to mobile or transportable units.

6.1.4.7 Rooms and cabins containing sauna heaters

In zone 3 of a room (or cabin) containing a sauna heater, the insulation and sheaths of cables shall be capable of withstanding a minimum temperature of 170°C.

 See BS 7671:2018 Chapter 713 for more detailed requirements relating to rooms and cabins containing sauna heaters.

6.1.4.8 Solar and photovoltaic power supply systems

- Where an electrical installation includes a photovoltaic (PV) power supply system without at least simple separation between the a.c. side and the d.c. side, an RCD shall be installed so as to provide either fault protection by automatic disconnection of supply or additional protection.
- On the a.c. side, the PV supply cable shall be connected to the supply side of the protective device for automatic disconnection of circuits supplying current-using equipment.
- PV string cables, PV array cables and PV d.c. main cables shall be selected and erected so as to minimise the risk of Earth faults and short-circuits.
- Overload protection may be omitted to the:

  - o PV main cable if the continuous current-carrying capacity is equal to (or greater than) 1.25 times I_{sc} STC of the PV generator;
  - o PV string and PV array cables when the continuous current-carrying capacity of the cable is equal to (or greater than) 1.25 times I_{sc} STC at any location.

- The PV supply cable on the a.c. side shall be protected against fault current by an overcurrent protective device installed at the connection to the a.c. mains.
- Where protective bonding conductors are installed, they shall be parallel to, and in as close contact as possible with, d.c. cables and a.c. cables and accessories.
- Wiring systems shall be capable of withstanding expected external influences such as wind, ice formation, temperature and solar radiation.

 See BS 7671:2018 Chapter 712 for more detailed requirements relating to rooms and cabins containing sauna heaters.

6.1.4.9 Swimming pools and other basins

- In zones 0, 1 and 2 of a swimming pool (or similar basin) any metallic sheath or metallic covering of a wiring system shall be connected to the supplementary equipotential bonding.
- Cables should preferably be installed in conduits made of insulating material.

For a fountain, the following additional requirements shall be met:

- a cable for electrical equipment in zone 0 shall be installed as far outside the basin rim as is reasonably practicable and run to the electrical equipment inside zone 0 by the shortest possible route;
- in zone 1, a cable shall be selected, installed and provided with mechanical protection to medium severity and the expected submersion in water depth.

See BS 7671:2018 Chapter 702 for more detailed requirements relating to swimming pools and other basins.

6.1.4.10 Temporary electrical systems in amusement parks, circuses and fairgrounds

All parts of these installations shall conform to the relevant standards:

- Conduit systems shall comply with the BS EN 61386 series.
- Cable trunking systems and cable ducting systems shall comply with part 2 of BS EN 50085.
- Tray and ladder systems shall comply with BS EN 61537.
- All cables shall meet the requirements of BS EN 60332-1-2.

Other wiring system requirements include:

- Armoured cables or cables protected against mechanical damage shall be used wherever there is a risk of mechanical damage due to external influence.
- Buried cables shall be protected against mechanical damage.
- Cables shall have a minimum rated voltage of 450/750 V.
- Joints shall not be made in cables except where necessary as a connection into a circuit.
- Supply cables shall be flexible and have adequate protection against mechanical damage.
- The routes of cables buried in the ground shall be marked at suitable intervals.
- Where strain can be transmitted to terminals, the connection shall incorporate some form of cable anchorage.
- Insulation-piercing lampholders shall not be used unless the cables and lampholders are compatible and the lampholders cannot be removed once fitted to the cable.
- Luminaires and floodlights shall be so fixed and protected that a focusing or concentration of heat is not likely to cause ignition of any material.

See BS 7671:2018 Chapter 740 for more detailed requirements relating to temporary electrical systems in amusement parks, circuses and fairgrounds.

6.1.5 Inspection of cables

As well as checking the routing of cables in safe zones (for protection against mechanical damage), where relevant to the installation, and where necessary, **all** cables should also be checked during erection.

Note: Inspection should always precede testing and shall normally be done with that part of the installation being inspected – ensuring that the power is switched off first. (Further details concerning maintenance and inspections are contained in Chapters 9 and 10.)

6.1.5.1 Mechanical stresses

Any wiring system, cable or conduit that is buried in a floor should be sufficiently protected to prevent damage caused by the intended use of the floor.

A wiring system:

- that is designed to draw conductors or cables in or out shall have adequate means of access to allow for this operation;
- shall be selected and erected to avoid damage to the sheath or insulation of cables and their terminations during installation, use, or maintenance;
- shall have a radius of every bend sufficient that the cables do not suffer damage and terminals are not stressed.

A cable:

- buried in the ground (as opposed to being installed in a conduit or duct) shall include an earthed armour or metal sheath (or both) as a protective conductor;
- shall be supported in such a way that it is not exposed to undue mechanical strain and so that there is no appreciable mechanical strain on the terminations of the conductors. If the conductors or cables are not supported continuously, they shall be borne by suitable means at appropriate intervals in such a manner that the conductors or cables do not suffer damage by their own weight;
- support and enclosure shall not have sharp edges that could damage the wiring system.
- shall not be damaged by the means of fixing;
- which passes across expansion joints shall be selected and/or erected so that any anticipated movement will not cause damage to the electrical equipment.

Buried cables need to be at a sufficient depth to avoid being damaged by any reasonably foreseeable disturbance of the ground and should be clearly marked with cable covers or a suitable marking tape.

A conduit system or cable ducting system that is going to be buried in the structure shall be completely erected between access points before any cable is drawn in.

See IEC 61386-24 for further details concerning underground conduits.

6.1.5.2 Vibration

Vibrations fall into two categories: free and forced. Free vibrations occur when the system is disturbed momentarily and then allowed to move without restraint. Forced vibrations occur if a system is continuously driven by an external agency.

For the purpose of the Wiring Regulations:

- stationary equipment which is moved temporarily for the purposes of connecting, cleaning etc, (e.g. cookers) shall be connected with flexible cable;
- the cables and cable connections of a wiring system supported by or fixed to a structure or equipment that is subject to vibration of medium or high severity shall be suitable for such conditions;
- the connection to suspended current-using equipment, (e.g. luminaires) shall be made by cable with flexible cores.

6.2 Conductors and conduits

As shown in Figure 6.3, there are a number of different types of conductors used in electrical installations.

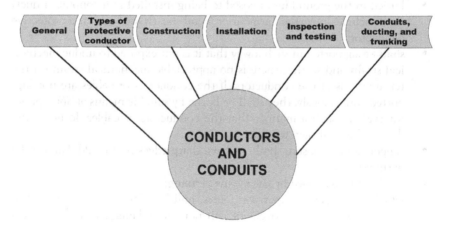

Figure 6.3 Conductors and conduits.

Conductors:

- intended to operate at temperatures above 70°C shall **not** be connected to switchgear, protective devices, accessories or other types of equipment;
- shall **not** be subjected to excessive mechanical stress;
- **shall** be capable of withstanding all foreseen electromechanical forces (including fault current) during service.

6.2.1 General

- The number of conductors to be considered in a circuit are those carrying load current less those conductors which only serve the purpose of protective conductors, and the conductor size shall be chosen on the basis of the highest line current.
- PEN conductors need to be taken into consideration in the same way as neutral conductors.
- Where there is a risk due to structural movement, the cable support and protection system employed shall be capable of permitting relative movement so that conductors are not subjected to excessive mechanical stress.

6.2.2 Types of protective conductor

As shown in Figure 6.4 there are a number of protective conductors that can be used for different situations and installations.

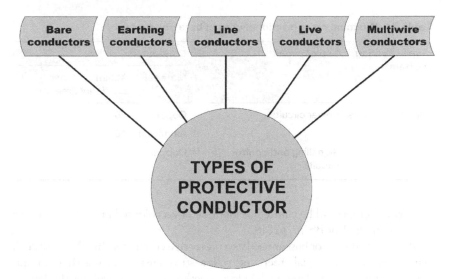

Figure 6.4 Types of protective conductor.

A protective conductor may consist of one or more of the following:

- a single-core cable;
- a conductor in a cable;
- an insulated or bare conductor in a common enclosure with insulated live conductors;
- a fixed bare or insulated conductor;
- a metal covering (e.g. the sheath, screen or armouring of a cable);
- a metal conduit or other enclosure or electrically continuous support system for conductors.

If a protective conductor is formed by a conduit, trunking, ducting or the metal sheath and/or armour of a cable, the earthing terminal of each accessory shall be connected by a separate protective conductor to an earthing terminal incorporated in the associated box or enclosure.

 A gas pipe, oil pipe, flexible or pliable conduit, support wires or other flexible metallic parts, or constructional parts that are subject to mechanical stress in normal service, shall **NOT** be selected as a protective conductor.

6.2.2.1 Bare conductors

Bare conductors shall be protected against corrosion at their supports and on their passage through walls.

 The minimum cross-sectional area of phase conductors in a.c. circuits and of live conductors in d.c. circuits shall be as shown in Table 6.1.

Table 6.1 Minimum cross-sectional area of bare conductors

Type of wiring system	Use of circuit	Conductor	
		Material	Minimum cross sectional area (mm²)
Bare conductors	Power circuits	Copper	10
		Aluminium	16
	Signalling and control circuits	Copper	4

 Bare conductors shall be painted or identified by a coloured tape, sleeve or disc as per Table 6.51 of BS 7671:2018.

A bare conductor or busbar used as a protective conductor shall be identified, where necessary, by equal green and yellow stripes (each not less than 15 mm and not more than 100 mm wide) close together, either throughout the length of the conductor or in each compartment and unit and at each accessible position. If adhesive tape is used, it shall also be bi-coloured.

6.2.2.2 Earthing conductors

Earthing conductors (see example shown in Figure 6.5) must be capable of being disconnected to enable the resistance of the earthing arrangements to be measured.

In every installation, a main earthing terminal shall be provided to connect the following to the earthing conductor:

- the circuit protective conductors;
- the protective bonding conductors;

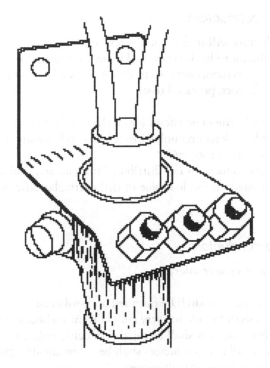

Figure 6.5 Typical earthing conductor. (Courtesy Stingray.)

- functional earthing conductors (if required);
- lightning protection system bonding conductor (if any).

If a protective conductor forms part of a cable, then this **shall** only be earthed in the installation containing the associated protective device with a cross-sectional area not less than that stated in BS 7671:2018 Table 52.3.

For further recommendations and guidance on meeting the requirements for earthing land-based electrical installations in and around buildings in the UK, see BS 7430 which includes the following requirements:

- The connection of an earthing conductor to an Earth electrode or other means of earthing shall be:

  o soundly made;
  o electrically and mechanically satisfactory;
  o labelled; and
  o suitably protected against corrosion.

- The earthing conductor of a street electrical fixture shall have a minimum copper equivalent cross-sectional area not less than that of the supply neutral conductor at that point or not less than 6 mm², whichever is the smaller.

6.2.2.3 Line conductors

All line conductors will include an overcurrent detection facility which will cause the conductor to be disconnected if any overcurrent is detected.

In a TN or TT system, overcurrent detection need not be provided for one of the line conductors, provided that:

- there exists, in the same circuit (or on the supply side), differential protection capable of detecting unbalanced loads and causing disconnection of all the line conductors; and
- the neutral conductor is not distributed from an artificial neutral point of circuits situated on the load side of this particular differential protective device.

6.2.2.4 Live conductors

From the point of view of safety:

- bare live conductors **shall** be installed on insulators;
- conductors **shall** be able to carry fault current without overheating;
- live supply conductors **shall** be capable of being isolated from circuits;
- the supply to all live conductors **shall** be automatically interrupted in the event of an overload or fault current;
- persons and livestock **shall** be protected against injury, and property shall be protected against damage, due to excessive temperatures (or electromechanical stresses) caused by any overcurrents likely to arise in live conductors.

6.2.2.5 Multi-wire, fine wire and very fine wire conductors

Terminals shall be used at the end, of a conductor to avoid the possible separation or spreading of individual wires of multi-wire, fine wire or very fine wire conductors.

 Soldering (tinning):

- of the whole conductor end of multi-wire, fine wire and very fine wire conductors is **not** allowed if screw terminals are used;
- conductor ends on fine wire and very fine wire conductors are **not** allowed at connection and junction points owing to the possible movement between the soldered and the non-soldered part of the conductor.

6.2.3 Construction and installation

This section describes the various elements of construction (see Figure 6.6) that need to be considered.

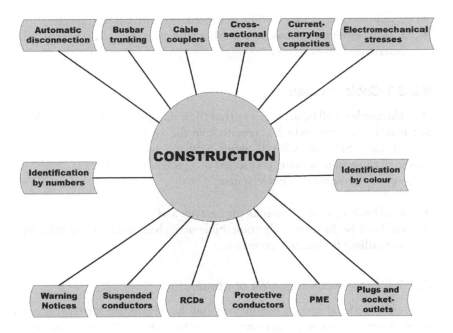

Figure 6.6 Construction.

6.2.3.1 Automatic disconnection in case of a fault

Automatic disconnection of supply (ADS) is the most widely used method for achieving protection against electric shock in an installation. As described in the Wiring Regulations, it works by limiting the size and duration of any voltages between exposed-conductive-parts to Earth by automatically interrupting the supply to the line conductor of a circuit, or equipment, in the event of a fault of negligible impedance between the line conductor and an exposed-conductive-part (or a protective conductor) in the circuit or equipment. (BS 7671:2018 Table 41.1 indicates the disconnection time of the circuit or equipment.)

 Disconnection is not required for protection against electric shock but may be required for other reasons, such as protection against thermal effects (see BS 7671:2018 Table 41.1), but where automatic disconnection cannot be achieved in the time required, supplementary equipotential bonding shall be provided.

In a TN system, a disconnection time not exceeding 5 s is permitted.

6.2.3.2 Busbar trunking

Where a busbar trunking system is used as a protective conductor:

- it shall be protected against insulation faults;
- its electrical continuity shall be assured, either by construction or by suitable connection, in such a way as to be protected against mechanical, chemical or electrochemical deterioration;

- its cross-sectional area shall be in accordance with BS EN 60439-1;
- it shall permit the connection of other protective conductors at every pre-determined tap-off point.

6.2.3.3 Cable couplers

A cable coupler shall be arranged so that the connector of the coupler is fitted at the end of the cable which is remote from the supply.

Except for a SELV or a Class II circuit, a cable coupler shall be non-reversible and shall be capable of being connected to a protective conductor.

Every cable coupler in a FELV system:

- shall have a protective conductor contact; and
- shall not be dimensionally compatible with those used for any other system utilised in the same premises.

6.2.3.4 Cross-sectional area of conductors

The cross-sectional area of conductors shall be determined for both normal operating conditions and, where appropriate, for fault conditions according to:

- the admissible maximum temperature;
- the admissible voltage drop limit;
- the electromechanical stresses likely to occur due to short-circuit and Earth fault currents;
- other mechanical stresses to which the conductors are likely to be exposed;
- the maximum impedance for correct operation of short-circuit and Earth fault protection;
- the method of installation;
- harmonics;
- thermal insulation.

The cross-sectional area of a phase conductor in an a.c. circuit or of a live conductor in a d.c. circuit shall be as shown in Tables 52.3 and 55.2 contained in BS 7671:2018.

The neutral conductor shall have a cross-sectional area greater than that of the line conductor.

For a polyphase circuit where each line conductor has a cross-sectional area greater than 16 mm² for copper or 25 mm² for aluminium, the neutral conductor can have a smaller cross-sectional area than that of the line conductors, providing that:

- the expected maximum current including harmonics (if any) in the neutral conductor during normal service is not greater than the current-carrying capacity of the reduced cross-sectional area of the neutral conductor; and

- the neutral conductor is protected against overcurrents; and
- the size of the neutral conductor is at least equal to 16 mm² for copper or 25 mm² for aluminium.

In caravans and motor caravans, the cross-sectional area of every conductor shall be not less than 1.5 mm².

6.2.3.5 Current-carrying capacities of conductors

The current (including any harmonic current) to be carried by any conductor for sustained periods during normal operation varies according to the type of insulation used – for example thermoplastic insulation's temperature limit would be 70°C at the conductor, whilst mineral-based insulation's would be 105°C, dependent on whether or not it was exposed to touch.

6.2.3.6 Electromechanical stresses

Every conductor or cable shall have adequate strength and be installed so as to withstand the electromechanical forces that may be caused by any current, including fault current, that it may have to carry whilst in service.

6.2.3.7 Identification of conductors – by colour

Whilst Table 6.51 of BS 7671:2018 provides full details of conductor colours, the main ones to remember are:

- Identification by colour or marking is not required for:
  o bare conductors where permanent identification is not practicable;
  o concentric conductors of cables;
  o extraneous-conductive-parts used as a protective conductor;
  o exposed-conductive-parts used as a protective conductor;
  o metal sheath or armour of cables when used as a protective conductor;
- conductors with **green-and-yellow** colour identification shall **not** be numbered other than for the purpose of circuit identification;
- clear marking shall be provided at the interface between conductors.

 The single colour **green** shall **not** be used for live conductors in power circuits, protective conductors, functional earthing or bonding conductors.

6.2.3.7.1 Bare conductors

A bare conductor shall be identified by the use of tape, sleeve or disc of the appropriate colour prescribed in BS 7671:2018 Table 6.51 or by painting it with such a colour.

Colour or marking is not required for bare conductors (where permanent identification is not practicable).

6.2.3.7.2 Neutral or midpoint conductors

Where a circuit includes a neutral or midpoint conductor, the colour used shall be blue.

6.2.3.7.3 PEN conductor

A PEN conductor shall be marked by one of the following methods:

- **green-and-yellow** throughout its length and with **blue** markings at the terminations;
- **blue** throughout its length with **green-and-yellow** markings at the terminations.

6.2.3.7.4 Other conductors

All other conductors shall be identified by colour in accordance with BS 7671:2018 Table 6.51.

6.2.3.7.5 Main protective bonding conductors

Unless protective multiple earthing (PME) conditions apply, a main protective bonding conductor shall have a cross-sectional area not less than half the cross-sectional area required for the earthing conductor of the installation, and not less than 6 mm².

 The cross-sectional area need not exceed 25 mm² if the bonding conductor is of copper or of a cross-sectional area affording equivalent conductance in other metals.

Except for highway power supplies and street furniture (where PME conditions apply), the main protective bonding conductor shall be selected in accordance with the neutral conductor of the supply and BS 7671:2018 Table 54.8 (which provides details of the minimum cross-sectional area of the main protective bonding conductor in relation to the neutral of the supply).

6.2.3.7.6 Non-standard colours

If wiring additions or alterations are made to an installation so that some of the wiring complies with the current Regulations but there is also wiring to previous versions of these Regulations, a warning notice (see Figure 6.7) shall be affixed at or near the appropriate distribution board with the following wording.

Figure 6.7 Warning notice – non-standard colours.

6.2.3.8 Identification of conductors by letters and/or numbers

The lettering or numbering system applied to identification of individual conductors and of conductors in a group:

- shall be clear, legible and durable;
- all numerals shall be in strong contrast to the colour of the insulation;
- shall be given in letters or Arabic numerals (in order to avoid confusion, unattached numerals 6 and 9 shall be underlined).

6.2.3.8.1 Numerical

Conductors may be identified by numbers, the number 0 being reserved for the neutral or midpoint conductor.

6.2.3.9 Plugs and socket outlets

Except for SELV, every plug and socket outlet shall be of the non-reversible type, with provision for the connection of a protective conductor.

6.2.3.10 Protective multiple earthing

Where protective multiple earthing (PME) exists, the minimum cross-sectional area of the main equipotential bonding conductor in relation to the

Table 6.2 Cross-sectional area of the main equipotential bonding conductor

Copper equivalent cross-sectional area of the supply neutral conductor	Minimum copper equivalent cross-sectional area of the main equipotential bonding conductor
35 mm² or less	10 mm²
over 35 mm² up to 50 mm²	16 mm²

neutral shall be in accordance with BS 7671:2018 Table 54.8 and as shown in Table 6.2.

In some cases the local distributor's network conditions may require a larger conductor.

6.2.3.11 Protective conductors

The Wiring Regulations emphasise that:

- a gas pipe, oil pipe, flexible or pliable conduit, support wires or other flexible metallic parts (or constructional parts subject to mechanical stress in normal service) shall **not** be selected as a protective conductor;
- a protective conductor shall be suitably protected against mechanical and chemical deterioration and electrodynamic effects;
- exposed-conductive-parts of equipment shall not be used as a protective conductor for other equipment;
- in installations and locations where the risk of an electric shock is increased by a reduction in body resistance and/or by contact with Earth potential, all plugs, socket outlets and cable couplers of a reduced low-voltage system shall have a protective conductor contact.

If the protective conductor:

- is not an vital part of a cable; or
- is not formed by conduit, ducting or trunking; or
- is not contained in an enclosure formed by a wiring system;

then the cross-sectional area shall be not less than:

- 2.5 mm² copper equivalent if protection against mechanical damage is provided; and
- 4 mm² copper equivalent if mechanical protection is not provided.

Where a protective conductor is common to two or more circuits, its cross-sectional area shall be:

- calculated for the most onerous of the values of fault current and operating time encountered in each of the various circuits; or

- selected so as to correspond to the cross-sectional area of the largest line conductor of those circuits.

A protective conductor with a cross-sectional area up to and including 6 mm² shall be protected throughout by a covering at least equivalent to that provided by the insulation of a single-core non-sheathed cable having a voltage rating of at least 450/750 V unless it is:

- a protective conductor forming part of a multi-core cable;
- cable trunking or conduit used as a protective conductor.

6.2.3.11.1 Types of protective conductor

The metal covering (including the sheath – bare or insulated) of a cable, trunking, ducting or metal conduit, may be used as a protective conductor for the associated circuit.

A protective conductor may consist of one or more of the following:

- a single-core cable;
- a conductor in a cable;
- a fixed bare or insulated conductor;
- a metal covering (e.g. the sheath, screen or armouring of a cable);
- a metal conduit, metallic cable management system or other enclosure or electrically continuous support system for conductors;
- an insulated or bare conductor in a common enclosure with insulated live conductors;
- an extraneous-conductive-part, provided that:
  - o electrical continuity can be assured;
  - o it is either constructed or connected so that it is protected against mechanical, chemical or electrochemical deterioration;
  - o it has been suitably adapted, if necessary, to suit its use;
  - o precautions have been taken against its removal.

 If the cross sectional area of the protective conductor of the types listed above is 10mm² or less, it shall be of copper.

If a metal enclosure or frame of a low-voltage switchgear or controlgear assembly or busbar trunking system is used as a protective conductor:

- its electrical continuity shall ensure that it is adequately protected against mechanical, chemical or electrochemical deterioration;
- its cross-sectional area shall be in accordance with BS EN 60439-1;
- it shall permit the connection of other protective conductors at predetermined tap-off point.

A protective conductor shall be identified by a bi-colour combination of **green-and-yellow**. One of the colours shall cover at least 30% (and at most 70%) of the surface being coloured, while the other colour shall cover the remainder of the surface.

If a bare conductor or busbar is used as a protective conductor, it will be identified by equal green and yellow stripes between 15 and 100 mm wide, close together and either throughout the length of the conductor or at each accessible position.

6.2.3.12 RCDs

An RCD shall be selected and erected to limit the risk of unwanted tripping, and should be capable of disconnecting all the line conductors of the circuit at substantially the same time.

6.2.3.13 Suspended conductors

Suspension devices and supporting conductors for extra-low-voltage luminaires shall be capable of carrying five times the mass of the luminaires)including their lamps) intended to be supported, but not less than 5 kg.

The suspended system shall be fixed to walls or ceilings by insulated distance cleats and terminations and connections of conductors shall be made by screw terminals or screwless clamping devices.

6.2.3.14 Warning notices

6.2.3.14.1 Safety Earth

A warning notice (as shown in Figure 6.8) shall be permanently fixed at the point of connection of every earthing conductor to an Earth electrode and every bonding conductor to an extraneous-conductive-part, as well as the main Earth terminal (when separated from the main switchgear). Protective bonding conductors shall also have a warning notice as shown in Figure 6.9.

6.2.3.14.2 Electrical separation

Where electrical separation to the supply to more than one current using equipment item is used the warning notice shall read as follows:

6.2.3.14.3 Alternative supplies

Where an installation includes alternative or additional sources of supply, warning notices shall be affixed at the following locations in the installation:

- the origin of the installation;
- the meter position, if remote from the origin;

Figure 6.8 Warning notice – earthing and bonding.

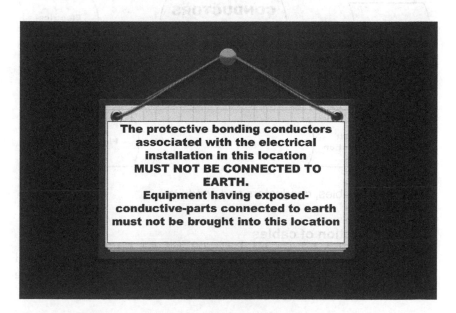

Figure 6.9 Warning notice – protective bonding conductors.

- the consumer unit or distribution board to which the alternative or additional sources are connected;
- at all points that isolate the supply source.

6.3 Installation

Some of the most common electrical installation issues we encounter today are caused by poor installation. For example:

- insecure wiring;
- wrong wiring sizes;
- improper wire length;
- unprotected wiring;
- connections.

Figure 6.10 shows the main areas that need particular attention.

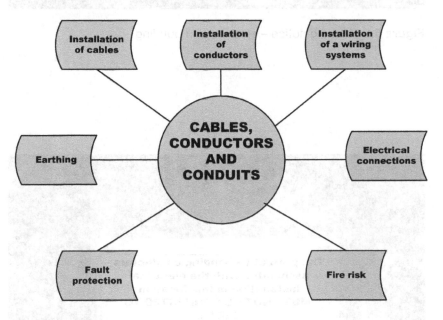

Figure 6.10 Cables, conductors and conduits.

6.3.1 Installation of cables

Every cable should have adequate strength and be so installed as to withstand the electromechanical forces that may be caused by any current, including fault current, it may have to carry in service.

 All bare, live, cables must be installed on insulators.

Non-sheathed cables:

- for fixed wiring shall be enclosed in conduit, ducting or trunking;
- are permitted in a cable trunking system which provides a minimum of IP4X or IPXXD protection, but only if the cover can only be removed.

6.3.1.1 Cable couplers

A cable coupler (a small device, usually a metal port or plastic box with two female inputs which fuses together two cables without soldering or crimping) is fitted at the end of the cable that is remote from the supply.

Every cable coupler of a reduced low-voltage system shall:

- have a protective conductor contact; and
- be dimensionally incompatible with all other couplers used for other systems in that particular premises.

On a FELV system the coupler:

- shall have a protective conductor contact; and
- shall not be dimensionally compatible with those used for any other system in use in the same premises.

For a SELV system, or a Class II circuit, a cable coupler shall be non-reversible and enable the connection of a protective conductor, and shall comply (where appropriate) with BS 6991, BS EN 61535, BS EN 60309-2 or BS EN 60320-1.

6.3.1.2 Current-carrying capacities of cables

The current-carrying capacity of an insulated cable is the maximum current that it can continuously carry without exceeding its temperature rating. It is also known as *ampacity*.

In compliance with the Wiring Regulations, the cross-sectional areas of the fixed wiring should be such that the permissible voltage drop is not exceeded.

6.3.1.3 Groups containing more than one circuit

Groups containing non-sheathed or sheathed cables which have different maximum operating temperatures, the current-carrying capacity of all cables in this group shall be based on the lowest maximum operating temperature of any one cable in the group together with the appropriate group rating factor.

 If a non-sheathed or sheathed cable is expected to carry a current less than 30% of its grouped current-carrying capacity, it may be ignored for the purpose of obtaining the rating factor for the rest of the group.

6.3.2 Installation of conductors

6.3.2.1 Electrical connections to bare connectors and/or busbars

Where a cable is to be connected to a bare conductor or busbar, the type of insulation and/or sheath shall be chosen with respect to the maximum operating temperature of the bare conductor or busbar.

A busbar trunking or a powertrack system:

- shall comply with BS EN 60439-6;
- shall comply with the appropriate part of the BS EN 61534 series;
- shall take account of external influences. (See Appendix 8 of BS 7671:2018 for further details.)

6.3.2.2 Bare conductors

If the nominal voltage does not exceed 25 V a.c. or 60 V d.c., bare conductors may be used for extra-low-voltage lighting installations, provided that:

- that the risk of a short-circuit is reduced to a minimum;
- the conductors have a cross-sectional area of at least 4 mm^2;
- the conductors are not placed directly on combustible material.

For suspended bare conductors, at least one conductor and its terminals shall be insulated for that part of the circuit between the transformer and the short-circuit protective device.

6.3.2.3 Bonding conductors

In agricultural and horticultural premises, protective bonding conductors shall be protected against mechanical damage and corrosion and shall have been selected to avoid electrolytic effects.

In solar photovoltaic (PV) power supply systems (where protective bonding conductors are installed) they shall be parallel to and in as close contact as possible with d.c. cables and a.c. cables and accessories.

A permanent label/warning notice, with the words shown in Figure 6.11, shall be permanently fixed at or near the bonding conductor's connection point to an extraneous part.

6.3.2.4 Connecting conductors

Connecting conductors shall meet the following requirements:

- all bare live conductors shall be installed on insulators;

Figure 6.11 Warning notice – where protective bonding conductors are installed.

- a circuit protective conductor shall be run to (and terminated at) each point in a wiring circuit supplying one or more items of Class II equipment;
- an insulation monitoring device (IMD) shall be connected between Earth and a live conductor of the monitored equipment;
- the *'line'* terminal(s) of an IMD shall be connected as close as practicable to the origin of the system, to either:

  o the neutral point of the power supply; or
  o an artificial neutral point with impedances connected to the line conductors; or
  o a line conductor or two or more line conductors.

 For d.c. installations, the *'line'* terminal(s) of the IMD shall be connected either to the midpoint or to one or all of the supply conductors.

- In some particular d.c. IT two-conductor installations, a passive IMD that does not inject current into the system may be used, provided that:

  o the insulation of all live distributed conductors is monitored;
  o all exposed-conductive-parts of the installation are interconnected;
  o circuit conductors are selected and installed so as to reduce the risk of an Earth fault to a minimum.

6.3.2.5 Conductors in parallel

Where two or more live conductors or PEN conductors are connected in parallel in a system, they shall be of the same material and correctional area and either:

- measures shall be taken to achieve equal load current sharing between them; or
- the conductors in parallel shall be:

  o multi-core cables; or
  o twisted single-core cables; or
  o non-sheathed cables; or
  o non-twisted single-core cables or non-sheathed cables in flat formation and where the cross-sectional area is greater than 50 mm^2 in copper or 70 mm^2 in aluminium.

 Notes:

1. This Regulation does not preclude the use of ring final circuits with or without spur connections.
2. Where adequate current sharing is not possible (or where four or more conductors have to be connected in parallel), consideration shall be given to the use of busbar trunking.

6.3.2.6 Connection of multi-wire, fine wire and very fine wire conductors

To avoid undesirable separation or spreading of individual wires, care should be taken to ensure that suitable terminals are used (and that conductor ends are suitably treated) for all connections of multi-wire, fine wire or very fine wire conductors. To achieve these conditions, BS EN 7671:2018 stipulates that:

- non-sheathed cables found at the termination of a conduit, ducting or trunking, shall be enclosed;
- soldering (tinning) the whole of the conductor end of multi-wire, fine wire and very fine wire conductors is not permitted if screw terminals are used.

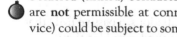 Soldered (tinned) conductor ends on fine wire and very fine wire conductors are **not** permissible at connection and junction points which (whilst in service) could be subject to some form of movement between the soldered and the non-soldered part of the conductor.

6.3.3 Installation of a wiring systems

The installation of wiring systems will depend on the type of conductor or cable used (see Section 521 of BS 7671:2018 for complete details) and:

- the rated voltage of the cable(s) shall not be less than the nominal voltage of the system and shall be at least 300/500 V; and
- adequate mechanical protection of the basic insulation shall be provided by (one or more) of the following:

  o the non-metallic sheath of the cable;
  o non-metallic trunking or ducting (complying with the BS EN 50085);
  o a non-metallic conduit.

The minimum cross-sectional area of the extra-low-voltage conductors shall normally be 1.5 mm² copper, but:

- for flexible cables with a maximum length of 3 m, a cross-sectional area of 1 mm² copper may be used;
- for suspended flexible cables (and especially for mechanical reasons) 4 mm² copper should be used;
- for composite cables consisting of a braided tinned copper outer sheath (with a material of high tensile strength inner core) 4 mm² copper should be used.

A voltage Band I circuit shall not be contained in the same wiring system as a Band II circuit, unless:

- every cable is insulated against the highest voltage present;
- each conductor of a multi-core cable is insulated for the highest voltage present;
- the cables are insulated for their system voltage and installed in a separate compartment to that for a cable ducting or cable trunking system;
- the cables are installed on a cable tray system where physical separation is provided by a partition;
- for a multi-core cable or cord, the cores of the Band I circuit are separated from the cores of the Band II circuit by an earthed metal screen of equivalent current-carrying capacity of the largest core of a Band II circuit.

6.3.4 Earthing

6.3.4.1 Earth electrodes

Earth electrodes are specifically designed and installed to improve a system's earthing and, as previously covered in Chapter 3 of this book ('*Earthing*'), the following types of Earth electrode are recognised as being suitable for the purposes of these Regulations:

- earth rods or pipes;
- Earth tapes or wires;
- Earth plates;
- underground structural metalwork embedded in foundations;

- welded metal reinforcement of concrete (except pre-stressed concrete) embedded in the Earth;
- lead sheaths and other metal coverings of cables;
- other suitable underground metalwork.

If the lead sheath or other metal covering of a cable is used as an Earth electrode, it will be subject to all of the following conditions:

- precautions have been taken to prevent excessive deterioration by corrosion;
- the sheath or covering shall be in effective contact with Earth;
- the consent of the owner of the cable has been obtained.

Further information on Earth electrodes can be found in BS 7430.

6.3.4.2 Earthing arrangements for protective conductors

Where the heat dissipation differs from one part of a route to another, the current-carrying capacity at each part of the route should be appropriate for that part of the route.

The earthing arrangements of protective conductors shall be such that:

- the value of impedance from the consumer's main earthing terminal to the earthed point of the supply meets the protective and functional requirements of the installation, and is considered to be continuously effective;
- Earth fault currents and protective conductor currents which may occur are carried without danger;
- the protective conductors are sufficiently robust (or have additional mechanical protection) against external influences.

For a TN-S system, the main earthing terminal of the installation needs to be connected to the earthed point of the source of energy.

For a TN-C-S system, where protective multiple earthing is provided; the main earthing terminal of the installation needs to be connected by the distributor to the neutral of the source of energy.

For a TT or IT system, the main earthing terminal shall be connected via an earthing conductor to an Earth electrode.

Where the supply to an installation is at high voltage, protection against faults between the high-voltage supply and Earth **shall** be provided.

Where a number of installations have separate earthing arrangements, any protective conductor common to any of these installations shall either:

- be capable of carrying the maximum fault current likely to flow through them; or

- be earthed within one installation only and insulated from the earthing arrangements of any other installation.

 Precautions should be taken against the risk of damage to other metallic parts through electrolysis and it is not recommended that an exposed-conductive-part of equipment is used to form a protective conductor for another equipment.

6.3.4.2.1 Earthing terminal

Where the protective conductor is formed by metal conduit, trunking, ducting or the metal sheath and/or armour of a cable, the earthing terminal of each accessory shall be connected by a separate protective conductor to an earthing terminal incorporated in the associated box or other enclosure.

Unless the circuit protective conductor is formed by a metal covering (or enclosure containing all of the conductors of the ring), the circuit protective conductor of every ring final circuit shall also be run in the form of a ring having both ends connected to the earthing terminal at the origin of the circuit.

The earthing conductor needs to be easily disconnected (possibly by means of a specialised tool) so as to allow the resistance of the earthing arrangements to be checked during maintenance and/or repair.

6.3.4.2.2 Earthing requirements for the installation of equipment having high protective conductor currents

Equipment having a protective conductor current exceeding 3.5 mA but less than 10 mA may either be permanently connected to the fixed wiring of the installation or connected by means of a plug and socket outlet complying with BS EN 60309-2.

Equipment having a protective conductor current exceeding 10 mA shall be connected via a flexible cable to the supply, permanently or via a protective conductor with an earth monitoring system.

If the total protective conductor current is likely to exceed 10 mA, then the wiring of every final circuit and distribution circuit shall have a high-integrity protective connection complying with one or more of the following:

- a single protective conductor with a cross-sectional area greater than 10 mm²;
- a single copper protective conductor having a cross-sectional area of not less than 4 mm²;
- two individual protective conductors;
- an Earth monitoring system which, in the event of fault occurring in the protective conductor, automatically disconnects the supply to the equipment;
- connection of the equipment to the supply is by means of a double-wound transformer or equivalent unit, such as a motor-alternator set.

Where two protective conductors are used, their ends shall be terminated independently of each other and at all connection points throughout the circuit.

 information shall be provided indicating which circuits have a high protective conductor current at the distribution board.

6.3.5 Electrical connections

Connections between conductors or between a conductor and other equipment should provide robust electrical continuity and ample mechanical strength and protection.

The type of connector chosen will need to take account of:

- the cross-sectional area of the conductor;
- the material of the conductor and its insulation;
- the number and shape of the wires forming the conductor;
- the number of conductors to be connected together;
- the provision of adequate locking arrangements in situations subject to vibration or thermal cycling.

Where a soldered connection is used, it shall be completed so that it is unaffected by creep, mechanical stress or any temperature rises under fault conditions.

Every electrical connection and joint shall be accessible for inspection, except for:

- a joint designed to be buried in the ground;
- an encapsulated joint;
- a connection between a cold tail and the heating element of equipment such as a floor heating system;
- a joint made by welding, soldering, brazing or an appropriate compression tool;
- joints or connections made in the equipment by the manufacturer that are not intended to be inspected or maintained;
 - equipment complying with the requirements of BS 5733 for a maintenance-free accessory and marked with the symbol.

Where necessary, precautions shall be taken so that the temperature of a connection whilst in normal service does not affect the conductors or any insulating material used to support the connection.

Where a cable is to be connected to a bare conductor or busbar, it should be capable of working within the maximum operating temperature of the bare conductor or busbar.

Every termination and joint in a live conductor or a PEN conductor shall be made within one (or a combination) of the following:

- a suitable accessory;
- an equipment enclosure;

- an enclosure partially formed or completed with non-combustible building material.

Where a connection is made in an enclosure, the enclosure shall provide adequate mechanical protection and protection against relevant external influences.

There shall be no appreciable mechanical strain on the connections of conductors.

6.3.5.1 Temperature

In electrical installations, risk of injury may result from excessive temperatures, which are likely to cause burns, fires and other injurious effects. To guard against this happening:

- cables and wiring accessories should only be installed or handled at temperatures within the limits stated in the relevant product specification or as given by the manufacturer;
- cables or flexible cords within an accessory, appliance or luminaire need to be suitable for the temperatures likely to be encountered, or provided with additional insulation suitable for those temperatures.

6.3.5.2 Thermally insulating materials

The current-carrying capacity of a cable is dependent on a number of factors, including the impact from thermally insulating materials on the current-carrying capacity of flat twin and Earth cables that are installed in buildings and for this reason the Regulations state that:

- a cable should preferably **not** be installed in a location where it is liable to be covered by thermal insulation, such as in a loft;
- if a cable has to be run in a space which is likely to become thermally insulated then, wherever practicable, the cable should be fixed in a position such that it will **not** be covered by the thermal insulation;
- if a single cable is likely to be totally surrounded by thermally insulating material for over 0.5 m, the current-carrying capacity shall be assumed to be at least 0.5 times the current-carrying capacity for that cable clipped directly to a surface and open;
- where a cable is to be totally surrounded by thermal insulation for less than 0.5 m the current-carrying capacity of the cable shall be reduced appropriately depending on the size of cable, length in insulation and thermal properties of the insulation.

6.3.5.3 Electric floor heating systems

The load of every floor-warming cable under operation shall be limited to a value such that the manufacturer's stated conductor temperature is not exceeded.

In locations containing a bath or a shower, only heating cables or thin-sheet flexible heating elements shall be provided which have either a metal sheath, metal enclosure or a fine-mesh metallic grid connected to the protective conductor of the supply circuit.

 Compliance with the latter requirement is not required if the protective measure SELV is provided for the floor heating system.

For electric floor heating systems in locations containing a bath or shower, the protective measure *'protection by electrical separation'* is not permitted.

6.3.5.4 Electrode water heaters and boilers

An electrode water heater or boiler is a device which heats the water it contains by having two or three electrodes immersed in the water and a single (or three-phase) supply connected to them. There is no element, the water being heated by the current which flows through it between the electrodes, and for these reasons:

- the shell of the electrode water heater or electrode boiler shall be bonded to the metallic sheath and armour (if any) of the incoming supply cable; and
- the protective conductor shall be connected to the shell of the electrode water heater or electrode boiler.

If an electrode water heater or electrode boiler is connected to a three-phase low-voltage supply, the equipment's shell shall be connected to the neutral of the supply as well as to the earthing conductor.

If the supply to an electrode water heater or electrode boiler is single-phase and one electrode is connected to a neutral conductor earthed by the distributor, the equipment's shell shall be connected to the neutral of the supply as well as to the earthing conductor.

If the electrode water heater or electrode boiler is not piped to a water supply or in physical contact with any earthed metal, a fuse in the line conductor may be substituted for the circuit-breaker and the equipment's shell need not be connected to the neutral of the supply.

6.3.5.4.1 Water heaters having immersed and uninsulated heating elements

All metal parts of the heater or boiler which are in contact with the water (other than current-carrying parts) shall be solidly and metallically connected to a metal water pipe through which the water supply to the heater or boiler is provided and that water pipe shall be connected to the main earthing terminal by means independent of the circuit protective conductor.

Single-phase water heaters and boilers with an uninsulated heating element immersed in the water shall not have a single-pole switch, non-linked circuit-breaker or fuse fitted in the neutral conductor, in any part of the circuit between the heater or boiler, or in the origin of the installation.

6.3.5.5 Enclosures

Conductive parts that are within an insulating enclosure shall be connected to a protective conductor (all conductive parts separated from live parts by basic insulation only shall be contained in an insulating enclosure).

No exposed-conductive-part or intermediate part shall be connected to a protective conductor, unless the specification for the equipment concerned allows for this to happen.

Where it is necessary to remove a barrier or to open an enclosure or remove parts of enclosures, this shall be possible only:

- by the use of a key or tool; or
- after disconnection of the supply to live parts against which the barriers or enclosures afford protection.

Enclosures containing rectifiers and transformers shall be adequately ventilated and the vents shall not be obstructed when in use.

6.3.6 Fault protection

Fault protection by automatic disconnection of supply shall be provided either by an overcurrent protective device (OCPD) in each line conductor or by an RCD.

Live parts of the separated circuit shall not be connected at any point to another circuit or to Earth or to a protective conductor (for separated circuits the use of separate wiring systems is recommended).

No exposed-conductive-part of the separated circuit shall be connected either to the protective conductor or exposed-conductive-parts of other circuits, or to Earth.

The exposed-conductive-parts of the equipment of the FELV circuit shall be connected to the protective conductor of the primary circuit of the source, **provided** that the primary circuit is subject to protection by automatic disconnection of supply.

Where OCPDs are used for fault protection, the protective conductor shall be incorporated in the same wiring system as the live conductors or in their immediate proximity.

6.3.6.1 Protection against fault current

 Any fault occurring at any point in a circuit it is quickly interrupted so that the fault current does not cause the agreed limiting temperature of any conductor or cable to be exceeded.

A device protecting a conductor may be installed on the supply side of the point where a change occurs, provided that it possesses an operating characteristic that can protect the wiring situated on the load side against fault current.

A single protective device may protect conductors in parallel against the effects of fault currents, provided that the device operates as soon as a fault occurs in one of the parallel conductors.

Conductors (other than live conductors), and any other parts intended to carry a fault current, shall be capable of carrying that current without attaining an excessive temperature.

A device for protection against fault current need not be provided for:

- a conductor connecting a generator, transformer, rectifier or an accumulator battery to the associated control panel where the protective device is placed in the panel;
- a circuit where disconnection could cause danger for the operation of the installation concerned;
- certain measuring circuits;
- the origin of an installation where the distributor installs one or more devices that provide protection against fault current; and

provided that both of the following conditions are simultaneously fulfilled:

- the wiring is carried out in such a way as to reduce the risk of a fault occurring to a minimum; and
- the wiring is installed in such a manner as to reduce to a minimum the risk of fire or danger to persons.

A device providing protection against fault current shall be installed at the point where a reduction in the cross-sectional area (or other change) causes a reduction in the current-carrying capacity of the conductors of the installation.

Note: Conductors should be provided with protection against electromechanical stresses of fault currents as necessary to prevent injury or damage to persons, livestock or property.

6.3.6.2 Ferromagnetic enclosures: electromagnetic effects

The conductors of an a.c. circuit that are installed in a ferromagnetic enclosure must be arranged so that the line conductors, the neutral conductor (if any) and the appropriate protective conductor are all contained in the same enclosure.

Where the conductors enter a ferrous enclosure, they should be arranged so that the conductors are only collectively surrounded by ferrous material.

6.3.6.3 *Final circuits*

The following arrangements of the final circuit are acceptable:

- a ring final circuit with a ring protective conductor;
- a radial final circuit with a single protective conductor.

If a final circuit contains a number of socket outlets or connection units that are intended to supply two or more items of equipment (and it is known that the total protective conductor current will exceed 10 mA) they shall be provided with a high-integrity protective conductor connection.

In a caravan and/or motor caravan, each final circuit forming part of an electrical installation shall be protected by an OCPD which disconnects all live conductors of that circuit.

6.3.7 Fire risk

A fire risk assessment is a process involving the systematic evaluation of the factors that determine the hazard from fire, the likelihood that there will be a fire and the consequences if one were to occur. There are both qualitative and quantitative methods of risk assessment (see Figure 6.12) that can be used, but broadly speaking, assessments are conducted in five key steps:

1. Identify the fire hazards.
2. Identify equipment and/or systems at risk.
3. Evaluate, remove or reduce the risks.
4. Record your findings, prepare an emergency plan and provide training.
5. Review and update the fire risk assessment regularly.

 It is recommended that, wherever possible, every termination of a live conductor or connection or joint between live conductors shall be contained within an enclosure.

Within the Wiring Regulations, all of these potential fire risks are covered in different ways depending on the situation, equipment or system involved. For example, the following are just a few of the most relevant stated requirements to be found in BS 7671:2018:

- In locations where a particular risk of fire exists, preventive protection measures against the risk of fire occurring must be investigated at the earliest opportunity.
- An appropriate evaluation of the risk should be carried out by a person who is experienced in fire risk assessments and if the evaluation indicates a possible fire risk, some form of safety precaution should be installed at the origin of the circuit to be protected. These could include:
  - o a residual current device (RCD) for overall protection against the risk of fire, or for protection against the risk of fire in IT systems;

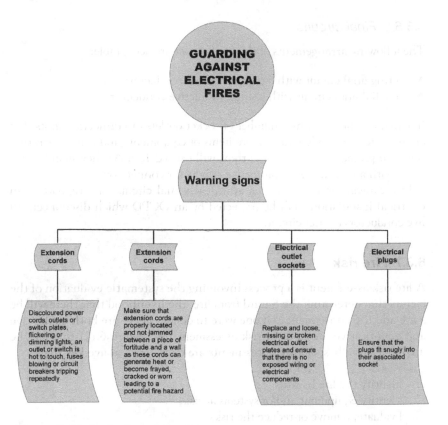

Figure 6.12 Basic fire precautions.

- o   a residual current monitoring device (RCM); or
- o   an insulation monitoring device (IMD);
- o   an arc fault detection device (AFDD).

- Where practicable, the passage of circuits through locations presenting a fire risk should be avoided.
- In locations where a fire risk exists, conductors of circuits supplied at extra-low voltage should be protected by either barriers or enclosures.
- Luminaires and floodlights should be fixed and protected so that a focusing or concentration of heat is not likely to cause ignition of any material.
- A wiring system which passes through the location but is not intended to supply electrical equipment within that location shall:

  - o   have no connection or joints in the location, unless the connection or joint is installed in an enclosure; and
  - o   is protected against overcurrent; and
  - o   does not use bare live conductors.

- A PEN conductor shall not be used unless it is a circuit traversing the location.

BS 7671:2018 Requirement

As stated in BS 7671:2018 para 422.3.13, every circuit shall be provided with a means of isolation from all live supply conductors by a linked switch or a linked circuit- 'breaker'.

6.3.7.1 Fire precautions

Fire precautions within the England and Wales are covered by the Regulatory Reform (Fire Safety) Order 2005, whilst in Scotland, the requirements on general fire safety are covered in Part 3 of the Fire (Scotland) Act 2005 – which is very similar to the Regulatory Reform (Fire Safety) Order 2005.

The aims of these pieces of legislation are to reduce the likelihood of a fire occurring in a building, and to prevent its spread if a fire incident does occur. Specifically, however, they are aimed at protecting people, buildings and their contents from damage.

Other chapters in this book have explained in detail how BS 7671:2018 addresses the necessary precautions that need to be taken for different aspects of the Building and Wiring Regulations, but for the installation of cables and conductors the following situations also need to be considered:

- installing cables with improved fire-resisting characteristics in case of a fire hazard;
- installing of cables in non-combustible solid walls, ceilings and floors;
- installing cables in areas with constructional partitions which have a fire-resisting time capability of 30 minutes or 90 minutes;
- installing mineral-insulated cables according to BS EN 60702.

Where these measures are not practicable, improved fire protection may be possible by the use of reactive fire protection systems.

6.3.7.1.1 Precautions within a fire-segregated compartment

Although cables that meet the requirements of BS EN 60332-1-2 can be installed without special precautions, cables that do not comply with this standard must be limited to short lengths for connection of appliances to the permanent wiring system and shall not pass from one fire-segregated compartment to another.

In installations where a definite fire risk is identified, the likelihood of a fire happening shall be minimised by ensuring that **all** cables meet the flame propagation requirements given in the relevant part of the BS EN 50266 series.

6.3.7.1.2 Locations with risks of fire due to the nature of processed or stored materials

Where there is a risk of fire due to the manufacture, processing and/or storage of flammable materials (usually referred to as BE2 conditions) in areas such as barns (due to the accumulation of dust and fibres), or in woodworking facilities or paper mills and textile factories (due to the storage and processing of combustible materials), then cables shall, as a minimum, satisfy the fire conditions specified in BS EN 60332-1-2.

Cables that are not completely embedded in a non-combustible material such as plaster or concrete or otherwise protected from fire shall still need to meet the flame propagation characteristics specified in BS EN 60332-1-2, and:

- a cable trunking system or cable ducting system needs to satisfy the fire conditions specified in BS EN 50085;
- precautions need to be taken to ensure that a cable or wiring system cannot propagate flame;
- a cable tray system or cable ladder shall meet the requirements of BSEN 61537.

Except for mineral-insulated cables, a wiring system shall be protected against insulation faults:

- in a TN or TT, system by an RCD having a rated residual operating current (IA) not exceeding 300 mA;
- in an IT system, by an insulation monitoring device with audible and visual signals.

Flexible cables and flexible cords shall be either:

- of heavy-duty type (with a voltage rating of not less than 450/750 V); or
- suitably protected against mechanical damage.

6.4 Safety precautions

Electrical energy is extremely dangerous if installed and/or used incorrectly and there are a numerous methods (see examples shown in Figure 6.13 and this section below) for ensuring of safety.

6.4.1 Fuses

A fuse is an electrical safety device that operates to provide overcurrent protection of an electrical circuit. Its essential component is a metal wire or strip that melts when too much current flows through it, thereby interrupting the current.

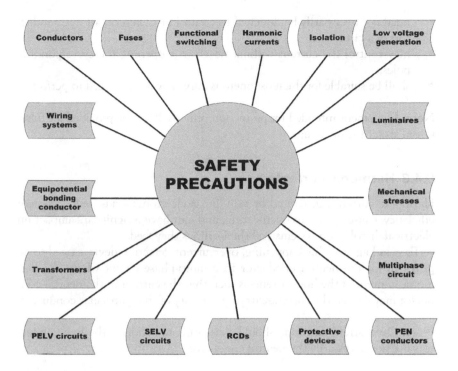

Figure 6.13 Safety precautions.

Fuse links shall preferably be of the cartridge type and shall either:

- have marked on or adjacent to them the type of fuse link that should to be used; or
- be of a type such that there is **no possibility** of replacing the existing fuse link with one that has a higher fusing factor than that intended.

a fuse carrier shall be arranged so that there is no contact between conductive parts belonging to two adjacent fuse bases.

a fuse base using screw-in fuses shall be connected so that the centre contact is connected to the conductor from the supply and the shell contact is connected to the conductor to the load.

6.4.2 Functional switching

Functional switching and its devices are used in the normal operation of circuits to switch them on and off and:

- shall be provided for every part of a circuit that needs to be controlled independently from other parts of the installation; but
- may control several items of current-using equipment intended to operate simultaneously;

- need not necessarily be designed to control or switch off **all** live conductors of a circuit;
- may control the current without necessarily opening the corresponding poles;
- shall be suitable for the most onerous duty they are intended to perform.

Note: Equipment intended for protection only shall not be provided for functional switching of circuits.

6.4.3 Harmonic currents

Harmonic currents, generated by non-linear electronic loads, reduce system efficiency, cause apparatus overheating, and can have a significant impact on electrical distribution systems and the facilities they feed.

To guard against this happening, overcurrent detection devices need to be provided for the neutral conductor in a multi-phase circuit where the harmonic content of the line currents is such that the current in the neutral conductor may exceed the current-carrying capacity of that particular conductor.

Note: Overcurrent detection should cause disconnection of the line conductors but not necessarily the neutral conductor.

6.4.4 Isolation

Isolation is intended (for reasons of safety) to make dead a circuit by separating an installation or section from every source of electric energy.

Every circuit shall be capable of being isolated from each of the live supply conductors.

Provision shall be made for disconnecting the neutral conductor.

In a TN-S or TN-C-S system, it is not necessary to isolate or switch the neutral conductor if it is reliably connected to Earth by suitably low impedance.

Semiconductor devices shall **not** be used as isolating devices.

Where necessary, suitable means shall be provided for the discharge of stored electrical energy.

- Where an isolating device for a particular circuit is remote from the equipment to be isolated, the means of isolation should always be secured in the open position.
- If the means of isolation is a lock or removable handle, then the key or handle shall be non-interchangeable with any other used for a similar purpose within the premises.
- If a switch is provided for this purpose:

  o   it shall be capable of cutting off the full load current of the relevant part of the installation; but o   if used as a device for switching off for mechanical maintenance, the switch need not necessarily interrupt the neutral conductor.

- In temporary electrical installations, marinas and similar locations, this switching device shall disconnect all live conductors including the neutral conductor.

6.4.4.1 Devices for isolation

In accordance with the Building Regulations, *"the electrical installation of each building or part of a building shall be isolated by a single isolation device"*, and the Wiring Regulations state that *"every circuit shall be provided with a means of isolation from all live supply conductors by a linked switch or a linked circuit-breaker"*. These requirements shall be achieved as follows:

- A means of isolation shall be installed on both sides of a static convertor.
- An isolation device shall isolate all live supply conductors from the circuit concerned.
- If a link is inserted in the neutral conductor, the link shall:

    o   not be capable of being removed without the use of a tool;
    o   only be accessible to skilled persons.

- All fixed electric motors shall be provided with a readily accessible and efficient means of switching off that is easy to operate and placed so as to prevent danger.
- Each device used for isolation shall be clearly identified by position or permanent marking to indicate the installation or circuit it isolates.
- In TN-S and TN-C-S systems, isolation or switching of the neutral conductor is not required if protective equipotential bonding is installed and either:

    o   the neutral conductor is reliably connected to Earth by a low resistance to meet the disconnection times of the protective devices accordingly; or
    o   the distributor declares that either the PEN or the neutral conductor of the supply is reliably connected to Earth.

- Isolation devices shall be designed and installed so as to prevent unintentional or inadvertent closure. For example, they shall be:

    o   located within a lockable space or lockable enclosure;
    o   located adjacent to the associated equipment.

- Isolation devices shall not be erected in agricultural and horticultural premises where they could be accessible to livestock or installed in a position where access could be impeded by livestock.
- Means of isolation shall preferably be provided by a multi-pole switching device.

Semiconductor devices shall **not** be used as isolating devices and no means of isolation or switching shall be inserted in the outer conductor of a concentric cable.

BS 7671:2018 Table 537.4 provides detailed guidance on the selection of protective, isolation and switching devices.

6.4.5 Live conductors

6.4.5.1 Background reminder

The live wire carries the current to the appliance, while the neutral wire carries it back. The Earth wire is for our protection, in case the live wire makes a contact with metal casing of an appliance. When this happens, the current will pass to the Earth instead of our body, thus saving us.

6.4.5.2 Some of BS 7671:2018's requirements and advice

The Regulations are full of advice about live wire conductors and the following is (in the author's opinion) probably the most important:

- A main switch that is intended to be operated by ordinary persons (such as a householder or similar individual) shall interrupt both live conductors of a single-phase supply.
- Bare live conductors shall be installed on insulators.
- Conductors shall be able to carry fault current without overheating.
- In agricultural and horticultural premises:

  o   the electrical installation of each building or part of a building shall be isolated by a single isolation device;
  o   means of isolation of all live conductors, including the neutral conductor, should also be provided for circuits that are used occasionally (such as during harvest time).

- In solar photovoltaic (PV) power supply systems, earthing of one of the live conductors of the d.c. side is permitted if there is at least simple separation between the a.c. side and the d.c. side.
- In zone 1 of swimming pools (and other basins) the supply circuit of the equipment shall be protected by:

  o   SELV; or
  o   an RCD; or
  o   electrical separation.

- Live supply conductors **shall** be capable of being isolated from circuits.
- Persons and livestock **shall** be protected against injury, and property shall be protected against damage due to excessive temperatures or electromechanical stresses caused by any overcurrents likely to arise in live conductors.

- The supply to all live conductors **shall** be automatically interrupted in the event of overload current and fault current.
- Where an installation is supplied from more than one source of energy (one of which requires a means of earthing independent of the means of earthing of other sources, and it is necessary to ensure that no more than one means of earthing is applied at any time), a switch may be inserted in the connection between the neutral point and the means of earthing, **provided** that the switch is a linked switch that has been arranged to disconnect and connect the earthing conductor for the appropriate source, at substantially the same time as the related live conductors.
- Where an IT system is designed not to disconnect in the event of a first fault, the occurrence of the first fault shall be indicated by either:

  o   an insulation monitoring device (IMD) combined with an insulation fault location system (IFLS); or
  o   a residual current monitor (RCM), provided the residual current is sufficiently high to be detected.

 Where no neutral point or midpoint exists, a line conductor may be connected to Earth.

6.4.6 Low voltage generating sets

Where a generating set is intended to operate in parallel with a system that distributes electricity to the public (or where two or more generating sets operate in parallel), circulating harmonic currents shall be limited so that the thermal rating of conductors is not exceeded.

6.4.7 Luminaires

Any flexible cable between the fixing means and the luminaire should be installed so that any expected stresses in the conductors, terminals and terminations will not affect the safety of the installation.

In circuits of a TN or TT system (except for E14 and E27 lampholders complying with BS EN 60238), the outer contact of every Edison screw or single-centre bayonet-cap type lampholder shall be connected to the neutral conductor.

 This regulation also applies to track-mounted systems!

Groups of luminaires divided between the three line conductors of a three-phase system with only one common neutral conductor shall be provided with at least one device that simultaneously disconnects all line conductors.

Luminaires and floodlights shall be so fixed and protected that a focusing or concentration of heat is not likely to cause ignition of any material.

Extra-low-voltage luminaire suspension devices and supporting conductors shall be capable of carrying five times the mass of the luminaires, including their lamps intended to be supported, but not less than 5 kg. The suspended system shall be fixed to walls or ceilings by insulated distance cleats and terminations and connections of conductors shall be made by screw terminals or screwless clamping devices.

6.4.8 Mechanical stresses

- Cable shall not be installed across a site road or a walkway unless adequate protection of the cable against mechanical damage is provided, and every conductor and wiring system shall be supported in such a way that it is not exposed to undue mechanical strain.
- A wiring system:
  - o shall be selected and erected to avoid, during installation, use or maintenance, damage to the sheath or insulation of cables and their terminations;
  - o shall ensure that the radius of every bend in a wiring system is such that conductors do not suffer damage and terminals are not stressed;
  - o intended for the drawing in or out of conductors shall have adequate means of access to allow this operation.
- Cables, busbars and other electrical conductors which pass across expansion joints shall be so selected or erected that anticipated movement does not cause damage to the electrical equipment.
- A cable buried in the ground (that is not installed in a conduit or duct) shall incorporate an earthed armoured or metal sheath (or both) suitable for use as a protective conductor.

6.4.9 Multi-core cables, conduits, ducting systems, franking systems or tray or ladder systems

Where multi-core cables are installed in parallel, each cable shall contain one conductor of each line.

The line and neutral conductors of each final circuit shall be electrically separate from those of every other final circuit, so as to prevent the indirect energising of a final circuit intended to be isolated.

Each part of a circuit shall be arranged such that the conductors are not distributed over different multi-core cables, conduits, ducting, trunking, tray or ladder systems.

6.4.10 Multi-phase circuit

In multi-phase (and single-phase) circuits, an independently operated single-pole switching device or protective device shall not be inserted in the neutral conductor.

6.4.11 Neutral conductor

Consideration should be given to the fact that:

- if the neutral conductor in a three-phase TN or TT system is interrupted, basic, double and reinforced insulation can be temporarily stressed with the line-to-line voltage (to guard against this, the neutral conductor shall be protected against short-circuit current);
- if a line conductor of an IT system is earthed accidentally, insulation or components can be temporarily stressed with the line-to-line voltage;
- if a short-circuit occurs in the low-voltage installation between a line conductor and the neutral conductor, the voltage between the other line conductors and the neutral conductor can reach the value of $1.45 \times U_o$ for a time up to 5s.

To guard against the above problems, the following Regulations should be observed:

- Owing to the possibility of increased harmonic distortion, the neutral conductor shall not be smaller than the line conductors.
- In an IT system, the neutral conductor shall not be distributed unless:

  o overcurrent detection is provided for the neutral conductor of every circuit; or
  o the neutral conductor is effectively protected against short-circuit by a protective device installed on the supply side; or
  o the circuit is protected by an RCD with a rated residual operating current not exceeding 0.2 times the current-carrying capacity of the corresponding neutral conductor.

- In a TN-S or TN-C-S system the neutral conductor need not be isolated or switched where it can be regarded as being reliably connected to Earth by suitably low impedance.
- The neutral conductor shall not be disconnected before the line conductors and shall be reconnected at the same time as (or before) the line conductors.
- In temporary electrical installations, the neutral conductor of the star-point of the generator shall, except for an IT system, be connected to the exposed-conductive-parts of the generator.
- If a link is inserted in the neutral conductor for isolating purposes, the link:

  o should not be capable of being removed without a tool; and
  o should be accessible to only one or more skilled persons.

- If an autotransformer is connected to a circuit having a neutral conductor, the common terminal of the winding shall be connected to the neutral conductor.

6.4.12 Non-conducting location

In a non-conducting location there is no need for a protective conductor.

6.4.13 Non-earthed equipotential bonding conductor

- Source supplies may supply more than one item of equipment, provided that:

    o   all exposed-conductive-parts of the separated circuit are connected together by an insulated and non-earthed equipotential bonding conductor;
    o   the non-earthed equipotential bonding conductor is not connected to a protective conductor (or to an exposed-conductive-part of any other circuit or to any extraneous-conductive-part).

6.4.14 Overhead conductors

- In caravan and camping parks; marinas and similar locations, all overhead conductors:

    o   shall be insulated;
    o   shall be at a height above ground of not less than 6 m in all areas subject to vehicle movement and 3.5 m in all other areas;
    o   shall have poles and other supports for overhead wiring located or protected so that they are unlikely to be damaged by any foreseeable vehicle movement.

See BS 7671:2018 Chapter 709 for more detailed requirements relating to marinas and similar locations.

6.4.15 PEN conductors

The following conditions are applicable to all combined protective and neutral (PEN) conductors:

- In Great Britain, Regulation 8(4) of the Electricity Safety, Quality and Continuity Regulations 2002 prohibits the use of PEN conductors in consumers' installations.
- Automatic disconnection using an RCD shall **not** be applied to a circuit incorporating a PEN conductor.
- A separate metal cable enclosure shall **not** be used as a PEN conductor.
- PEN conductors shall **not** be isolated or switched.

PEN conductors may **only** be used within an installation:

- where any necessary authorisation for use of a PEN conductor has been obtained and where the installation complies with the conditions for that authorisation; or

- where the installation is supplied by a privately owned transformer or converter, provided there is no metallic connection (except for the earthing connection) with the distributor's network; or
- where the supply is obtained from a private generating plant.

For a fixed installation, a PEN conductor can be used, provided that the part of the installation concerned is not supplied through an RCD.

Other than a cable conforming to BS EN 60702-1, all PEN conductors shall be insulated or have an insulating covering suitable for the highest voltage to which they may be subjected.

 Note: If, from any point of the installation, the neutral and protective functions are provided by separate conductors, those conductors should not then be reconnected beyond that point.

6.4.15.1 Marinas, caravans and camping parks

For electrical installations in marinas (and similar locations), and in caravans and camping parks, ESQCR Regulations prohibit the connections of a PME earthing facility to any metalwork of a boat or caravan etc., and in caravan and camping parks, socket outlet protective conductors shall not be connected to any PEN conductor of the electricity supply.

6.4.15.2 Medical locations

PEN conductors shall not be used in medical locations and medical buildings downstream of the main distribution board.

6.4.15.3 Temporary electrical installations

In temporary electrical installations (where the type of system earthing is TN) a PEN conductor shall not be used downstream of the origin of the temporary electrical installation.

6.4.16 Protection against overcurrent

A protective device shall be provided to break any overcurrent in the circuit conductors before such a current could cause a danger due to thermal or mechanical effects detrimental to insulation, connections, joints, terminations or the surroundings of the conductors.

 Protection of conductors according to these Regulations does not necessarily protect the equipment connected to the conductors.

6.4.17 Protection against overload current

Overload protection devices shall be installed at any point where a reduction would affect the value current-carrying capacity of the conductors of the installation.

Note: An overcurrent may be an overload current or a fault current.

The rated current or current setting of an overload protection device shall not be less than the design current of the circuit, and:

- shall not exceed the lowest of the current-carrying capacities of any of the circuit's conductors; and
- the current causing effective operation of the protective device shall not exceed 1.45 times the lowest of the current-carrying capacities of any of the circuit's conductors.

Note: A reduction in current-carrying capacity may be due to a change in cross-sectional area, method of installation, type of cable or conductor, or in environmental conditions.

An overload protection device may be installed along the run of that conductor, provided:

- it is protected against fault current; or
- its length does not exceed 3 m;
- it is installed so as to reduce the risk of fault to a minimum; and
- it is installed in so as to reduce to a minimum the risk of fire or danger to persons.

Overload protection devices need not be provided:

- for a conductor situated on the load side of the point where a reduction occurs in the value of current-carrying capacity; or
- where the conductor is effectively protected against overload by a protective device installed on the supply side of that point; or
- for a conductor which, because of the characteristics of the load or the supply, is not likely to carry overload current; or
- at the origin of an installation where the distributor provides an overload device.

Where a single protective device protects two or more conductors in parallel there shall be no branch circuits or devices for isolation or switching in the parallel conductors.

6.4.18 Protection by Earth-free local equipotential bonding

Earth-free local equipotential bonding shall connect together every simultaneously accessible exposed-conductive-part so as to prevent the occurrence of dangerous touch voltages and shall **only** be used in special circumstances.

Unless protection by automatic disconnection of supply can be applied, local protective bonding conductors shall **not:**

- be in direct electrical contact with Earth; nor
- pass through exposed-conductive-parts; nor
- pass through extraneous-conductive-parts.

All socket outlets shall be provided with a protective conductor contact that is connected to the equipotential bonding system.

All flexible cables (unless they supply equipment with double or reinforced insulation) shall include a protective bonding conductor.

If two faults affect two exposed-conductive-parts at the same time (and these parts are fed by conductors with a different polarity), a protective device shall disconnect the supply in a disconnection time described in BS 7671:2018 Table 41.1.

 In order to meet the requirements of BS 7671:2018, protective equipotential bonding shall be applied to any metallic sheath of a telecommunication cable.

6.4.19 Protective conductors

All exposed-conductive-parts of the installation shall be connected by a protective conductor to the main earthing terminal of the installation, which shall then be connected to the earthed point of the power supply system.

- For an outdoor lighting installation, a protective conductor shall not be provided and the conductive parts of the lighting column shall not be intentionally connected to the earthing system.
- In caravans and motor caravans, all circuit protective conductors shall be incorporated in a multi-core cable or in a conduit together with the live conductors.
- In a fixed installation, a single conductor may serve both as a protective conductor and as a neutral (PEN) conductor.
- If an RCD is used for fault protection in a TN system, the circuit should also incorporate a PEN conductor (in the case of a TN-C-S system, the PEN conductor shall be on the source side of the RCD).

6.4.19.1 Preservation of electrical continuity of protective conductors

A protective conductor:

- shall be suitably protected against mechanical, chemical deterioration and electrodynamic effects;
- with a cross-sectional area up to and including 6 mm² shall be protected by a covering equivalent to that provided by the insulation of a single-core non-sheathed cable.

Where the sheath of a cable incorporating an uninsulated protective conductor of cross-sectional area up to and including 6 mm^2 is removed adjacent to joints and terminations, the protective conductor shall be protected by insulating sleeving complying with the BS EN 60684 series.

Joints intended to be disconnected for test purposes are permitted in a protective conductor circuit.

Where electrical monitoring of earthing is used, no dedicated devices (e.g. operating sensors, coils etc.) shall be connected in series with the protective conductor.

A switching device shall not be inserted in a protective conductor, unless:

- the switch has been inserted in the connection between the neutral point and the means of earthing;
- the switch is a linked switch arranged to disconnect and connect the earthing conductor;
- it is a multi-pole linked switch or plug-in device in which the protective conductor circuit has not been interrupted before the live conductors and re-established not later than when the live conductors are reconnected.

Every connection and joint shall be accessible for inspection, testing and maintenance and every joint in a metallic conduit shall be mechanically and electrically continuous.

6.4.20 Protective devices and switches

Single-pole fuses, switches or circuit-breakers shall only be inserted in the line conductor and no switch, circuit-breaker (except where linked) or fuse shall be inserted in an earthed neutral conductor.

Any linked switch or linked circuit-breaker inserted in an earthed neutral conductor shall be capable of breaking all of the related line conductors.

6.4.21 Protective earthing

Exposed-conductive-parts shall be connected to a protective conductor which shall comply with Chapter 54 of BS 7671:2008 and run to and be terminated at each point in wiring and at each accessory (except a lampholder having no exposed-conductive-parts and suspended from such a point).

Simultaneously accessible exposed-conductive-parts may be connected to the same earthing system individually, in groups or collectively.

6.4.22 Protective equipotential bonding

For each installation, a main protective bonding conductor will be used to connect all extraneous-conductive-parts to the main earthing terminal. Such installations include the following:

- central heating and air-conditioning systems;
- exposed metallic structural parts of the building;
- gas installation pipes;
- water installation pipes;
- other installation pipework and ducting.

Where an installation serves more than one building, the above requirement needs to be applied to each building.

6.4.22.1 Exhibitions and showgrounds

In exhibitions and showgrounds, structural metallic parts which are accessible from within the stand, vehicle, wagon, caravan or container shall be connected through the main protective bonding conductors to the main earthing terminal within the unit.

6.4.22.2 Mobile or transportable units

In mobile or transportable units:

- accessible conductive parts of the unit, such as the chassis, shall be connected through the main protective bonding conductors to the main earthing terminal within the unit;
- the main protective bonding conductors shall be finely stranded.

6.4.22.3 Caravans and motor caravans

In caravans and motor caravans, where protection by automatic disconnection of supply is used, an RCD and the wiring system must include a circuit protective conductor which shall be connected to:

- the protective contact of the inlet; and
- the exposed-conductive-parts of the electrical equipment; and
- the protective contacts of the socket outlets.

 Structural metallic parts which are accessible from within the caravan shall be connected through main protective bonding conductors to the main earthing terminal within the caravan.

6.4.23 RCDs

Where it is necessary to limit the consequence of fault currents in a wiring system from the point of view of fire risk, the circuit shall:

- be protected by an RCD for fault protection; and
- have the RCD installed at the origin of the circuit to be protected; and
- the RCD shall switch all live conductors.

An RCD:

- ensures that any protective conductor current which occurs during normal operation of the connected load(s) will be unlikely to cause unnecessary tripping of the device;
- shall be capable of disconnecting all the line conductors of the circuit at substantially the same time;
- associated with a circuit that would normally be expected to have a protective conductor shall not be considered sufficient for fault protection if there is no such conductor.

In an IT system, where protection is provided by an RCD, the non-operating residual current of the device shall be at least equal to the current which circulates on the first fault to Earth of negligible impedance affecting a line conductor.

6.4.24 SELV and PELV circuits

The earthing of PELV circuits may be achieved by a connection to Earth or to an earthed protective conductor within the source itself.

Protective separation of the wiring systems of SELV or PELV circuits from the live parts of other circuits shall be achieved by one of the following arrangements:

- SELV and PELV circuit conductors;
- in addition to their basic insulation, enclosing them in a non-metallic sheath or insulating enclosure;
- using an earthed metallic sheath or earthed metallic screen.

Plugs and socket outlets in a SELV system shall not have a protective conductor contact.

Exposed-conductive-parts of a SELV circuit shall not be connected to protective conductors.

6.4.25 Supplementary equipotential bonding

The use of supplementary bonding does not exclude the need to disconnect the supply for other reasons, for example protection against fire, thermal stresses in equipment etc.

A supplementary equipotential bonding system:

- is provided by a supplementary conductor, a conductive part of a permanent and reliable nature, or by a combination of these;
- is connected to the protective conductors of all equipment items, including those of socket outlets;

- that is connected to two exposed-conductive-parts (or an extraneous-conductive-part) shall have a conductance not less than that of the smaller protective conductor connected to the exposed-conductive-parts;
- that connects two extraneous-conductive-parts shall have a cross-sectional area not less than 2.5 mm² if sheathed, or 4 mm² if mechanical protection is not provided.

Where supplementary bonding is to be applied to a fixed appliance, the circuit protective conductor within the flexible cord shall be deemed to provide the supplementary bonding connection to the exposed-conductive-parts of the appliance.

6.4.25.1 Locations containing a bath and/or a shower

In locations containing a bath and/or a shower, local supplementary equipotential bonding shall connect the terminals of the protective conductor of each circuit supplying Class I and Class II equipment to the accessible extraneous-conductive-parts such as:

- metallic pipes supplying services and metallic waste pipes (e.g. water, gas);
- metallic central heating pipes and air-conditioning systems.

 Supplementary equipotential bonding may be installed outside or inside rooms containing a bath or shower, preferably close to the point of entry of extraneous-conductive-parts into such rooms. It should be noted, however, that accessible metallic structural parts of the building (metallic door architraves, window frames and similar parts) are not considered to be extraneous-conductive-parts unless they are connected to metallic structural parts of the building.

6.4.25.2 Swimming pools and/or other basins

All extraneous-conductive-parts in zones 0, 1 and 2 of a swimming pool (and/or other basins) shall be connected by supplementary protective bonding conductors to the protective conductors of exposed-conductive-parts of equipment situated in these zones.

6.4.26 Suspended conductors

Extra-low-voltage luminaire suspension devices and supporting conductors shall be capable of carrying five times the mass of the luminaires (including their lamps) intended to be supported, but not less than 5 kg.

Insulation-piercing connectors and termination wires which rely on counterweights hung over suspended conductors to maintain the electrical connection should not be used.

The suspended system should be fixed to walls or ceilings by insulated distance cleats which are continuously accessible throughout the route and the terminations and connections of conductors should be made by screw terminals or screwless clamping devices.

6.4.27 Transformers

Where an autotransformer is connected to a circuit that has a neutral conductor, the common terminal of the winding will be connected to the neutral conductor.

Where a step-up transformer is used, a linked switch shall be provided for disconnecting the transformer from all live conductors of the supply.

 A step-up autotransformer shall **not** be connected to an IT system.

6.4.28 Wiring systems

 Examples of wiring systems are shown in Table 6.4A2 of the Wiring Regulations.
A voltage Band I circuit shall **not** be part of the same wiring system as a Band II circuit, unless:

- every cable or conductor is insulated for the highest voltage present;
- each conductor of a multi-core cable is insulated for the highest voltage present in the cable.

Metallic structural parts of buildings (such as pipe systems or parts of furniture) shall not be used as live conductors.
The minimum cross-sectional area of the extra-low-voltage conductors shall normally be 1.5 mm² copper, **but** in the case of:

- flexible cables with a maximum length of 3 m, a cross-sectional area of 1 mm² copper may be used;
- suspended flexible cables or insulated conductors, for mechanical reasons, 4 mm² copper should be used;
- composite cables consisting of braided tinned copper outer sheath, having a material of high tensile strength inner core, 4 mm² copper should be used.

6.4.28.1 Marinas and similar locations

In marinas and similar locations, the following wiring systems are suitable for distribution circuits:

- overhead cables or overhead insulated conductors;
- cables with copper conductors and thermoplastic or elastomeric insulation and its sheath installed within an appropriate cable management system;

- cables with armouring and serving of thermoplastic or elastomeric material;
- mineral-insulated cables with a PVC protective covering.

6.4.28.2 Caravans and motor caravans

In caravans and motor caravans, the wiring systems shall be installed using one or more of the following:

- insulated single-core cables, with flexible class 5 conductors, in a non-metallic conduit;
- insulated single-core cables, with stranded class 2 conductors (minimum of seven strands), in a non-metallic conduit;
- sheathed flexible cables.

6.5 Inspection and testing

 Note: Inspection shall precede testing and shall normally be done with that part of the installation under inspection disconnected from the supply.

As indicated in Figure 6.14, every electrical installation should be subjected to periodic inspection and testing (combined with any maintenance and repair that is required) and a notice to that effect shall be fixed on or nearby to that installation as shown in Figure 6.15.

During erection and on completion of an installation (or an addition or alteration to an installation), and before it is put into service, appropriate

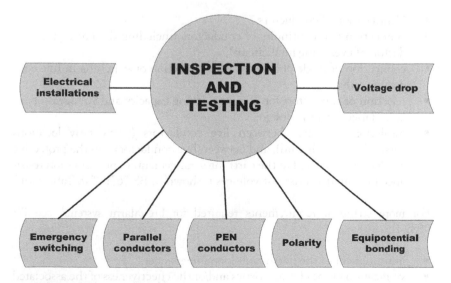

Figure 6.14 Inspection and testing.

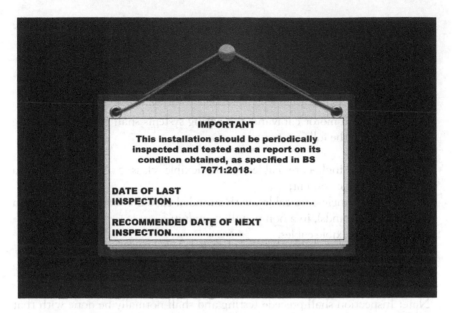

Figure 6.15 Warning notice for periodic electrical installation inspections and testing.

inspection and testing shall be carried out by skilled persons competent to verify that the requirements of this Standard have been met.

The inspection shall include at least the checking of the following items, where relevant to the installation and, where necessary, during erection:

- identification of conductors;
- connection and continuity of conductors (including the protective conductor, of every ring final circuit);
- connection of single-pole devices for protection or switching in line conductors only;
- selection of conductors for current-carrying capacity and voltage drop, in accordance with the design;
- insulation resistance between live conductors (particularly locations exposed to a fire hazard), and between live conductors and the protective conductor connected to the earthing arrangement. (The insulation resistance measured against test voltages is shown in BS 7671:2018 Table 6.61.)

 For more stringent requirements required for fire alarm systems, see BS 5839-1.

- measurement of the Earth fault loop impedance;
- verification of the characteristics and/or the effectiveness of the associated protective device.

6.5.1 Certification for initial verification

On completion of the verification of a new installation (or an addition or alteration to an existing installation), an Electrical Installation Certificate based on the example contained in BS 7671 Appendix 6 shall be issued to the person ordering the work. This certificate shall include details of any defect or omission revealed (and corrected) during the inspection and testing.

6.5.2 Periodic inspection and testing

Periodic inspection and testing of every electrical installation shall be carried out, on an as required basis, in order to determine, so far as is reasonably practicable, whether the installation is in a satisfactory condition for continued service.

A full report (based on Appendix 6 of BS 7671:2018) needs to be produced showing:

- details of those parts of the installation that have been inspected and tested;
- any limitations of the inspection and testing;
- details of any damage, deterioration, defects or dangerous conditions that have been found during the inspection;
- any non-compliance with the requirements of BS 7671 which may give rise to danger.

6.5.2.1 Electrical installations

 The characteristics of the electrical equipment shall **not** be impaired by the process of erection.

Every installation shall be divided into circuits, as necessary, to reduce the possibility of unwanted tripping of RCDs due to excessive protective conductor (PE) currents that are **not** due to a fault.

The number of final circuits required, and the number of points supplied by any final circuit, shall enable compliance with the requirements for overcurrent protection, isolation and switching (with due regard to the current-carrying capacities of conductors).

6.5.2.2 Emergency switching

 Emergency switching off is classified as an emergency operation that is intended to dislocate the supply of electrical energy to all or part of an installation where a risk of electric shock or another risk of electrical origin is involved.

Selection and erection of devices for emergency switching off shall be in accordance with the following regulations and shall comply with Regulation 537.2.

Means for emergency switching off may consist of:

- an isolating device that will interrupt all live conductors;
- one switching device capable of directly cutting off the appropriate supply; or
- a combination of devices activated by a single action for the purpose of cutting off the appropriate supply;
- hand-operated switching devices for direct interruption of the main circuit, which shall be selected where practicable;
- remote-control switching of circuit-breakers, control and protective switching devices or RCDs.

Plugs and socket outlets shall **not** be provided for use as means for emergency switching off.

The means of operating (handles, push-buttons, etc.) devices for emergency switching off shall be:

- clearly identified, preferably by colour;

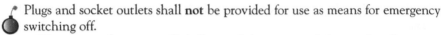

If a colour is used for identification, this shall be **red** with a contrasting background (e.g. yellow).

- readily accessible, particularly at places where a danger might occur;
- capable of latching in the 'OFF' position (unless both the means of operation for emergency switching off and for re-energising are under the control of the same person).

The operation of the emergency switching device shall have priority over any other function relative to safety and shall not be inhibited by any other operation of the installation.

The release of an emergency switching device operated remotely shall not re-energise the relevant part of the installation.

6.5.2.3 Protective bonding conductor

Where an installation has more than one source of supply to which PME conditions apply, the main protective bonding conductor shall be selected according to the largest neutral conductor of the supply.

6.5.2.4 Parallel conductors

Except for a ring final circuit, the value of the current-carrying capacity of the conductor shall be the sum of the current-carrying capacities of the parallel conductors. It should be noted, however, that although a single protective device may protect conductors in parallel against the effects of fault currents, problems can occur with the sharing of the fault currents between the parallel conductors, as a fault can be fed from both ends of a parallel conductor.

Where the use of a single conductor is impractical and the currents in the parallel conductors are unequal, the design current and requirements for overload protection for each conductor shall be considered individually.

 Currents in parallel conductors are considered to be unequal if the difference between the currents is more than 10% of the design current for each conductor.

6.5.2.5 PEN conductors

Where any necessary authorisation for use of a PEN conductor has been obtained:

- the outer conductor of a concentric cable shall not be common to more than one circuit;
- the conductance of the outer conductor of a concentric cable (measured at a temperature of 20°C) shall:
  - o for a single-core cable not be less than that of the internal conductor;
  - o for a multi-core cable serving a number of points contained within one final circuit not be less than that of the internal conductors connected in parallel;
- the continuity of every joint in the outer conductor of a concentric cable (and at a termination of that joint) may be supplemented by an additional conductor used for sealing and clamping the outer conductor.

Where two or more PEN conductors are connected in parallel in a system, either:

- there should be equal load current sharing between them; or
- the conductors shall be multi-core cables or twisted single-core cables or non-sheathed cables; or
- the conductors shall be non-twisted single-core cables or non-sheathed cables in trefoil or flat formation with a cross-sectional area greater than 50 mm^2 in copper or 70 mm^2 in aluminium.

 No means of isolation or switching shall be inserted in the outer conductor of a concentric cable.

6.5.2.6 Polarity

A test of polarity shall be made and it shall be verified that:

- every fuse and single-pole control and protective device is connected in the line conductor only; and

- circuits (except for E14 and E27 lampholders to BS EN 60238) which have an earthed neutral conductor, and centre-contact-bayonet and Edison-screw lampholders, have the outer or screwed contacts connected to the neutral conductor; and
- wiring has been correctly connected to socket outlets and similar accessories;
- all polarity-sensitive appliances have terminals clearly marked '−' and '+', or have two conductors, indicating polarity by colour or by identification tags or sleeves marked '−' or '+'.

6.5.2.7 Protective equipotential bonding conductors

Equipotential bonding is basically an electrical connection maintaining various exposed-conductive-parts and extraneous-conductive-parts at substantially the same potential, and is very important for ensuring that:

- all socket outlets are provided with a protective conductor contact (that is connected to the equipotential bonding conductor;
- all flexible equipment cables (other than Class II equipment) shall have a protective conductor for use as an equipotential bonding conductor.

 Such parts are bonded to the electrical service Earth point of the building to ensure the safety of building occupants.

6.5.2.8 Verification of voltage drop

When required, compliance to the Regulations may be confirmed by evaluating the voltage drop:

- by measuring the circuit impedance;
- by using calculations (e.g. by diagrams or graphs showing maximum cable length v load current for different conductor cross-sectional areas with different percentage voltage drops for specific nominal voltages, conductor temperatures and wiring systems).

 Note: Verification of voltage drop is not normally required during initial verification.

6.5.3 Conduits, cable ducting, cable trunking, busbar or busbar trunking

As shown in Figure 6.16, there are a lot of different methods of ensuring the continued safety of electrical sources to various locations.

6.5.3.1 Bonding conductors

In each installation, main protective bonding conductors shall connect extraneous-conductive-parts to the main earthing terminal.

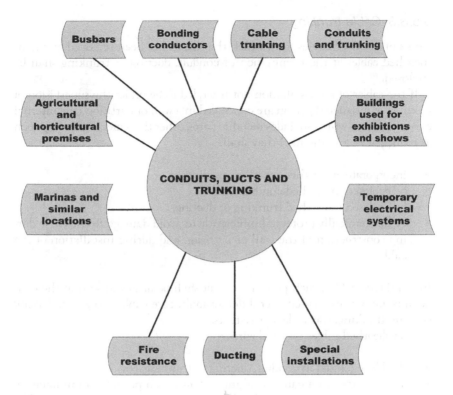

Figure 6.16 Conduits, ducting and trunking.

6.5.3.2 Busbars

A busbar trunking system shall comply with BS EN 60439-2 (powertrack with the BS EN 61534 series) and should always be installed in accordance with the manufacturer's instructions taking account of external influences, particularly with respect to the following:

- busbars passing across expansion joints shall be selected and erected so that any movement does not cause damage to the electrical equipment;
- any busbar trunking system used as a protective conductor should ensure that:
  - o its electrical continuity is guaranteed (either by construction or by suitable connection) and that it is protected against mechanical, chemical or electrochemical deterioration;
  - o its cross-sectional area is in accordance with BS EN 60439-1;
  - o the connection of other protective conductors at every predetermined tap-off point is always possible.

6.5.3.3 Cable trunking

Cores of sheathed cables from which the sheath has been removed and non-sheathed cables at the termination of conduit, ducting or trunking shall be enclosed.

If the cables of an installation not intended to be under the supervision of a skilled or instructed person are concealed in a wall or partition (the internal construction of which includes metallic parts, other than metallic fixings such as nails, screws and the like) they shall:

- incorporate an earthed metallic covering; or
- be enclosed in earthed conduit: or
- be enclosed in earthed trunking or ducting; or
- be mechanically protected sufficiently to avoid damage to the cable during construction of the wall or partition and during Installation of the cable.

In multi-core cables, each part of a circuit shall be arranged so that the conductors are not distributed over different multi-core cables, conduits, ducting systems, trunking, tray or ladder systems.

Non-sheathed cables for fixed wiring:

- shall be enclosed in conduit, ducting or trunking;
- are permitted in a cable trunking system which provides a minimum of IP4X or IPXXD protection, **but only** if the cover can only be removed by means of a tool or a deliberate action.

The metal covering (including the sheath – bare or insulated) of a cable, trunking, ducting or metal conduit, may be used as a protective conductor for the associated circuit.

Where the protective conductor is formed by metal conduit, trunking, ducting or the metal sheath and/or armour of a cable, the earthing terminal of each accessory shall be connected by a separate protective conductor to an earthing terminal incorporated in the associated box or other enclosure.

Wherever equipment is fixed on or in cable trunking, skirting trunking or in mouldings it should not be fixed on covers which can be inadvertently removed.

6.5.3.4 Conduits and trunking

- Conduit and trunking **shall** comply with the resistance to flame propagation requirements of BS EN 50085 or BS EN 50086.
- Flexible or pliable conduit shall **not** be selected as a protective conductor.

- Conduits and conduit fittings **shall** comply with the appropriate British Standard shown in Table 6.3.

Table 6.3 Conduits and conduit fittings

Conduit fitting	BS Standard	BS EN Standard
Steel conduit and fittings	BS 31	BS EN 60423 and BS EN 50086-1
Flexible steel conduit	BS 731-1	BS EN 60423 and BSEN50086-1
Steel conduit fittings with metric threads	BS 4568	BS EN 60423 and BS EN 50086-1
Non-metallic conduits and fittings	BS 4607	BS EN 60423 and BS EN 50086-2-1

6.5.3.5 Ducting

 A cable trunking system or cable ducting system shall satisfy the test under fire conditions specified in BS EN 50085.

A cable concealed in a wall or partition at a depth of less than 50 mm from a surface of the wall or partition shall:

- incorporate an earthed metallic covering; or
- be enclosed in earthed conduit; or
- be enclosed in earthed trunking or ducting; or
- be mechanically protected against damage sufficient to prevent penetration of the cable by nails, screws etc.; or
- form part of a SELV or PELV circuit.

 Note: Two or more circuits are allowed in the same conduit, ducting or trunking.

6.5.3.6 Fire resistance

Any wiring system (such as a conduit system, cable ducting system, cable trunking system, busbar or busbar trunking system) which penetrates elements of building construction that have specified fire resistance shall be internally sealed to the degree of fire resistance required before penetration as well as being externally sealed.

 Conduit and trunking systems shall be in accordance with BS EN 61386-1 and BS EN 50085-1 respectively and shall meet the fire-resistance tests within these Standards.

6.5.4 Special installations

A cable installed under a floor or above a ceiling shall be run in such a position that it is not liable to be damaged by contact with the floor or ceiling or their fixings. The best method of achieving this regulation is to:

- enclose the cable in earthed conduit, earthed trunking or ducting; or
- mechanically protect it against damage sufficiently to prevent penetration of the cable by nails, screws etc.

A conduit system, cable trunking system or cable ducting system:

- which is going to be buried in the structure shall be completely erected between access points before any cable is drawn in;
- classified as non-flame propagating according to the relevant product standard and having a maximum internal cross-sectional area of 710 mm^2 need not be internally sealed, provided that:

  o the system satisfies the test of BS EN 60529 for IP33; and
  o any termination of the system in one of the compartments, separated by the building construction being penetrated, satisfies the test of BS EN 60529 for IP33.

A voltage Band I circuit shall not be contained in the same wiring system as a Band II circuit, unless:

- every cable or conductor is insulated for the highest voltage present;
- the cables are insulated for their system voltage and installed in a separate compartment of a cable ducting or cable trunking system;
- the cables are installed on a cable tray system where physical separation is provided by a partition;
- a separate conduit, trunking or ducting system is employed.

6.5.4.1 Agricultural and horticultural premises

In agricultural and horticultural premises (where the wiring system may be exposed to impact and mechanical shock due to vehicles and mobile agricultural machines etc.), the external influences shall be classified AG3 and:

- conduits shall provide a degree of protection against impact of 5 J according to BS EN 61386-2;
- cable trunking and ducting systems shall provide a degree of protection against impact of 5 J according to BS EN 50085-2-1.

6.5.4.2 Buildings used for exhibitions and shows etc.

Where no fire alarm system is installed in a building used for exhibitions and shows etc., cable systems shall be either:

- flame retardant to BS EN 60332-1-2 and low smoke to BS EN 61034-2; or
- single-core or multi-core unarmoured cables enclosed in metallic or non-metallic conduit or trunking, providing a degree of fire protection of at least IP4X.

6.5.4.3 Marinas and similar locations

In marinas and similar locations, the following wiring systems shall not be used on or above a jetty, wharf, pier or pontoon:

- cables in free air suspended from, or incorporating, a support wire;
- non-sheathed cables in cable management systems;
- cables with aluminium conductors;
- mineral-insulated cables.

6.5.4.4 Temporary electrical systems in amusement parks, circuses and fairgrounds

The following apply to all temporary electrical systems in amusement parks, circuses and fairgrounds:

- Cable trunking systems and cable ducting systems shall comply with the relevant part 2 of BS EN 50085.
- Conduit systems shall comply with the BS EN 61386 series.

Author's end note

Now that we have seen how the selection, installation and use of cables, conductors and conduits plays such an important part in the structure and safety of electrical wiring systems, the next thing is see if there are any 'external influences' that will have an effect on these products once they are installed.

In the following chapter we will see how equipment items and their interconnections can be simultaneously exposed to a large number of external influences (such as temperature, humidity, weather, pollutants, fire, electromagnetic and mechanical etc.), and why a combination of environmental factors has become so important and needs to be considered during manufacture, operation, and maintenance and repair of electrical installations.

7
External influences

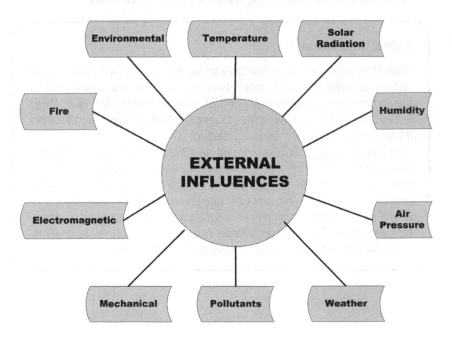

Figure 7.1 External influences.

BS 7671:2018 Requirement

512.2 External influences
512.2.1 Equipment shall be of a design appropriate to the situation in which it is to be used or its mode of installation shall take account of the conditions likely to be, encountered.
512.2.2 If the equipment does not, by its construction, have the characteristics relevant to the external influences of its location, it may nevertheless be used be used on condition that it is provided with appropriate additional protection in the erection of the installation. Such protection shall not adversely affect the operation of the equipment thus protected.
512.2.3 Where different external influences occur simultaneously, they may have independent or mutual effects and the degree of protection shall be provided accordingly.
512.2.4 The selection of equipment according to external influences is necessary not only for proper functioning, but also to ensure the reliability of the measures of protection for safety complying with these Regulations generally. Measures of protection afforded by the construction of the equipment are valid only for the given conditions of external influence if the corresponding equipment specifications are made in these conditions of external influence.

It has also been determined that for the purpose of these Regulations, the following classes of external influence are conventionally regarded as normal (i.e. that the requirement must generally satisfy applicable standards).

AA	Ambient temperature	AA4
AB	Atmospheric humidity	AB4
AC to AS	Other environmental conditions	XX1 of each parameter
B and C	Utilisation and construction of buildings	XXI of each parameter, except XX2 for the parameter BC

Note: A list of all external influences and their characteristics have been included as an appendix to BS 7671:2018 (i.e. Appendix 5) and the following notes concerning external influences are offered as guidance.

7.1 Environmental factors and influences

The actual environment to which equipment is likely to be exposed is normally complex and comprises a number of conditions. When defining the conditions for a certain application it is, therefore, necessary to consider **all** environmental influences which may be as a result of:

- conditions from the surrounding medium;
- conditions caused by the structure in which the equipment is situated or attached;
- influences from external sources or activities.

7.1.1 Combined environmental factors

Equipment may, of course, be simultaneously exposed to a large number of environmental factors and corresponding parameters. Some of the parameters are statistically dependent (e.g. low air velocity and low temperature; sun radiation and high temperature). Other parameters are statistically independent (e.g. vibration and temperature). The effect of a combination of environmental factors is, therefore, extremely important and has to be considered during manufacture, operation as well as maintenance and repair.

7.1.2 Sequences of environmental factors

Certain effects of exposing electrical equipment and electrical installations to environmental conditions are a direct result of two or more factors or parameters, happening either simultaneously or after each other (e.g. thermal shock caused when exposing equipment to a high temperature immediately after exposing it to a low temperature). These possibilities must always be taken into account when designing and installing electrical equipment.

7.1.3 Environmental application

Although the conditions affecting electrical equipment mainly consist of the environment (ambient and created), consideration must also have to be given to where the equipment will be operating from and how it will be used.

For simplicity this can be broken down into two basis categories:

Conditions	The environmental conditions that have been identified as having an effect on equipment (Table 7.1)
Situations	The main uses of electronic equipment (Table 7.2).

7.1.4 Environmental conditions

Table 7.1 (and Figure 7.2) shows the eight basic conditions that have a direct effect on electrical equipment and electrical installations.

Table 7.1 Basic conditions

Climatic	Externally generated influences
• altitude ambient temperature • ambient temperature • atmospheric pressure • condensation • precipitation (i.e. rain, snow and hail) • relative atmospheric humidity • solar radiation • wind	• air movement • dust • temperature • precipitation (e.g. water spray) • pressure changes (e.g. tunnels)
Mechanical	**Ergonomic aspects**
• shock (sinusoidal and random) • vibration	• achieving maximum task effectiveness • protecting the health of the engineer and end user • the comfort of the operator and end user
Electrical	**Chemical**
• earthing and bonding • electromagnetic environment (EMC and EMI) • power supplies • susceptibility and generation • transients (spikes and surges)	• corrosion • dangerous substances • pollution and contamination • resistance to solvents
Biological	**General**
• animals • humans (vandalism) • vegetation	• components • design of equipment • earthquakes • flammability and fire hazardous areas • maintainability • safety • reliability • waste

7.1.5 Equipment situations

Obviously not all equipment will be fully operational, all of the time, and so various equipment 'situations' also have to be considered.

 As a reference on the whole topic of environmental requirements, the following book (by the same author no less!) is recommended. This book (see overleaf) has become known as the definitive reference for designers and manufacturers of electrical and electromechanical equipment worldwide.

Table 7.2 Environmental conditions

Operational	Storage	Transportation
• when installed and operational • when installed and not in use	• when in storage	• when being transported

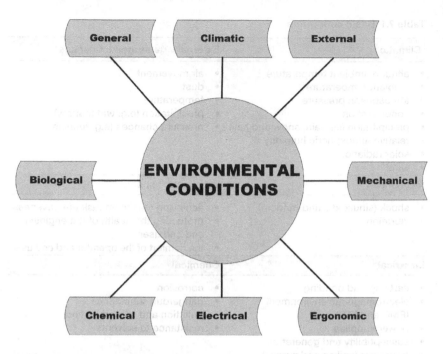

Figure 7.2 Environmental conditions.

 *'Environmental requirements for the electromechanical and electronic equipment'*, which has been written in the form of a reference book, contains background guidance concerning the environment and environmental requirements. It also provides a case study of typical requirements, values and ranges currently being requested from industry in today's contracts. Test specifications aimed at proving conformance to these requirements are also listed and the most used national, European and international standards and specifications (e.g. BS, EN, CEN, CENELEC, IEC and ISO) are described in relative detail. ISBN: 0-7506-3902-4

7.1.6 Requirements from the Regulations – environmental factors and influences

7.1.6.1 General environmental conditions

Electrical equipment shall be selected that can safely withstand the stresses, environmental conditions and characteristics of its location. (A coating of paint, varnish or similar product is generally not considered to comply with these requirements!).

The design of electrical equipment shall:

• take in to account the environmental conditions to which it will be subjected;

- be appropriate to the external influences foreseen;
- have an adequate level of immunity against electromagnetic disturbances and emissions.

Equipment that is located within or near an area susceptible to the risk of fire or explosion shall be constructed or protected to safely prevent danger,

7.1.6.2 Types of wiring and methods of installation

The choice of the type of wiring system and the method of installation shall include consideration of the following:

- the nature of the location;
- the nature of the structure supporting the wiring;
- accessibility of wiring to persons and livestock;
- voltage;
- the electromechanical stresses likely to occur due to short-circuit and earth fault currents;
- electromagnetic interference;
- other external influences (e.g. mechanical, thermal and those associated with fire) to which the wiring is likely to be exposed during the erection of the electrical installation or in service.

7.1.6.2.1 Electrical connections

If an electrical connection is made within an enclosure, then that enclosure shall provide adequate mechanical protection and protection against relevant external influences.

7.1.6.2.2 Sealing of wiring system penetrations

Any sealing arrangement for the installation shall resist external influences to the same degree as the wiring system with which it is used (or the building construction element in which it has been installed) and, in addition, it shall:

- be resistant to the products of combustion;
- be protected from water penetration;
- be compatible with the wiring system's material that it is in contact with;
- permit thermal movement of the wiring system without a reduction in sealing quality;
- withstand (through its own mechanical stability) to the stresses which may arise through damage to the support of the wiring system due to fire.

7.2 Air pressure and altitude

Air pressure frequently referred to as atmospheric pressure is 'the force exerted on a surface of a unit area caused by the Earth's gravitational attraction on the *air vertically above that area'*. Air pressure varies with altitude (i.e. elevation above mean sea level) and location (see Table 7.3). For instance at the equator where the trade winds of both hemispheres converge, there is a low pressure zone, (known as the ITCZ or International Conveyance Zone), which is characterised by high humidity.

Table 7.3 Air pressure and altitude

Air Pressure		Approximate altitude above sea level
(kPa)	(mbar)	(m)
1	10	31,200
2	20	26,600
4	40	22,100
8	80	17,600
15	150	13,600
25	250	10,400
40	400	7,200
55	550	4,850
70	700	3,000

It is not widely appreciated that the location of equipment, especially with respect to its altitude above sea level, can affect the working of that equipment (see Figure 7.3). But it is not just the height above sea level that has the most effect! Even air pressure variations at ground level have to be considered.

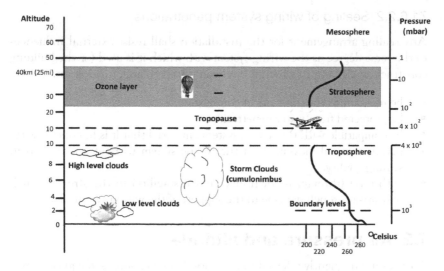

Figure 7.3 Atmospheric structure.

7.2.1 Low air pressure

At altitudes above sea level, low air pressure can cause:

- change of physical or chemical properties;
- decreased efficiency of heat dissipation by convection and conduction in air, that will effect equipment cooling (e.g. an air pressure decrease of 30% has been found to cause an increase of 12% in temperature);
- erratic breakdown or malfunction of equipment from arcing or corona;
- leakage of gases or fluids from gasket sealed containers;
- ruptures of pressurised containers;
- speeding up effects essentially due to temperature, (e.g. volatilisation of plasticisers evaporation of lubricants etc.).

7.2.2 Typical Requirements – air pressure and altitude

Table 7.4 shows the most common environmental requirements concerning air pressure and altitude.

Table 7.4 Air pressure and altitude

High air pressure	High air pressure occurring in natural depressions and mines can have a mechanical effect on sealed containers and should always be borne in mind when designing and installing electrical equipment
Installations up to 2000m above sea level	Electrical equipment must be capable of working to an altitude between -120m to 2000m above sea level - which corresponds to an air pressure range from 110.4 to 74.8 kPa

7.3 Ambient temperature

Temperature, humidity, rainfall, wind velocity and the duration of sunshine all affect the climate of an area. These elements are in turn the result of the interaction of a number of determining causes, such as latitude, altitude, wind direction, distance from the sea, relief, and vegetation. Elements and their determining causes are similarly inter-related which also contribute to temperature changes, for example, the length of day is a factor which helps to determine temperature, however, the duration of actual sunshine is an element with far-reaching effects on plant and animal life.

Of all the elements that have an effect on man and equipment none is more vital to living organisms than temperature. Temperature has a large influence on where humans live and in the areas where they work. Protective housing or artificial heat sources may overcome low temperatures (and high altitudes), similarly cooling devices and reflective coatings protect equipment from high temperatures. Temperature is therefore a particularly important aspect of the environment and its accurate measurement and definition requires careful consideration.

The ambient temperature at any given time is the temperature of the air measured under standardised conditions and with certain recognised precautions against errors introduced by radiation from the sun or other heated body. Temperature figures with respect to climate are generally 'shade' temperatures (i.e. the temperature of the air measured from a location that excludes the influence of the direct rays of the sun) and it is usual for the temperature to be much higher in the direct sunshine. Many mountain areas have air temperatures in the region of zero in winter but the presence of bright sunshine will produce a feeling of warmth and permits the wearing of light clothing (as shown in Figure 7.4).

Figure 7.4 Enjoying the environment. (Courtesy of HEC Ltd.)

Seasonal fluctuations in temperature do not pass below ground deeper than 60-80ft. Below that depth, borings and mine-shafts show that the temperature increases (downwards) depending on the geographical position, location and depth (See Figure 7.5).

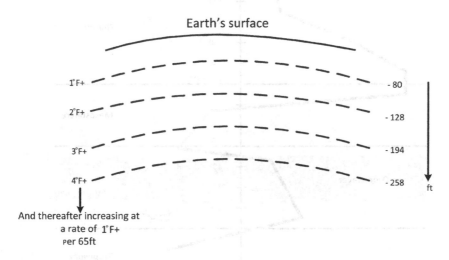

Figure 7.5 Temperature changes below the Earth's crust.

On average, however, a rise of about 1° C may be taken for each 64ft of descent. Assuming that this rate of increase is maintained, it stands to reason that the interior of the Earth must be excessively hot and, therefore, it must warm the surface to some extent.

It is not possible to determine the precise influence of this temperature increase but it has an effect on tunnels at a depth greater than 80ft. As the heat from the core is virtually negligible on the surface of the Earth, (compared with that of the sun), it has not been considered in this book and the only source of heat that has been taken into account, is the sun.

The difference between summer and winter temperatures for any locality is known as the *'annual range of temperature'* or the *'absolute range of temperature'* of that particular locality and it is the difference between the highest and lowest temperatures ever experienced at the place in question. The maximum and minimum temperatures are obviously not the same every year and, should their average over a series of years be taken, it would be known as the *'mean annual extreme range'*.

Although air near the surface (especially at night) may be cooler than the air just above it there is, generally speaking, a gradual falling off of temperature from the ground level up. (See Figure 7.6). Over thousands of feet this cooling averages 1° C per 300ft and thus at approx. 25-55,000 ft. (five to ten miles) above the ground the temperature will be down to 55-60° C below zero.

Above this the air temperature ceases to fall off regularly, in fact it may even rise for a bit. Usually, however, it remains fairly constant and because of this

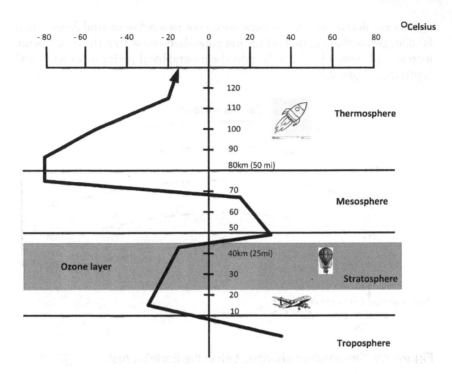

Figure 7.6 Temperature changes above the Earth's surface.

it is sometimes referred to as the isothermal layer – but to meteorologists and airmen it is known as the stratosphere. The lower layers of the atmosphere (i.e. where temperature falls off with height) is known as the troposphere and the boundary layer between the two is called the *'tropopause'*.

7.4 Weather and precipitation

Water is one of the most remarkable substances on the Earth. It is the substance that we most often see in all its three states: liquid (water), solid (ice) and gas (steam). No living organism can exist without water and as much as half the weight of plants and animals is made up of water. Water in the oceans makes up approximately eleven times the volume of the solid part of the Earth in addition, that is, to water frozen in ice flows, in lakes, rivers, within the ground and in living plants and animals.

Water is a constantly moving cycle. As the sun beats down, some of the surface water is evaporated; this water vapour rises as part of the air and is moved along by the wind. Should it pass over a land mass it may become a cloud and as more moisture is attracted to the cloud or the clouds pass over rising ground the water particles become larger and fall as rain sleet or snow. See Figures 7.7 and 7.8.

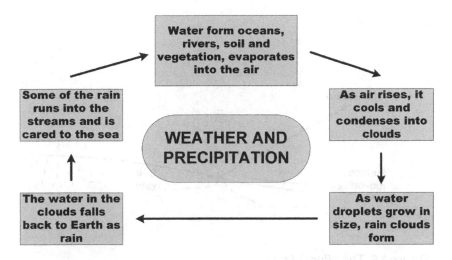

Figure 7.7 The hydrological cycle.

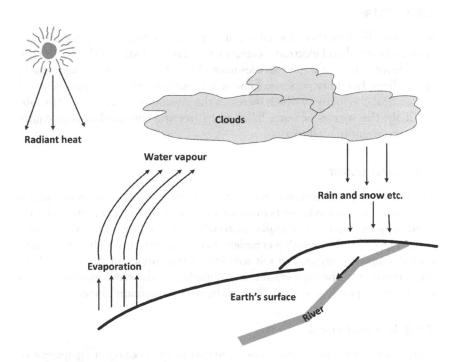

Figure 7.8 Simplified water cycle.

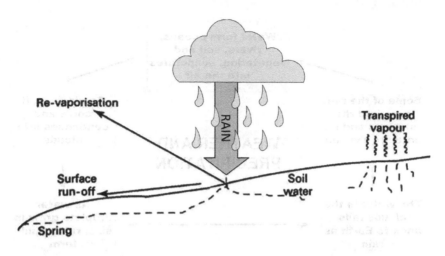

Figure 7.9 The effect of rain.

As the rain comes into contact with the ground (se Figure 7.9) there are several avenues open to it; it may be re-vaporised and return to the atmosphere, be absorbed by the ground or will remain on the surface of the ground and run downhill forming streams and rivers which run into the sea and the cycle begins once more.

7.4.1 Water

Water (in all of its three forms) is a major cause of failure in every application of electrical and electronic components. The humidity of the air and the possible formation of water particles must always be taken into consideration, especially as humidity possesses (almost without exception) a certain amount of electrical conductivity which increases the possibility of corrosion of metals. Similarly, the ingress of water followed by freezing within electronic equipment can result in malfunction.

7.4.2 Salt water

Salt has an electrochemical effect on metallic materials (i.e. corrosion) which can damage and degrade the performance of equipment and/or parts that have been manufactured from metallic materials. Non-metallic materials can also be damaged by salt through a complex chemical reaction which is dependent on the supply of oxygenated salt solution to the surface of the material, its temperature and the temperature and humidity of the environment. This is particularly a problem in areas close to the sea or mountain ranges.

7.4.3 Ice and snow

Water in the form of ice can cause problems in the cooling of equipment or freezing and thawing, which will result in cracks occurring, breaking cases, etc.

Powdered snow can easily be blown through ventilation ducts and then melt down in equipment compartments and cubicles that can cause damp problems in critical systems if not prevented in the original construction.

7.4.4 Weathering

As shown in the following paragraphs, there are several types of 'weathering' (which is the collective term for the processes by which rock at or near the earth's surface is disintegrated and decomposed by the action of atmospheric agents, water, and living things).

7.4.5 Exfoliation

During the day rocks are warmed by the sun and at night the surface can cool more rapidly than the underlying rocks. The outer skin of the rock then becomes tight and cracks, thus layers of the rock peel off and the mountain becomes rounded or dome shaped. This exfoliation can have an effect on equipment that is sunk into rock faces or mounted on the surface of equipment.

7.4.6 Freeze thaw

When water freezes it turns to ice, expanding by about one twelfth of its volume. If this water is in the joint or a crack in a casing then the space will become enlarged and the casing on either side will be forced apart. When the ice eventually thaws more water will penetrate into the crack and the cycle repeats itself with the crack constantly enlarging.

7.4.7 Chemical weathering

Water can pick up quantities of sulphur dioxide from the atmosphere and will form a weak solution of acid. This acid can attack certain equipment housings and the process whereby the housing is worn away is known as chemical weathering.

7.4.8 Erosion

As the wind blows over dry ground it collects grit and 'throws' it vigorously against the surfaces nearby. This grit acts in a similar way as sandpaper and gradually wears away the surface with which it comes into contact.

7.4.9 Mass movement

Once solids, such as sedimentary rocks have been broken up, there is often a downwards movement of the particles which have been broken off. This 'soil creep' can gather momentum and can, in certain circumstances, submerge equipment.

7.4.10 Requirements from the Regulations – weather and precipitation

Table 7.5 shows the most common environmental requirements concerning weather and precipitation:

Table 7.5 Typical requirements – solar radiation

Environment	Protection
Operation	All equipment should be capable of operating during rain, snow and hail, and be unaffected by ice, salt and water.
Rain	All equipment should be capable of operating in rain and be capable of preventing the penetration of rainfall at a minimum rate of 13 cm/hour and an accompanying wind rate of 25 m/s.
Snow and hail	Consideration needs to be given to the effect of all forms of snow and/or hail. The maximum diameter of the hailstones is conventionally taken as 15 mm, but larger diameters can occur on occasion.
Salt water	Equipment should be capable of operating in (or be protected from) heavy salt spray, as would be experienced in seacoast areas and in the vicinity of salted roadways.
Weather protection	Equipment that is operated adjacent to the sea shore or on mountain ranges, and therefore subject to water and precipitation, must be able to function equally well as the same equipment housed in arid deserts.

7.4.10.1 Equipment

> **BS 7671:2018 Requirement**
>
> *"The design of the electrical installation shall take into account the environmental conditions to which it will be subjected."*

7.4.10.2 Water

Wiring systems must be selected and erected so that no damage is caused by condensation or ingress of water during their installation, use and/or maintenance.

Special consideration needs to be given to wiring systems that are liable to frequent splashing, immersion or submersion, and if a wiring system is likely to be exposed to waves (AD6) it shall be protected from mechanical damage caused by impact, vibration and other physical stresses.

Where a wiring system passes through floors, walls, roofs, ceilings, partitions or cavity barriers, the openings remaining after its passage should be sealed so as to provide the same degree of protection from water penetration as the building construction element in which it has been installed.

In situations where water might collect or condensation could form in a wiring system, or where that system is routed below services that are liable to cause condensation (such as water, steam or gas services), the wiring system needs to be protected.

Wiring systems that are likely to be affected by corrosion or deterioration through being in the vicinity of corrosive or polluting substances (including water) shall be protected by protective tapes, paints or grease and/or manufactured from a material that is resistant to such substances.

7.4.10.3 Ice

Wiring systems for solar photovoltaic power supplies (whether they be PV string cables, PV array cables, PV power supplies or PV d.c. main cables) shall withstand the expected external influences such as ice formation.

7.4.10.4 Marinas

In marinas, equipment installed on or above a jetty, wharf, pier or pontoon shall be selected according to the external influences (e.g. water splashes, water jets or water waves) which may be present.

For marinas, particular attention must always be given to the likelihood of corrosive elements, structural movement, mechanical damage, presence of flammable fuel and the increased risk of electric shock owing to:

- the existence of water;
- a reduction in body resistance;
- bodily contact with an Earth potential.

 Note: See BS 7671:2018 Chapter 709 for more detailed requirements relating to marinas and similar locations.

7.4.10.5 Electrical installations in caravan and camping parks

Socket outlets of wiring systems shall be placed at a height of 0.5 to 1.5 m from the ground, as measured from the lowest part of the socket outlet.

In special cases, due to environmental conditions such as risk of flooding or heavy snowfall, the maximum height of socket outlets is permitted to exceed 1.5 m.

 Note: See BS 7671:2018 Chapter 708 for more detailed requirements relating to electrical installations in caravan camping parks and similar locations.

7.5 Humidity

The atmosphere is normally described as "a shallow skin or envelope of gases surrounding the surface of the Earth which is made up of nitrogen, oxygen and a number of other gases which are present in very small quantities". While the ratio of these components shows no appreciable variation either with latitude or altitude, the water vapour content of the atmosphere is subject to extremely wide fluctuations and the amount of water present in the air is referred to as 'humidity'.

When air and water come into contact, they will exchange particles with each other (i.e. the air particles will pass into the water and water particles will pass into the air) in the form of vapour. There is always a certain amount of water vapour present in the air, and a certain amount of air present in water, and there is always a constant movement between the two mediums. If there is only a small amount of water vapour present in the air, then more particles of water will pass from the water into the air than from the air into the water and so the water will gradually dry up or evaporate. Conversely, if the amount of water vapour in the air is large, then as many particles of vapour will pass from the air into the water as from the water into the air, and the water will not evaporate. In such a case the air is said to be 'saturated' – or, to put it another way, it holds as much water vapour as it can possibly contain.

Water vapour is collected in the air above the oceans and is carried by the wind towards the land masses. The amount of water vapour in the air varies greatly depending on the place and season, but, in general, evaporation is most rapid at high temperature and slower at lower temperatures. We may, therefore, expect to find the greatest amount of water vapour over the oceans near the Equator and the smallest amount over the land in a cold region such as North-East Asia in winter. However, even when the surface is covered with snow and ice, water evaporation may take place, and, occasionally during a long frost, the snow will gradually disappear without melting.

Except for the water vapour being present, the composition of the atmosphere near the surface of the Earth up to a height of some 2000 ft is practically uniform throughout the globe. However, at greater altitudes (i.e. above the atmospheric boundary in the tropopause), there is practically no water vapour, or water, in any form.

7.5.1 What is humidity?

Temperature and the relative humidity of air (in varying combinations) are climatic factors which act upon electrical equipment and installations during storage, transportation or operation – see Figure 7.10. Humidity, and the electrolytic damage resulting from moisture, mostly affects plug points, soldered joints (in particular dry joints), bare conductors, relay contacts and switches. Humidity also promotes metal corrosion (see Section 7.7 – Pollutants and Contaminants) owing to its electrical conductivity.

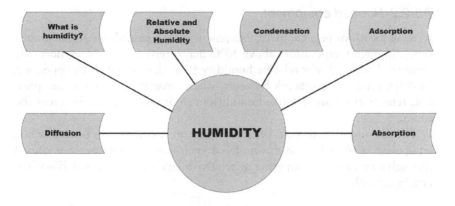

Figure 7.10 The effects of humidity.

In many cases, however, environmental influences such as mechanical and thermal stresses are merely the forerunner of the impending destruction of components by humidity – especially as the majority of electronic component failures are caused by water!

 Note: Humidity (in the context of this book) has been taken to cover relative humidity, absolute humidity, condensation, adsorption, absorption and diffusion, and details of these '*subsets*' are provided in the following paragraphs.

7.5.2 Relative and absolute humidity and their effect on equipment performance

The performance of virtually all electrical equipment is influenced and limited by its internal temperature which, in turn, is dependent on the external ambient conditions and on the heat generated within the device itself. Fortunately, most electrical and electronic components (especially resistors) will normally remain dry when under load owing to the amount of internal/external heat dissipation. Indeed, many components either have to be de-rated in order to improve their reliability or, for reasons of circuit function, only energised intermittently.

7.5.2.1 Externally mounted equipment

Equipment and components that are mounted in external cabinets run the risk of coming into contact with water or water vapour (e.g. drifting snow, fog, dew, rain, spray water or water from hoses) and the equipment must, therefore, be adequately protected from such humidity in order to prevent the ingress of vapour into the system within the casing.

7.5.2.2 Housed equipment

In most locations (e.g. cabinets, equipment rooms, workshops and laboratories), although temperatures above 30°C may often occur, they are normally combined with a lower relative humidity than that found in the open air. In other rooms (e.g. offices), however, where several heat sources are present, temperatures and relative humidities can differ dramatically across the room.

The sun also plays its part, because in certain circumstances (such as when equipment is placed in an unventilated enclosure) the intense heat caused through solar radiation can generate relative humidities in excess of 95% when combined with:

- high relative humidity caused by the release of moisture from hygroscopic materials;
- the breathing and perspiration of human beings;
- open vessels containing water or other sources of moisture.

7.5.3 Condensation

Condensation occurs when the surface temperature of an item is lower than that of the dew point (i.e. the temperature with a relative humidity of 100% at which condensation occurs) and which can change electrical characteristics (e.g. decrease surface resistance, increase loss angle) between the absolute point at which atmospheric vapour condenses into droplets (i.e. the dew point), absolute humidity and vapour pressure.

For example, if a piece of equipment has a low thermal time constant, condensation (normally found on the surface of the equipment) will occur only if the temperature of the air increases very rapidly, or if the relative humidity is very close to 100%. Sudden changes in temperature may cause water to condense on parts of equipment, and leakage currents can occur.

7.5.4 Absorption

The quantity of water that can be absorbed by a material depends largely on the water content of the ambient air, and the speed of penetration of the water molecules generally increases with the temperature.

7.5.5 Adsorption

Adsorption is the amount of humidity that may adhere to the surface of a material and depends on the type of material, the surface structure and the vapour pressure. This layer of water (no matter how small) can cause electrical short-circuits and material distortion etc.

7.5.6 Diffusion

Water vapour can penetrate encapsulations of organic material (e.g. into a capacitor or semiconductor) via the sealing compound, and enter the casing. This factor is frequently overlooked and can become a problem, especially as the moisture absorbed by an insulating material can cause a variation in a number of electrical characteristics (e.g. reduced dielectric strength, reduced insulation resistance, increased loss angle, increased capacitance etc.).

7.5.7 Protection

The effects of humidity mainly depend on temperature, temperature changes and impurities in the air. As shown in Table 7.6, there are three basic methods of protecting the active parts of equipment and components from humidity.

Table 7.6 Protective methods – humidity

Heating the surrounding air so that the relative humidity cannot reach high values	This method normally requires a separate heat source which (especially in the case of equipment mounted in external cabinets) usually means a separate power supply must be provided. This method is disadvantaged by the reliability of the circuit being dependent on the efficiency of the heating.
Hermetically sealing components or assemblies using hydroscopic materials	This is an extremely difficult process, as the smallest crack or split can allow moisture to penetrate the component, particularly in the area of connecting-wire entry points. Metal, glass and ceramic encapsulation do nevertheless produce some very satisfactory results.
Ventilation and the use of moisture-absorbing materials	Most water-retaining materials and paint etc. are suitable for the temporary absorption of excessive high air humidity in the casing, which, because of the risk of pollution and dust penetration, cannot be fully ventilated (i.e. air exchange with the outside temperature is not possible).

7.5.8 Typical requirements – humidity

Table 7.7 gives the most common environmental requirements concerning humidity, with Table 7.8 showing the levels of external humidity.

7.5.9 Requirements from the Regulations – humidity

Note: For the purpose of these Regulations, the AB2 and AB4 classes of ambient temperature (between 5% and 100%) are generally recommended.

A wiring system shall be selected so that no damage is caused by condensation or ingress of water during installation, use and maintenance.

Table 7.7 Typical requirements – humidity

Equipment in cubicles and cases	The design of equipment should take into account temperature rises within cubicles and equipment cases in order to ensure that the components do not exceed their specified temperature ranges.
Equipment interoperability	Equipment that is operated adjacent to the sea shore (and, therefore, subject to extreme humidity) must be able to function equally well as the same equipment housed in the low humidity of (for example) the desert.
External humidity levels	Equipment should be designed and manufactured to meet external humidity levels (limit values), over the complete range of ambient temperature values anticipated.
	Note: Meteorological measurements made over many years have shown that, within Europe, a relative humidity greater than 95% combined with a temperature above 30°C does not occur over long periods in free air conditions.
Condensation	Operationally caused infrequent and slight moisture condensation should not lead to malfunction or failure of the equipment.
Indoor installations	In all indoor installations, provision must be made for limiting the humidity of the ambient air to a maximum of 75% at -5°C.
Product configuration	All proposed all in date equipment, components or other articles must be tested in their production configuration without the use of any additional external devices that have been added expressly for the purpose of passing humidity testing.
Peripheral units	For peripheral units (e.g. measuring transducers etc.) or equipment employed in a decentralised configuration (i.e. where ambient temperature ranges are exceeded) the actual temperature occurring at the location of the equipment concerned should be utilised when designing equipment.

Table 7.8 Humidity – external humidity levels

Duration	Limit value
Yearly average	75% relative humidity
On 30 days of the year, continuously	95% relative humidity
On the other days, occasionally	100% relative humidity
On the other days, occasionally	30 g/m^3 occurring in tunnels

Special consideration needs to be given to wiring systems that are liable to frequent splashing, immersion or submersion.

It shall be ensured that any condensation which might form in a wiring system or where water might collect is swiftly eliminated.

Wiring systems that could be subjected to waves (AD6) shall be protected from mechanical damage (e.g. impact, vibration and other mechanical stresses).

Where corrosive or polluting substances (including water) could cause corrosion and/or deterioration, the parts of the wiring system that are likely to be affected shall be suitably protected (e.g. by protective tapes, paints or grease) and/or manufactured from a material resistant to such substances.

If a wiring system is routed below services that are liable to cause condensation (such as water, steam or gas services), precautions shall be taken to protect the wiring system from harmful effects.

7.6 Solar radiation

Author's note

In this age of global warming and climate change, it is interesting to note the effects the sun's radiation has, not only on plant life, but also on electrical engineering.

Of all the factors that control the weather, the sun is by far the most powerful, and practically everything that occurs on the Earth is controlled, directly or indirectly, by it. The sun affects the places humans inhabit, the kind of homes that are built, the work that is done and the equipment that is used.

Less than one-millionth of the energy emitted from the sun's surface travels the ninety-odd million miles to reach this planet. The sun's energy crosses those miles in the form of short electromagnetic radio waves, identical in nature to those used in broadcasting, which pass through the atmosphere and are absorbed by the Earth's surface. These waves warm the Earth's surface and are then re-radiated back to space.

The wavelength of the energy emitted by the Earth is very much longer than that emitted by the sun (because the Earth is much cooler than the sun), and these longer waves are not able to pass through the atmosphere as freely as short waves. For this reason, a large proportion of the energy emitted by the Earth is absorbed by the water vapour and water droplets in the lower atmosphere, and this energy in turn is re-radiated back to Earth (Figure 7.11). Thus the Earth plays the part of a receiving station absorbing short electromagnetic waves and converting them into longer electromagnetic waves, while the atmosphere acts as a trap containing most of the longer electromagnetic waves before they are lost to space.

Radiation from the sun consists of rays of three differing wavelengths: heat rays, actinic rays and light rays. Heat rays and actinic rays are intercepted by

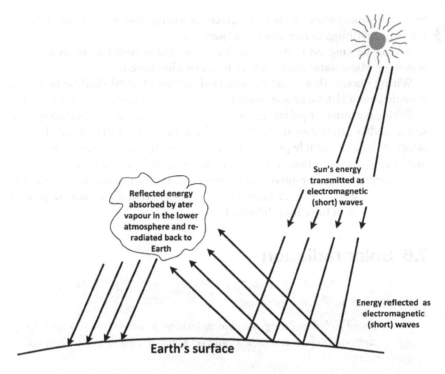

Figure 7.11 Solar radiation – energy.

solid bodies and produce peculiar effects in varying degrees according to the nature of the surface on which they fall. The light rays are responsible for daylight and both light rays and actinic rays are necessary for the life processes of plants. The heat rays' most important aspect is temperature, and the amount of sunshine (and therefore the temperature) will depend on latitude and the length of day (Figure 7.12).

Radiant energy can be reflected from solid surfaces and intensified by that reflection. For example, reflection from walls is frequently used for ripening peaches and pears. Reflection from bare ground can also assist in the ripening of melons and other creeping plants, while reflection from water surfaces enhances the 'climatic reputation' of waterside resorts. However, radiant energy can also cause damage to equipment as heat rays can warm the material and/or its environment to dangerous levels, and photochemical degradation of materials can be caused by the ultraviolet content of solar radiation.

7.6.1 What are the effects of solar radiation?

On cloudless nights when atmospheric radiation is very low, objects exposed to the night sky will attain surface temperatures below that of the surrounding air temperature. For example (and by experiment), a horizontal disk thermally

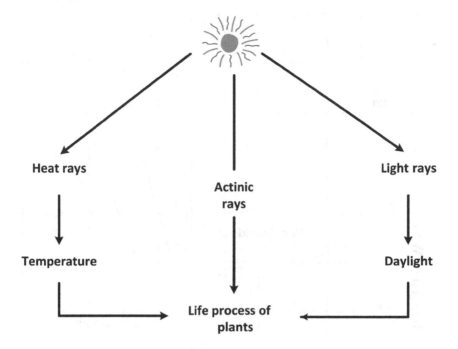

Figure 7.12 Sun's radiation.

insulated from the ground and exposed to the night sky during a clear night can attain a temperature of -14°C when the air temperature is 0°C and the relative humidity is close to 100%, and these values are of assistance when determining the *'under temperature'* of components (see Figure 7.13).

The sun's electromagnetic radiation consists of a broad spectrum of light ranging from ultraviolet to near infrared. Owing to the distance of the sun from the Earth, solar radiation appears on the Earth's surface as a parallel beam and the highest (maximum) level of radiation occurs at noon, on a cloudless day, at a surface perpendicular to the sun.

Most of the sun's energy reaches the surface of the Earth in the 0.3 –0.4 μm range and the density of the solar radiated power (or irradiance – expressed in watts per square metre) is dependent on the content of aerosol particles, ozone and water vapour in the air. The actual amount of irradiance will vary considerably with geographical latitude and type of climate (i.e. temperature, humidity, air velocity etc.).

Having said that, an object subjected to solar radiation will obtain a temperature depending on the surrounding ambient air temperature, the intensity of radiation, the air velocity, the incidence angle of the radiation on the object, the duration of exposure and the thermal properties of the object itself (e.g. surface reflectance, size, shape, thermal conductance and specific heat), together with other factors such as wind and heat conduction to mountings and surface absorbency etc.

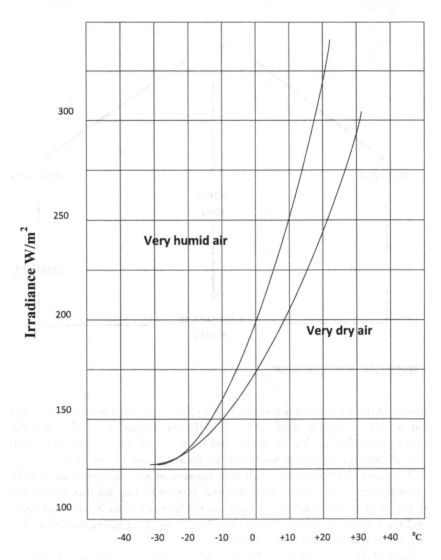

Figure 7.13 Lowest values of atmospheric radiation during clear nights.

7.6.2 Photochemical degradation of material

One of the biggest problems caused by solar radiation is the photochemical degradation of most organic materials, which in turn causes the elasticity and plasticity of certain rubber compounds and plastic materials to be affected and can, in exceptional cases, make optical glass opaque. Although solar radiation can bleach out colours in paints, textiles, paper, etc. (a major consideration when trying to read the colour coding of components!!) by far the most important effect is the heating of materials.

The combined effect of solar radiation, atmospheric gases, temperature and humidity changes etc., are often termed '*weathering*' and result in the '*ageing*' and ultimate destruction of most organic materials e.g. plastics, rubbers, paints, timber etc. Typical defects caused by weathering are:

- bleaching out colours in paints, textiles and paper;
- cracking and disintegration of cable sheathing;
- fading of pigments;
- rapid deterioration and breakdown of paints.

7.6.3 Effects of irradiance

To guard against the effects of irradiance, the following guidelines should be considered when locating electrical equipment:

- the sun should be allowed to shine only on the smallest possible casing surfaces;
- windows should be avoided on the sunny side of rooms housing electronic equipment;
- heat-sensitive parts must be protected by heat shields made, for instance, of polished stainless steel or aluminium plate;
- air-conditioning plant and cooling fans (when used) in rooms housing electronic equipment should be efficient and reliable;
- convection flow should sweep across the largest possible surfaces of materials with good conduction properties.

7.6.4 Heating effects

As noted previously, probably the most important effect of solar radiation heating is mainly caused by the short-term, high-intensity radiation around noon on cloudless days. Typical peak values of irradiance are shown in Table 7.9.

Table 7.9 Typical peak values of irradiance from a cloudless sky

Area type	Irradiance (W/m²)		
	Large cities	Flat land	Mountainous areas
Subtropical climates and deserts	700	750	1180
Other areas	1050	1120	1180

As equipment (if fully exposed to solar radiation) in an ambient temperature (e.g. 35–40°C) can attain temperatures in excess of 60°C, one has to consider the outside surface of the equipment. To a major extent the surface reflectance of an object affects its temperature rise from solar heating, and changing the finish from a dark colour to a gloss white can cause a considerable reduction in temperature.

On the downside, a pristine finish designed specifically to reduce temperature can be expected to deteriorate in time and result in an increase in temperature.

Another problem found in most of today's materials is that they are also selective reflectors (i.e. their spectral reflectance factor changes with wavelength). For example, paints are, in general, poor infrared reflectors although they may be very efficient as a visible warning. Care should, therefore, be taken when selecting materials and finishes for equipment casings.

7.6.5 Requirements from the Regulations – solar radiation

Table 7.10 lists the most common environmental requirements concerning solar radiation.

Table 7.10 Typical requirements – solar radiation

Survivability	Equipment that is exposed to the effect of solar radiation should remain unaffected.
Exposure	The sun should be allowed to shine only on the smallest possible casing surfaces and the convection flow should sweep across the largest possible surfaces of materials with good conduction properties.
Heat shields	Heat-sensitive parts shall be protected by heat shields made of (for instance) polished stainless steel or aluminium plate.
Windows	Windows should be avoided on the sunny side of rooms housing electronic equipment.
Air conditioning	Air-conditioning plant and cooling fans (when used) in rooms housing electronic equipment should be efficient and reliable.

Wiring systems shall be selected, erected and (where necessary) shielded where and whenever significant solar radiation (AN2) or ultraviolet radiation is experienced or anticipated. (Special precautions may need to be taken for equipment that is subject to ionising radiation).

Solar photovoltaic (PV) modules must be installed so that there is adequate heat dissipation when the site is subject to conditions of maximum solar radiation.

7.7 Pollutants and contaminants

Over the last twenty years environmental matters have become an area of widespread public concern, particularly those concerning the issues of pollution and contaminants. Pollutants and contaminants come in many forms and can have an effect on the air, land or water courses. As pollutants move from one medium to another they may be deposited on equipment and equipment housing – and they can cause extensive damage.

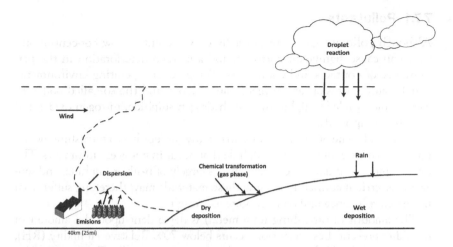

Figure 7.14 Process between the emissions of air pollutants and them being deposited on the ground.

Pollution of the air can occur in both the troposphere and stratosphere as shown in Figure 7.14. In the troposphere pollutants from chimneys (for example) are carried by the air and can be deposited over time and distance, thus having a limited life span before they are washed out or deposited on the ground. If pollutants are injected straight into the stratosphere (as with a volcanic eruption) they will remain there for some time and result in noticeable effects over the whole region. On the other hand, the roughness of the ground will produce air turbulence which will itself promote the mixing of pollutants and even low wind speeds will result in high pollutant concentrations.

Sources of natural pollutants include:

- sulphur – emitted by volcanoes and from biological processes;
- nitrogen – from biological processes in soil and lightning, and biomass burning;
- hydrocarbons – methane from fermentation of rice paddies, fermentation of the digestive tract of ruminants (e.g. cows). Also released by insects, coal mining and gas extraction.

Sources of man-made pollutants include:

- carbon dioxide and carbon monoxide, produced during the burning of fossil fuels;
- soot formation accompanied by carbon monoxide, generally due to inadequate or poor air supply;
- hydrocarbons – most boilers and central heating units burning fossil fuels result in very low emissions of gaseous hydrocarbons or oxygenated hydrocarbons such as aldehydes.

7.7.1 Pollutants

Although pollutant gases are normally only present in low concentrations, they can cause significant corrosion and a marked deterioration in the performance of contacts and connectors. The gases in operating environments which cause corrosion are oxygen, water vapour and the so-called pollutant gases, which include sulphur dioxide, hydrogen sulphide, nitrogen oxides and chlorine compounds.

Silver and some of its alloys are particularly susceptible to tarnishing by the minute quantities of hydrogen sulphide that occur in many environments. The tarnish product is dark in colour, consists largely of b-silver sulphide, and separable electrical connections using these materials may, therefore, suffer from increased resistance and contact noise as a result.

The amount of tarnishing (of a metal) is dependent upon the amount of humidity present. Less corrosion occurs below 70% Relative Humidity (RH), but above 80% RH the rate of tarnishing increases rapidly. Temperature also has an effect on the amount of tarnishing as the nature of the corrosion mechanism has a tendency to change at temperatures above 30°C.

7.7.1.1 Sulphur dioxide

Sulphur dioxide is the pollutant gas most commonly found in the atmosphere and is usually present in high concentrations in urban and industrial locations. In combination with other pollutants and moisture (e.g. humidity) it is responsible for the formation of high-resistance, visible corrosion layers on all but the most noble metals (e.g. silver and gold) and alloys.

Although sulphur dioxide alone is less corrosive than other gases (such as sulphur trioxide, nitrogen oxides and chlorine compounds), the most extensive corrosion occurs when combustion products are present together with sulphur dioxide.

7.7.1.2 Hydrogen sulphide

Hydrogen sulphide is caused by bacterial reduction of sulphates in vegetation, soil, stagnant water and animal waste on a worldwide basis. In the atmosphere, hydrogen sulphide is oxidised to sulphur dioxide, which in turn is brought to the ground by rain. In an aerobic soil, a bacterium turns the sulphur dioxide to sulphates. Sulphate-reducing bacteria complete the cycle and turn the sulphates to hydrogen sulphide – which is the principal natural sulphur input in the atmosphere and is, therefore, a widespread pollutant of air.

7.7.1.3 Nitrogen oxides

The production of nitrogen oxides is particularly significant because the rate of corrosion by sulphur dioxide is greatly accelerated in the presence of nitrogen dioxide, although the corrosion products have similar compositions.

7.7.2 Contaminants

Contaminants are composed of dust, sand, smoke and other particles that are contained within the air and these can have an effect on electrical equipment in various ways, especially:

- ingress of dust into enclosures and encapsulations;
- deterioration of electrical characteristics (e.g. faulty contact, change of contact resistance);
- seizure or disturbance in motion bearings, axles, shafts and other moving parts;
- surface abrasion (erosion and corrosion);
- reduction in thermal conductivity;
- clogging of ventilating openings, bushes, pipes, filters and apertures that are necessary for operation etc.

The presence of dust and sand in combination with other environmental factors such as water vapour can also cause corrosion and promote mould growth. Damp heat atmospheres cause corrosion in connection with chemically aggressive dust, and similar effects are caused by salt mist. Effects of ion-conducting and corrosive dusts (e.g. de-icing salts) need also to be considered.

7.7.2.1 Dust and sand

The concentration of dust and sand in the atmosphere varies widely with geographical locality, local climatic conditions and the type and degree of activity taking place. The amount of dust and sand found in the air is dependent on terrain, wind, temperature, humidity and precipitation. Under certain conditions enormous amounts of dust and sand may be temporarily released and this suspended dust will drift away with the wind (see Table 7.11) depending on its concentration and the size of the particles.

Table 7.11 Concentration of dust and sand

Atmospheric region	Dust and sand concentration ($\mu g/m^3$)
Rural and suburban	40–110
Urban	100–450
Industrial	500–2000

(Extracted with permission from a paper by Herne European Consultancy.)

Particles larger than 150 μm are generally confined to the air layer in the first metre above ground, and in this layer about half of the sand grains move within the first 10 mm above the surface.

The dust and sand appearing in enclosed and sheltered locations is generated by several sources (e.g. quartz, de-icing salts, fertilisers etc.

penetrating into locations via ventilating ducts or badly fitting windows). The dust may also come from cloth or carpets in normal use in the working environment.

7.7.2.2 Dust

Dust may be defined as "particulate matter of unspecified origin, composition and size ranging from 1 μm to 150 μm originating from quartz, flour, organic fibres etc.".

Particles of less than 75 μm can, because of their low terminal velocity, remain suspended in the atmosphere for very long periods through the natural turbulence of the air. In sheltered and enclosed locations, the maximum grain size tends to be smaller (e.g. less than 100 μm) than non-weather-protected locations due to the filtering effect of the shelter.

The dust found in and around electrotechnical products may be generated by several different sources. The dust may be quartz, coal, de-icing salts, fertilisers or small fibres from cotton or wool (real or artificial that has been generated from cloth or carpets by normal use in living rooms and offices), can penetrate into a piece of equipment by a number of mechanisms:

- carried in by forced air circulation, (e.g. for cooling purposes);
- carried in by the thermal motion of the air;
- pumped in by variations in the atmospheric pressure caused by temperature changes;
- blown in by wind.

Dust itself can act as a physical agent or chemical component (or both), and can cause one or more of the following harmful effects:

- seizure of moving parts;
- abrasion of moving parts;
- adding mass to moving parts, thereby causing unbalance;
- deterioration of electric insulation;
- deterioration of dielectric properties;
- clogging of air filters;
- reduction of thermal conductivity;
- interference with optical characteristics;

and also

- corrosion and mould growth;
- overheating and fire hazard.

Dust adhering to the surface of materials may contain organic substances that provide a source of food for micro-organisms.

7.7.2.3 Sand

Sand is the term applied to "segregated unconsolidated accumulation of detrital sediment, consisting mainly of tiny broken chips of crystalline quartz or other mineral, between 100 μm and 1000 μm in size".

Particles greater than 150 μm are unable to remain airborne unless continually subjected to strong winds, induced airflows or turbulence. Sand is generally harder than most fused silica glass compositions and can, quite naturally, scratch the surface of most glass optical devices. Pressure applied over trapped grains of sand can also cause fractures to occur in equipment.

The electrostatic charges produced by friction of the particles in sandstorms can interfere with the operation of equipment and sometimes be dangerous to personnel. The breakdown of insulators, transformers and lightning arresters and the failure of car ignition systems have also been known to occur as a result of such charges. The electrostatic voltages produced can be very large. Indeed, voltages as high as 150 kV have made telephone and telegraph communications inoperable during sandstorms.

Quartz, because of its hardness, can result in rapid wear or damage to products, particularly moving parts. However, erosion of material requires that the presence of dust and sand is combined with a high-velocity airstream over an extensive period of time.

Sand and the majority of dusts usually deposited on insulated surfaces are poor conductors in the absence of moisture. The presence of moisture, however, will result in the dissolving of the soluble particles and the formation of conducting electrolytes. For example, the leakage currents flowing over contaminated power line insulators can be of the order of one million times greater than those which flow through clean, dry insulators.

7.7.2.4 Smoke or fumes

Smoke or fumes are "dispersive systems in the air consisting of particles below 1 μm".

As the particles are so small they do not usually effect equipment, provided that the equipment is properly designed.

7.7.2.5 Fauna and flora

With a few exceptions, fauna (rodents, insects, termites, birds etc.) and flora (plants, trees, seeds, fruit, blossom, mould, bacteria and fungi etc.) may be present at all locations where equipment is stored, transported or used. Whilst fauna may be the cause of damage inside buildings as well as in open-air locations, damage by flora will predominantly occur in open-air conditions. Moulds and bacteria can, however, be present both inside buildings and in open-air conditions.

The frequency of this flora and fauna depends on temperature and humidity. In warm damp climates, fauna and flora, especially insects and micro-organisms such as mould and bacteria, will find conditions favourable to life. Humid or wet rooms in buildings (or rooms in which processes produce humidity) are suitable living spaces for rodents, insects and micro-organisms. The range of temperature in which moulds may grow is from 0–40°C, and the most favourable temperatures for many cultures is between 20°C and 30°C.

If the surfaces of the products carry layers of organic substances (e.g. grease, oil, dust), or animal/vegetable deposits, the surfaces will become ideal locations for the growth of moulds and bacteria.

7.7.2.5.1 Effects of flora and fauna

The functioning of equipment and materials can be affected by physical attacks of fauna. Small animals and insects that feed from, gnaw at, eat into and chew at materials are particular problems, as are termites cutting holes into material.

Materials, such as wood, paper, leather, textiles, plastics (including elastomers) and even some metals such as tin and lead are all susceptible to attack.

Larger animals can also cause damage by stroke, impact or thrust. These attacks can cause:

- physical breakdown of material, parts, units or devices;
- mechanical deformation or compression;
- surface deterioration;
- electrical failure caused by mechanical deterioration.

Deposits from fauna (especially insects, rodents, birds etc.) can be caused by the presence of the animal itself, nest building, deposited feed stocks, or metabolic products such as excrement and enzymes etc.

Deposits from flora may consist of detached parts of plants (leaves, blossom, seeds, fruits etc.) and growth layers of cultures of moulds or bacteria. Adverse effects from these include:

- deterioration of material;
- metallic corrosion;
- mechanical failure of moving parts;
- electrical failure due to:
  - o increased conductivity of insulators;
  - o failure of insulation;
  - o increased contact resistance;
  - o electrolytic and ageing effects in the presence of humidity or chemical substances;
  - o moisture absorption and adsorption;
  - o decreased heat dissipation.

These in turn can cause interruption of electrical circuits, malfunctioning of mechanical parts and clouding of optical surfaces (including glass).

7.7.2.6 Mould

Surface contamination in the form of dusts, splashes, condensed volatile nutrients or grease may be deposited on equipment. When that equipment is exposed (in use, storage or transportation) to the atmosphere, and without proper protective covering, mould growth will occur and mould can cause unforeseen damage to equipment, whether it is constructed from mould-resistant materials or not!

A fungus grows in soil and in, or on, many types of common material. It propagates by producing spores which become detached from the main growth and later germinate to produce further growth. The spores are very small and are easily carried by the wind (or moving air). They also adhere to dust particles carried in the air. Contamination can also occur due to handling. Spores may be deposited by the hands or in the film of moisture left by the hands.

7.7.2.6.1 Germination and growth

Moisture is essential in allowing the spores to germinate and when a layer of dust or other hydrophilic (i.e. moisture retaining) material is present on the surface; sufficient moisture may be abstracted by it, from the atmosphere. In addition to high humidity, spores require (on the surface of the specimen) a layer of material that will absorb the moisture. Mould growth is also encouraged by stagnant air spaces and lack of ventilation.

When the relative humidity is below 65%, no germination or growth will occur. The higher the relative humidity above this value, the more rapid the growth will be. Spores can survive long periods of very low humidity and even though the main growth may have died, they will germinate and start a new growth as soon as the relative humidity becomes favourable again (i.e. in excess of 65%). The optimum temperature of germination for the majority of moulds is between 20°C and 30°C.

7.7.2.6.2 Effects of mould growth

Moulds can live on most organic materials, but some of these materials are much more susceptible to attack than others. Growth normally only occurs on surfaces exposed to the air, and those which absorb or adsorb moisture will generally be more prone to attack.

Even where only a slightly harmful attack on a material occurs, the formation of an electrically conducting path across the surface due to a layer of wet mycelium (i.e. the vegetative part of a fungus) can drastically lower the insulation resistance between electrical conductors supported by an insulation material. When the wet mycelium grows in a position where it is within the electromagnetic field of a critically adjusted electronic circuit, it can cause a serious variation in the frequency-impedance characteristics of that circuit.

Among the materials that are very susceptible to attack are leather, wood, textiles, cellulose, silk and other natural resources. Most plastic materials, although less susceptible, are also prone to attack as they will probably contain oligomers (i.e. natural or synthetic compounds of usually high molecular weight that consist of millions of linked simple molecules), non-polymerised monomers (i.e. a molecule that can combine with others to form a polymer) and/or additives which may radiate to the surface and be a nutrient for fungi. Some plastic materials depend, for a satisfactory life span, on the presence of a plasticiser (i.e. substances added to plastics to make or keep them soft or pliable) which, if it is readily digested by fungi, will eventually give rise to failure of the main material.

Mould attack on materials usually results in a decrease of mechanical strength and/or changes in other physical properties, and the growing mould on the surface of a material can yield acid products and other electrolytes which will cause a secondary attack on the material. This attack can lead to electrolytic or ageing effects, and even glass can lose its transparency due to this process. Oxidation or decomposition may be facilitated by the presence of catalysts secreted by the mould.

7.7.2.6.3 Prevention of mould growth

All insulating materials used should be chosen to give as great a resistance to mould growth as possible, thus maximising the time taken for mycelium to grow and minimising any damage to the material consequent upon such growth. The use of lubricants during assembly (e.g. varnishes, finishes etc.) is frequently necessary in order to obtain the required performance or durability of a product. Such materials should be chosen with regard to their ability to resist mould growth for even though it can be shown the lubricants do not support mould growth, they may collect dust which in turn will support mould growth.

Moisture traps which could possibly be formed during the assembly of equipment and in which mould can grow should be avoided. Examples of such less obvious traps are between unsealed mating plugs and sockets, or between printed circuit cards and edge connectors. Other preventatives of mould growth include:

- complete sealing of the equipment in (and with) a dry, clean atmosphere;
- continuous heating within an enclosure, which can ensure a sufficiently low humidity;
- operation of equipment within a suitable controlled environment;
- regularly replaced desiccants (e.g. silica beads);
- periodic and careful cleaning of enclosed equipment.

Where the material and functioning of the equipment allows such treatment, ultraviolet radiation or ozone may be used for sterilisation. Air currents flowing over the parts can retard the development of mould growth and can be used to control the action of acarids (i.e. mites and ticks).

7.7.3 Requirements from the Regulations – pollutants and contaminants

Table 7.12 shows the most common environmental requirements concerning pollutants and contaminants.

Table 7.12 Typical requirements – pollutants and contaminants

Pollutants	• Although the severity of pollution will depend upon the location of the equipment, the effects of pollution must be considered in the design of equipment and components. • Means need to be provided to reduce pollution by the effective use of protective devices. • The requirements of ISO 14001 regarding environmental protection and the prevention of pollution have to be met.
Contaminants	The following should be considered: • chemical active substances; • biologically active substances; • flora and fauna; • dust; • sand.
Mould	• In an assembled state, equipment needs to operate when exposed to airborne mould spores and within climates that will be conducive to the growth of moulds. • Insulating materials should be chosen to provide as much resistance to mould growth as possible, and all materials used should be chosen with regard to their ability to resist mould growth.

7.8 Mechanical

As shown in Figure 7.15, there are a number of effects that need to be considered with regard to the effect of the mechanical environment on the continued reliability of equipment

Mechanics is the branch of physics concerned with the motions of objects and their response to forces. Modern descriptions of such behaviour begin with a careful definition of such quantities as displacement (distance moved), time, velocity, acceleration, mass and force.

There is often a tendency to underestimate the effect that the mechanical environment can have on the reliability of equipment, especially the effects of vibration and shock. Mechanical stresses are normally attributed to a moving mass and there is frequently a tendency to underestimate the effect of the mechanical environment on the reliability of static installations. Experience suggests, however, that vibrations and shocks are a significant '*reliability, availability and maintainability*' (RAM) factor, not only from the point of view of vehicle-mounted equipment, but also with respect to permanent installations.

If a spurious vibration acts on a printed circuit board (PCB), module or equipment, resonant oscillations will be induced in all components at their

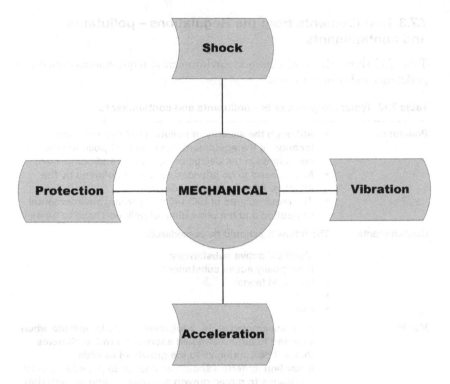

Figure 7.15 Effects of the mechanical environment.

natural frequency. If, however, the frequency spectrum has some more or less distinctive frequency bands, the elements will perform forced oscillations at the cyclic frequency of the interference and generally depend on both the characteristics of the oscillator (i.e. the component) and on the interference.

7.8.1 Shock

'*Shock*' is generally defined as "an impact shock characterised by a simple acceleration and free impact on a firm base", and is usually the result of a violent collision, or a heavy blow. Whilst it is difficult to design and install electrical equipment so that components and systems are completely immune to shock, precautions should, nevertheless, be taken to guard against potential problem areas.

7.8.2 Vibration

Components, equipment and other articles may during transportation or in service be subjected to conditions involving vibration of a harmonic pattern, generated primarily by rotating, pulsating or oscillating forces caused by machinery and seismic incidents. Adequate precautions need to be taken to prevent damage to electrical equipment and circuitry.

7.8.3 Acceleration

Equipment, components and electrotechnical products that are likely to be installed in moving bodies (e.g. rotating machinery) will be subjected to forces caused by steady accelerations. In general, the accelerations encountered in service will have different values along each of the major axes of the moving body, and also usually have different values in the opposite senses of each axis.

7.8.4 Protection

The resonant frequency of components is greatly influenced by the length of their connecting wires, and the actual length of these connecting wires may well be the decisive factor as to whether a component fails or remains functioning under given vibration and impact conditions. The amplitude of shocks on the equipment can be reduced by use of special mounting devices.

Shock absorption is based on storing the impact and releasing it at a retarded rate. The peak acceleration is reduced and the high frequencies damped, thus providing protection for the components with their relatively higher natural frequencies (of some hundred hertz).

Vibration dampers and shock absorbers are often used as a form of protection against mechanical stresses. The basic difference between vibration dampers and shock absorbers is that with the former the natural frequency lies below the interference frequency, whilst for the latter it is above it.

 Generally speaking, vibration dampers provide no protection whatsoever against shocks and similarly shock absorbers offer no protection against vibrations. Only in exceptional cases can vibration dampers, for high frequencies, be used as shock absorbers.

Elastic suspension of equipment can cause a critical increase in amplitude at certain frequencies and, where translatory and rotary displacements greater than six degrees of freedom are possible, very complex motions may arise. Wherever possible, therefore, one should try to ensure that none of the (possible) resonant frequencies fall within the range of the induced displacements.

Sheathing circuits by means of cast resins can, in most cases, be a very effective means of counteracting mechanical stresses combined with temperature humidity.

7.8.5 Requirements from the Regulations – mechanical

Table 7.13 gives the most common environmental requirements concerning solar radiation.

Table 7.13 Typical requirements – mechanical

Closed rooms	Equipment located in closed room installations must be capable of withstanding self-induced vibrations.
Encapsulated outdoor installations	Equipment contained in encapsulated outdoor installations must be capable of withstanding vibrations and shocks.
In service	Equipment should be capable of withstanding without deterioration or malfunction all mechanical stresses that occur in service.
Long-term exposure	Equipment must be capable of withstanding long-term exposure to shocks.
Mechanical shock	Equipment should be capable of withstanding shock pulses (e.g. a minimum of 20,000 shocks at a shock level of 20 g).
On or near the roadside	Equipment located on or near the roadside must be capable of withstanding vibrations and shocks.
Random vibration	Equipment should be capable of withstanding random vibration.
Vibrations and shocks	Any dampers or anti-vibration mountings must be integral so as to the equipment to prevent the unit being accidentally installed without them.

7.8.5.1 Mechanical and physical stresses on wiring systems

All wiring systems must be selected and erected to avoid (during installation, use or maintenance) damage to the sheath or insulation of its cables and their terminations.

A conduit system or cable ducting system (other than a pre-wired conduit assembly that has been specifically designed for the installation) that is going to be buried in the structure must be completely erected between access points before any cable is drawn in.

The radius of every bend in a wiring system shall be such that conductors or cables do not suffer damage and terminals are not stressed.

Cables and conductors shall be supported at appropriate intervals:

- so that they do not suffer any damage because of their own weight;
- so that they (and their terminations) are not exposed to any undue mechanical strain.

Cables and conductors shall:

- not be damaged by the means of fixing;
- have adequate means of access to allow them to be drawn in and out of a product wiring system.

The location of buried cables shall be marked by cable covers or a suitable marking tape.

Buried cables, conduits and ducts shall:

- be at a sufficient depth to avoid being damaged by any reasonably foreseeable disturbance of the ground;
- be suitably identified by cable covers or marking tape (Figure 7.16).

Figure 7.16 Buried electric cable marking note.

Other requirements include:

- a cable buried in the ground (that is not installed in a conduit or duct) shall incorporate an earthed armour or metal sheath, or both, suitable for use as a protective conductor;
- cable supports and enclosures shall not have sharp edges liable to damage the wiring system;
- cables, busbars and other electrical conductors which pass across expansion joints shall be selected and/or erected so that anticipated movement does not cause damage to the electrical equipment.

 See IEC 61386-24 for further details concerning underground conduits.

7.8.5.2 Shock

For locations where the wiring system may be exposed to impact and mechanical shock due to vehicles and mobile agricultural machines etc., the external influences shall be classified AG3 and:

- conduits shall provide a degree of protection against impact of 5 J according to BS EN 61386-2;
- cable trunking and ducting systems shall provide a degree of protection against impact of 5 J according to BS EN 50085-2-1.

7.8.5.3 Vibration

Stationary equipment which is moved temporarily for the purposes of connecting, cleaning etc., (e.g. a cooker) may be connected with a non-flexible cable. However, if they are subject to vibration whilst in use they shall be connected by flexible cables.

A wiring system supported by (or fixed to) a structure or equipment item that is subject to vibration of medium severity (AH2) or high severity (AH3) shall be suitable for such conditions, particularly where cables and cable connections are concerned.

Note: Where no vibration or movement can be expected, cable with non-flexible cores may be used.

7.8.5.4 Electrical connections

Connections between conductors or between a conductor and other equipment shall provide durable electrical continuity and adequate mechanical strength and protection, taking account of:

- the cross-sectional area of the conductor;
- the material of the conductor and its insulation;
- the number and shape of the wires forming the conductor;
- the number of conductors to be connected together;
- the temperature attained at the terminals in normal service;
- the provision of adequate locking arrangements in situations subject to vibration or thermal cycling.

7.8.5.5 Electrical connections in caravans and motor homes

In a caravan or motor caravan:

- as wiring will be subjected to vibration, all wiring needs to be protected against mechanical damage either by location or by enhanced mechanical protection;
- any wiring passing through metalwork must be protected by suitable bushes or grommets (which are securely fixed in position);
- precautions must be taken to avoid mechanical damage due to sharp edges or abrasive parts;
- all cables, unless enclosed in rigid conduit, and all flexible conduit shall be supported at intervals not exceeding 0.4 m for vertical runs and 0.25 m for horizontal runs.

See BS 7671:2018 Chapter 721 for more detailed requirements relating to electrical connections in caravans and motor homes.

7.8.5.6 Low voltage generating sets – generators

Electrical equipment associated with the generator shall be mounted securely and, if necessary, on anti-vibration mountings.

 See BS 7671:2018 Chapter 740 for more detailed requirements relating to generators.

7.8.5.7 Protection against overcurrent

Persons and livestock shall be protected against injury, and property shall be protected against damage, due to electromechanical stresses caused by any overcurrents likely to arise in live conductors.

7.8.5.8 Protection against fault current

Electrical equipment, including conductors, shall be provided with mechanical protection against electromechanical stresses of fault currents as necessary to prevent injury or damage to persons, livestock or property.

7.8.5.9 Cross-sectional area of conductors

The cross-sectional area of conductors shall be determined for both normal operating conditions and, where appropriate, for fault conditions according to:

- the electromechanical stresses likely to occur due to short-circuit and Earth fault currents;
- other mechanical stresses to which the conductors are likely to be exposed.

7.8.5.10 Waves

Wiring systems that may be subjected to the possibility of water waves at seashore locations (AD6) shall be protected from mechanical damage (i.e. impact, vibration and other mechanical stresses).

7.9 Electromagnetic compatibility

Most car owners normally accept that when they drive near electric pylons, their listening pleasure may be interrupted by loud crackles and/or buzzing noises. However, with the increased use of electronic equipment, the problem of interference has become one of our prime concerns.

Although most forms of interference are usually tolerated as being 'one of those things' (that you cannot do much about), the design of modern sophisticated equipment has become so susceptible to electromagnetic interference (EMI) that some form of regulation has had to be agreed.

Within Europe, this regulation is contained in the Electromagnetic Compatibility Directive 2004/108/EC (which repealed the original Directive 89/336/EEC), which clearly states that all electronic equipment shall be constructed so that:

- the electromagnetic disturbance it generates does not exceed a level allowing radio and telecommunications equipment and other apparatus to operate as intended;
- the apparatus has an adequate level of intrinsic immunity to electromagnetic disturbance.

Electromagnetic disturbances and EMI can seriously disrupt and even damage information technology (IT) systems and/or IT equipment, electronic components and circuits. Lightning, switching operations short-circuits and other electromagnetic phenomena can also cause overvoltages and electromagnetic interference.

These effects are potentially more severe:

- where large metal loops exist;
- where different electrical wiring systems are installed in common routes, such as power supply, signalling and/or data communication cables connecting IT equipment within a building.

 This is of particular relevance in (or near) rooms that are used for medical purposes, as electromagnetic disturbances can dramatically interfere with medical electrical equipment.

7.9.1 Electromagnetic interference

As previously noted, EMI is the disturbance caused by generated currents (typically due to lightning, switching operations, short-circuits and other electromagnetic phenomena) which cause overvoltages and electromagnetic interference, induction, coupling or conduction. This is a particular problem with sensitive equipment where signal transmission can become corrupted or distorted. Data transmission may also result in an increased error rate or total loss of data.

7.9.2 Sources of electromagnetic disturbances

Potential sources of electromagnetic disturbances within an installation typically include:

- electric motors;
- fluorescent lighting;
- frequency convertors/regulators including variable speed drives (VSDs);
- lifts;

- power distribution busbars and rectifiers;
- switchgear;
- switching devices for inductive loads;
- transformers;
- welding machines.

See also Figure 7.17.

 For further information regarding electromagnetic disturbances, please refer to the BS EN 50174 series of standards.

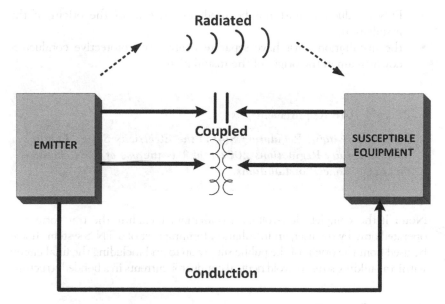

Figure 7.17 Electromagnetic sources.

7.9.3 Measures to reduce EMI

To reduce the effects of EMI, the following measures shall be considered:

- Ensuring that there is adequate separation between power and signal cables.
- Ensuring that the areas of all wiring loops are as small as possible in order to minimise voltages induced by lightning.
- Including surge protective devices and/or filters to improve the electromagnetic compatibility of electrical equipment sensitive to electromagnetic disturbances.
- Installing an equipotential bonding network.
- Installing power cables (i.e. line, neutral and any protective Earth conductors) close together in order to minimize cable loop areas.

- Limiting the amount of fault current from power systems flowing through the screens and cores of signal cables, or data cables, which are earthed.
- Using surge protective devices and/or filters to improve the electromagnetic compatibility (EMC) of electrical equipment sensitive to electromagnetic disturbances.

7.9.3.1 TN system

To minimize electromagnetic disturbances, the following requirements should be met:

- PEN conductors must not be used downstream of the origin of the installation;
- the installation must have separate neutral and protective conductors downstream of the origin of the installation.

> **BS 7671:2018 Requirement**
>
> *"In Great Britain, Regulation 8(4) of the Electricity Safety, Quality and Continuity Regulations 2002 prohibits the use of PEN conductors in consumers' installations."*

Note: If the complete low-voltage installation, including the transformer, is operated only by the user, an installation forming part of a TN-S system should be used from the origin of the public supply up to and including the final circuit within a building so as to avoid neutral conductor currents in a bonded structure.

7.9.3.2 TT system

If the installation forms part of a TT system, the possibility of overvoltages existing between live parts and any extraneous-conductive-parts of different buildings that are connected to different Earth electrodes needs to be considered. Using an isolating transformer, bypass equipotential bonding conductor or single-point bonding would also be beneficial.

If the live conductors of the supply into any of the buildings are less than 35 mm² in cross-sectional area, the main protective bypass conductor shall have a minimum cross-sectional area of 10 mm².

See BS 7671:2018 Table 54.8 for other sized supply neutral conductors.

7.9.3.3 Transfer of supply

To prevent electromagnetic fields due to stray currents in the main supply system of an installation, the transfer from one supply to an alternative supply for any installation forming part of a TN system shall be via a multi-pole switching device that switches both the line conductors and the neutral conductor (if any).

7.9.3.4 Separate buildings

Where different buildings have separate equipotential bonding systems, metal-free optical fibre cables or other non-conducting systems (such as a microwave signal transformer) should be used for signal and data transmission.

7.9.3.4.1 Earthing conductors

Within a single building, all protective and functional earthing conductors of an installation shall normally be connected to the main earthing terminal.

If a number of installations have separate earthing arrangements, then any protective conductor that is common to any of these installations shall either be

- capable of carrying the maximum fault current that is likely to flow through them; or
- earthed within one of the installations only.

Note: Where interconnection of the Earth electrodes is not possible or practicable, it is recommended that separation of communications networks is applied, for example by using optical or radio links.

7.9.3.5 Equipotential bonding networks

To avoid or reduce the possibility of electromagnetic disturbances affecting an installation (in particular IT systems):

- metal sheaths, screens or cables that have been armoured must be bonded to the common bonding network (CBN), unless such bonding is required to be omitted for safety reasons;
- screened signal or data cables that are earthed should be capable of limiting the fault current from power systems flowing through the screens and cores of signal and/or data cables;
- the impedance of equipotential bonding connections intended to carry functional Earth currents having high-frequency components shall be as low as practicable, and this should be achieved by the use of multiple, separated bonds that are as short as possible;
- the size and installation of an equipotential bonding ring network shall have the following minimum nominal dimensions:
  o flat cross-section: 25 mm × 3 mm;
  o round diameter: 8 mm.

Bare conductors shall be protected against corrosion at their supports and on their passage through walls.

The following parts shall be connected to the equipotential bonding network:

- metallic containment;
- conductive screens, sheaths or armoured data-transmission cables used for IT equipment;
- functional earthing conductors of antenna systems;
- the earthed pole's conductor of a d.c. supply used for IT equipment;
- functional earthing conductors;
- protective conductors.

Author's note

Parts of a cable that are inside an accessory, appliance or luminaire must be temperature resistant, or be provided with additional insulation suitable for all possible temperatures encountered.

For buildings with several floors, it is recommended that, on each floor:

- an equipotential bonding system be installed;
- the bonding systems of the different floors should be interconnected, at least twice, by protective conductors.

 See BS 7671:2018 Chapter 54 for more detailed requirements.

7.9.3.6 Earthing arrangements and equipotential bonding for IT installations

Information technology installations should:

- be connected to the main earthing terminal by the shortest practicable route from any point in the building;
- use one or more earthing busbars which should be accessible throughout its entire length.
- ensure that all bare conductors are protected to prevent corrosion.

 If the earthing busbar is used as part of a d.c. return current path, its cross-sectional area must be selected according to the expected d.c. return currents.

7.9.3.7 Segregation of circuits

Cables that are used at voltage Band II (low voltage) and Band I (extra-low voltage) which share the same cable management system or the same route shall be installed according to the following requirements:

- Each part of a circuit shall be arranged such that the conductors are not distributed over different multi-core cables, conduits, ducting systems, franking systems or tray or ladder systems.

- In order to prevent the indirect energising of an isolated final circuit, the line and neutral conductors of each final circuit must be electrically separate from those of every other final circuit.
- Whenever multi-core cables are installed in parallel, each cable should contain one conductor of each line.
- A voltage Band I circuit shall not be contained in the same wiring system as a Band II circuit unless:
  - each conductor of a multi-core cable is insulated for the highest voltage present in the cable;
  - for a multi-core cable, the cores of the Band I circuit are separated from the cores of the Band II circuit by an earthed metal screen of equivalent current-carrying capacity to that of the largest core of a Band II circuit.

- If underground telecommunication cables and underground power cables are liable to cross each other, a minimum clearance of 100 mm shall be maintained; and:
  - a fire-retardant partition shall be provided between the cables; and
  - mechanical protection between the cables shall be provided for the points where the cables cross over.

- The minimum distance between information technology cables and discharge, neon and mercury vapour (or other high-intensity discharge) lamps shall be 130 mm.
- Data wiring racks and electrical equipment shall always be separated.

7.9.3.8 Protection against voltage disturbances and measures against electromagnetic disturbances

Persons and livestock shall be protected against injury, and property shall be protected against any harmful effects, as a consequence of:

- a fault between live parts of circuits supplied at different voltages;
- overvoltages such as those originating from atmospheric events or from switching;
- undervoltage and any subsequent voltage recovery.

The installation shall have an adequate level of immunity against electromagnetic disturbances so as to function correctly in the specified environment.

The installation design shall take into consideration the anticipated electromagnetic emissions generated by the installation or the installed equipment.

See also the BS EN 62305 series of standards for further information concerning protection against lightning strikes.

7.9.3.9 Electrical installations

Electrical installations **shall** be arranged so that they do not mutually interfere (including electromagnetic interference) with other electrical installations and non-electrical installations in a building.

As stated in BS 7671:2018, *"all fixed installations shall be in accordance with the Electromagnetic Compatibility Regulations 2006 (SI 2006/3418)"*, which came into force on 20 July 2007 and which require that all electrical and electronic apparatuses marketed in the UK, including imports, satisfy the requirements of the EMC Directive.

Immunity levels of equipment shall be chosen taking into account:

- the electromagnetic influences that can occur when connected for normal use; and
- the intended level of continuity of service necessary for the application. (See BS EN 50082.)

Equipment shall be chosen with sufficiently low emission levels so that it cannot cause unacceptable electromagnetic interference with other electrical equipment (see BS EN 50081).

The area of all wiring loops shall be as small as possible so as to minimise voltages induced by lightning.

7.9.3.10 Wiring installations

The choice of the type of wiring system and the method of installation shall include consideration of the following:

- the nature of the location;
- the type of structure supporting the wiring;
- the accessibility of wiring to persons and livestock;
- voltage;
- the electromechanical stresses likely to occur due to short-circuit and Earth fault currents;
- electromagnetic interference;
- other external influences (e.g. mechanical, thermal and those associated with fire) to which the wiring is likely to be exposed during the erection of the electrical installation or in service.

Every installation shall be divided into circuits, as necessary, to:

- avoid danger and minimise inconvenience in the event of a fault;
- ensure safe inspection, testing and maintenance;
- prevent the indirect energising of a circuit that is intended to be isolated;
- reduce the possibility of unwanted tripping of RCDs;
- take account of hazards that may arise from the failure of a single circuit (e.g. a lighting circuit);

- take into account the possibility of EMI being present, and reduce its effects if it is;
- ensure that no mutual detrimental influence will occur between electrical installations and non-electrical installations.

7.9.3.11 Cables and conductors

The conductors of an a.c. circuit installed in a ferromagnetic enclosure shall be arranged so that the line conductors, the neutral conductor (if any) and the appropriate protective conductor are all contained in the same enclosure.

Where such conductors enter a ferrous enclosure, they shall be arranged such that the conductors are only collectively surrounded by ferrous material.

Single-core cables armoured with steel wire or steel tape shall **not** be used for an a.c. circuit.

 Notes:

1. The steel wire or steel tape armour of a single-core cable is regarded as a ferromagnetic enclosure.
2. For single-core armoured cables, the use of aluminium armour may be considered.

7.9.3.12 Medical locations

 Special considerations have to be made concerning EMI and electromagnetic compatibility (EMC) in medical locations.

It is recommended that radial wiring patterns are used to avoid 'Earth loops' that may cause electromagnetic interference.

In or near rooms for medical use, electromagnetic disturbances associated with electrical installations can interfere with medical electrical equipment.

7.9.4 Requirements from the Regulations – electromagnetic compatibility

Table 7.14 (overleaf) shows the most common environmental requirements concerning electromagnetic compatibility (EMC):

7.10 Fire

A fire will normally start when sufficient thermal energy, from, for example, an electric short-circuit or a burning cigarette, is supplied to a combustible material. Following ignition, the fire will then produce its own thermal energy, some of which will be used as feedback to maintain combustion, and some transferred via radiation and convection to other materials. These materials may also ignite and spread the fire.

Table 7.14 Typical requirements – electromagnetic compatibility

Equipment	The use of electronic equipment shall not interfere with the operation of other equipment.
	All active electronic devices shall comply with the EMC Directive.
CE Marking	Only CE (Conformity Europe) marked equipment may be offered for sale, and all active equipment connected to an electrical installation shall carry a CE mark.
Apparatus cases	Input/output connections from apparatus cases should always be of non-screened non-balanced signalling cable, and are normally restricted in length.
Atmospheric disturbances	To counteract the effects of storms, it is generally recommended that all equipment should be capable of withstanding (as a minimum) the following overvoltages:
	Magnitude 2000 V Rise time 1.2 µs Middle voltage time 50 µs
Equipment immunity levels	Equipment should be immune to induced common-mode voltages.
	Equipment should not experience a permanent loss of availability or suffer component damage for any induced common-mode voltage.
Magnetic field	As low-frequency fields can influence cathode ray tubes, equipment should be capable of withstanding the following intensities:
	Hz A/m 5 0.8 50 3.0 250 1.5
Power supply lines	Equipment should be immune to the following high-frequency bursts:
	Initial peak-to-peak voltage 1 kV Burst repetition rate 5 kHz
Transients	All electronic equipment should be capable of withstanding:
	• transients (either directly induced or indirectly coupled), so that no damage or failure occurs during operation; • without damage or abnormal operation, transient non-repetitive surges.

The environmental conditions relating to the occurrence, development and spread of fire within a building and its effect on electrotechnical products exposed to fire (see Figure 7.18) is primarily covered by Section 8 (Fire Exposure) of IEC 721.2. This section provides background information for selecting the appropriate parameters and severities related to exposure of

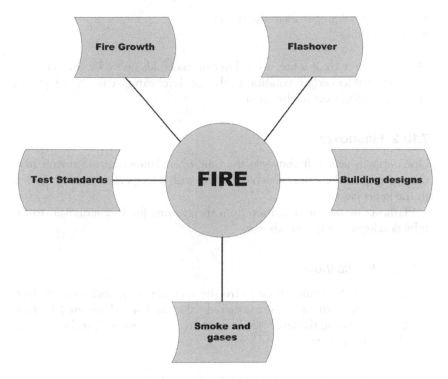

Figure 7.18 Fire exposure.

products to fire. More detailed information on fire condition characteristics and fire hazard testing is contained in specialist documentation.

The development of the fire generally consists of three processes:

- aerodynamic;
- chemical;
- thermal.

As a rule, radiation, convection and flame spread are the dominant physical factors.

7.10.1 Fire growth

Once a fire has started in a space (e.g. a room) its growth and spread is determined by:

- the aerodynamic conditions of the space;
- the arrangement of the fuel or fire load, its distribution, continuity, porosity and combustion properties;
- the shape and size of the space;
- the site;

- the thermal properties of the space;
- the volume.

During the growth of a fire, a hot layer of gas builds up under the ceiling of the space. Under certain conditions, this gas layer can give rise to a rapid fire growth, and flashover might occur.

7.10.2 Flashover

One normally defines flashover as the time when flames begin to emerge from the openings of the space, which correlates with a temperature of 500–600°C in the upper gas layer.

Flashover marks the transition from the growing fire (pre-flashover) to the fully developed fire (post-flashover).

7.10.2.1 Pre flashover

A pre-flashover fire primarily concerns the operation and function of products (e.g. detectors, alarm systems, associated cables and sprinklers etc.) that are vital to maintaining the level of safety required for escape and/or the rescue of people caught in a fire.

7.10.2.1.1 Characteristics of pre-flashover fire

The ignitability properties of exposed material will depend on:

- the exposure time;
- the heat supplied;
- the geometrical location;
- the presence or not of flames;
- the thermal data;

together with time variations such as:

- rate of heat release;
- rate of flame spread;
- gas temperature.

7.10.2.1.2 Fire hazard of a pre-flashover

The fire hazard of a pre-flashover situation is normally considered in terms of a series of probabilities, which depend on:

- the presence of ignition sources;
- the presence of products;
- the product fire performance properties;

- the environmental factors;
- the presence of people;
- the presence/operation of detection and suppression devices;
- the availability of escape.

See Figure 7.19.

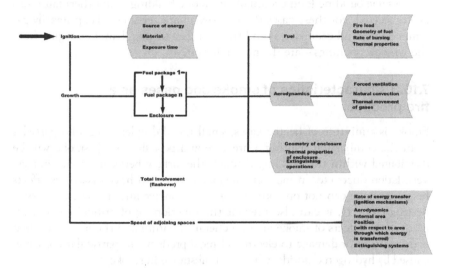

Figure 7.19 Factors affecting ignition, growth and spread of fire in a building.

7.10.2.2 Post flashover

Whilst most standards are normally concerned with conditions during the pre-flashover stage of a fire, conditions following flashover must also be considered. A post-flashover fire can seriously damage some of the structural and load-bearing elements of a building and the fire can then, quite easily, spread from one fire space to another via partitions and ventilation systems. This can, of course, seriously damage electrical equipment located in these voids. For example, in a large space it is quite possible that a fire, small in relation to that space, could be large enough to damage some of the structural elements in the post-flashover state. An often-overlooked but important factor of the post-flashover fire is the amount of smoke and toxic gases that can affect people in escape routes and remote safety areas in a building. Smoke and toxic gases can also significantly affect equipment.

7.10.2.2.1 Characteristics of post-flashover fire

The main characteristics of a post flashover fire are:

- the gas temperature;
- the geometrical and thermal data for external flames;

- the rate of heat release;
- the smoke and its optical properties;
- the composition of the combustion products, particularly corrosive and toxic gases.

The possibility of a large external fire spreading from one storey to another in the same building (and eventually from one building to another) must also be considered. For these cases the first two characteristics – i.e. primarily gas temperature, and geometrical and thermal data for the flames emerging from the window openings – are the most relevant.

7.10.3 Characteristics of smoke and gases as a fire product

Smoke is a mixture of heated gases, small liquid droplets and solid particles from the combustion. During a fire (pre and post flashover), smoke will be distributed within the building through the airflow between rooms and via ventilation ducts etc. In most circumstances this can have disastrous effects because smoke can not only kill people, and damage and in some cases even destroy property, it can also prevent the functioning of critical equipment. Most of the effects of smoke are of a chemical nature and the most prevalent is destruction or damage to electrotechnical products, in particular corrosion caused by hydrogen chloride, which is a substance in smoke.

Metal surfaces that are exposed to air under normal (non-fire) conditions often have a chloride deposit up to 10 mg/m^2. Such an amount is generally not harmful. However, after exposure to smoke from a fire involving polyvinyl chloride (PVC), a surface contamination of up to thousands of milligrams per square metre can be found, often causing significant damage. Chloride contamination of electrotechnical equipment can be removed by, for instance, detergents, solvents, neutralising agents, ultrasonic vibrations and clean air jets, but the procedures are not always effective, sometimes giving a temporary but not permanent cure.

Experiments, have proved that PVC coated electrical wires are sufficiently fire proof in most situations.

7.10.4 Building designs

In the design of buildings, the fire design of load-bearing structural elements and partitions is normally considered as a national problem and directly related to the results of standard national and (when available) international fire resistance tests. In such tests, the specimen is exposed, in a furnace, to a temperature rise, which is varied with time and within specified limits, according to the particular test being used.

Over the last few decades, rapid progress has been made in the development of analytical and computational methods for determining the fire design of load-bearing and separating structures and structural elements. In the long term, it is foreseeable that this will develop into an analytical and/or computational

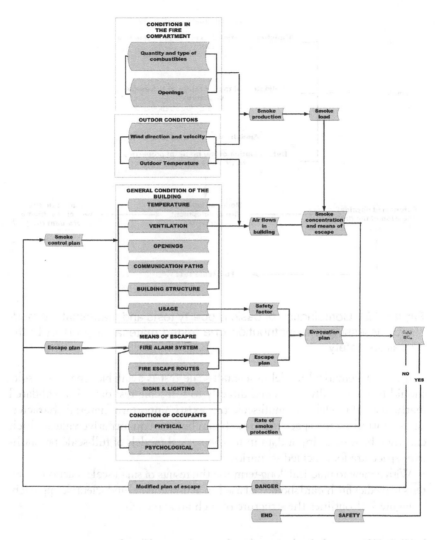

Figure 7.20 Flow diagram of a smoke-control design system in a building.

design, directly based on a natural fire exposure. The design will be specified with regard to the combustion characteristics of a fire load and the geometrical, ventilation and thermal properties of the fire space (see Figure 7.20).

7.10.5 Test standards

Fire tests on building materials, components and structures normally focus on the characteristics of pre-flashover fire. Simplified full-scale (i.e. room) tests for surface products' reaction to smoke, and in particular toxic combustion products, are already available, but considerable development work needs to be completed before a useful small-scale test is available.

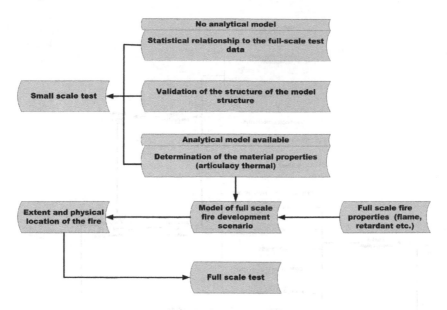

Figure 7.21 Combination of basic property tests and mathematical models for assessing the contribution of a tested material or product to the overall fire safety.

If no mathematical model of a small-scale test is available, the test results should be statistically correlated directly to full-scale test data. If a validated mathematical model of a small-scale test exists, important material characteristics controlling the space fire growth can be given quantitative values which can then be used as input data in mathematical models of full-scale pre-flashover space fire for specified scenarios.

With a view to practical, long-term use the results of small-scale reaction to fire tests to predict fire hazard should be based on a fundamental and scientific approach.

Figure 7.21 outlines the structure of such an approach.

7.10.6 Other related standards and specifications

IEC 60695 series	Fire hazard testing – Guidance, tests and specifications for assessing fire hazard of electrotechnical products
ISO 5657	Fire tests – Reaction to fire – Ignitability of building products
ISO 5658	Reaction to fire tests – Spread of flame on building products and vertical configuration
ISO 5660	Fire tests – Reaction to fire – Rate of heat release from building products
ISO 9705	Fire tests – Full-scale room test for surface products
ISO TR 5924	Fire tests – Reaction to fire – Smoke generated by building products (dual-chamber test)
ISO TR9112.1	Toxicity testing of fire effluents – General

7.10.7 Requirements from the Regulations – fire

In most contracts, reference is made to the IEC 60695 series of standards which cover the assessment of electrotechnical products against a nominated fire hazard.

The CENELEC (Comité Européen de Normalisation Électrotechnique) standards, on the other hand, show the requirement for equipment to operate in fire hazardous areas as three distinct clauses, as follows:

Class FO – no special fire hazard envisaged. This is considered a normal service condition and, except for the characteristics inherent to the design of the equipment, no special measures need to be taken to limit flammability.

Class F1 – equipment subject to fire hazard. This is considered an abnormal condition and restricted flammability is required. Self-extinction of fire shall take place within a specified time period. Poor burning is permitted with negligible energy consumption.

The emission of toxic substances shall be minimised. Materials and products of combustion shall, as far as possible, be halogen free and shall contribute a limited quantity of thermal energy to an external fire.

Class F2 – equipment subject to external fire. This is considered an abnormal condition and, in addition to the requirements of Class F1, the equipment shall (by means of special provisions) be able to operate for a given time period when subjected to an external fire.

Materials are normally expected to confirm to those requirements defined in EN 60721.3.3 and EN 60721.3.4.

BS 7671:2018 has been updated so as to maintain technical alignment with CENELEC harmonisation documents. One of the main changes concerned the requirements for safety services (e.g. emergency escape lighting, fire alarm systems, installations for fire pumps, fire rescue service lifts, smoke and heat extraction equipment) which now need to be observed.

Safety services have also been expanded in line with IEC standardisation.

7.10.7.1 Electrical installations

In electrical installations, risk of injury may result from excessive temperatures likely to cause burns, fires and other injurious effects. To guard against this happening:

- equipment in surroundings susceptible to risk of fire or explosion shall be constructed or protected so as to prevent danger;

- the choice of the type of wiring system and the method of installation shall include consideration of the following:
    - o   accessibility of wiring to persons and livestock
    - o   electromagnetic interference;
    - o   the electromechanical stresses likely to occur due to short-circuit and Earth fault currents;
    - o   the nature of the location;
    - o   the nature of the structure supporting the wiring;
    - o   voltage;
    - o   other external influences (such as mechanical, thermal and those associated with fire) to which the wiring is likely to be exposed during the erection of the electrical installation or in service.

Note: In structures where the shape and dimensions are such as will facilitate the spread of fire, precautions shall be taken to ensure that the electrical installation cannot propagate a fire (e.g. by the chimney effect).

7.10.7.2 Selection and erection of installations in locations of national, commercial, industrial or public significance

When selecting an erecting electromechanical equipment in locations of national, commercial, industrial or public significance, the following measures need to be considered:

- installation of cables in areas with constructional partitions having a fire-resisting capability for a time of 30 minutes or 90 minutes;
- installation of cables in non-combustible solid walls, ceilings and floors;
- installation of cables with improved fire-resisting characteristics in case of a fire hazard;
- installation of mineral-insulated cables according to BS EN 60702.

Note: Where these measures are not practicable, improved fire protection may be possible by the use of reactive fire protection systems.

7.10.7.3 Precautions where a particular risk of fire exist

In locations that include buildings or rooms with assets of significant value (such as national monuments, museums and other public buildings), in buildings such as railway stations and airports (that are generally considered to be of public significance) and/or in buildings or facilities such as laboratories, computer centres and certain industrial and storage facilities (which can be of commercial or industrial significance), the following needs to be taken into consideration:

- Electrical equipment shall be so selected and erected that its temperature in normal operation will not cause a fire.

A temperature cut-out device should always have a manual reset.

Where there is a risk of fire due to the manufacture, processing or storage of flammable materials (e.g. in barns, due to the accumulation of dust and fibres; and in woodworking facilities, paper mills and textile factories, due to the storage and processing of combustible materials), a fire risk will be present and the following precaution need be observed:

- A cable shall, as a minimum, satisfy the test under fire conditions specified in BS EN 60332-1-2.
- A cable that is not completely embedded in a non-combustible material (such as plaster or concrete) or otherwise protected from fire shall meet the flame propagation characteristics as specified in BS EN 60332-1-2.
- Cable tray systems or a cable ladder shall satisfy the requirements specified in BS EN 61537.
- Cable trunking or cable ducting systems shall fulfil the fire conditions specified in BS EN 50085.
- Conduit and trunking systems meet the fire conditions specified in BS EN 61386-1 and BS EN 50085-1 respectively and shall meet the fire-resistance tests within these standards.
- Powertrack systems shall satisfy the fire conditions specified in the BS EN 61534 series.
- Precautions shall be taken such that a cable or wiring system cannot propagate flame.
- Where the risk of flame propagation is high, the cable shall meet the flame propagation characteristics specified in the appropriate part of the BS EN 50266 series.

7.10.7.4 Protection against the risk of fire

Where it is necessary from the point of view of fire risk to limit the consequence of fault currents of higher frequencies, d.c. fault currents or increased leakage currents in a wiring system, the selection of protective and monitoring devices shall take into account the nature of the load and the likelihood of the device continuing to safely operate. Typical actions regarding protective devices include:

- installing an RCD (residual current device) at the origin of the circuit to be protected which will switch all live conductors (at a rated residual operating current not exceeding 300 mA);
- installing an RCM (residual current monitoring device) as an alternative to RCDs in IT systems, provided that the location is supervised by one or more skilled or instructed person(s), taking into consideration that RCMs shall:
  - o be in accordance with BS EN 62020;
  - o operate in conjunction with switchgear suitable for isolation;

- o be installed at the origin of final circuits;
- o not exceed 300 mA rated residual operating current;
- o provide audible and visual signals;

- installing an IMD (insulation monitoring device) to enable fault location on load;
- install AFDDs (arc fault detection devices), such as an optical detection system:

  - o at the origin of the final circuits to be protected;
  - o in a.c. single-phase circuits not exceeding 230 V;
  - o in switchgear and controlgear assemblies.

When selecting and erecting a luminaire, the thermal effects of radiant and convected energy on the surroundings shall be taken into account, including the fire resistance of adjacent material:

- at the point of installation; and
- in the thermally affected areas.

 Note: A device for protection against fault current need **not** be provided where the wiring is installed in such a manner as to reduce to a minimum the risk of fire or danger to persons; but the omission of devices for protection against overload **is** permitted for circuits supplying current-using equipment where unexpected disconnection of the circuit could cause danger or damage.

Examples of such circuits are:

- a circuit supplying a fire-extinguishing device;
- a circuit supplying a safety service, such as a fire alarm or a gas alarm.

 In such situations consideration should be been given to the provision of an overload alarm.

7.10.7.5 Protection against thermal effects

Persons, livestock and property (particularly agricultural and horticultural premises) **shall** be protected against harmful effects of heat or fire which may be generated or propagated in electrical installations. These effects include:

- failure of electrical equipment such as protective devices, switchgear, thermostats, temperature limiters, seals of cable penetrations and wiring systems;
- insulation faults or arcs, sparks and high-temperature particles;
- harmonic currents;
- heat accumulation, heat radiation, hot components or equipment;
- overcurrent;
- external influences such as lightning surge.

Note: The use of supplementary bonding does not exclude the need to disconnect the supply for other reasons, for example protection against fire, thermal stresses in equipment, etc.

Electrical heating appliances used for the breeding and rearing of livestock shall:

- comply with BS EN 60335-2-71; and
- be fixed at an appropriate distance from livestock and combustible material, to minimise any risks of burns to livestock and of fire.

 Note: See BS 7671:2018 Chapter 705 for more detailed requirements relating to agricultural and horticultural premises.

At exhibitions, shows and stands which normally use temporary electrical installations, the following apply:

- Lighting equipment and appliances with high-temperature surfaces (such as incandescent lamps, spotlights and small projectors) shall be suitably guarded and installed, and located in accordance with the relevant standard.
- Showcases and signs shall be made from material with heat-resistance, mechanical strength, electrical insulation and ventilation devices sufficient to overcome the combustibility of exhibits with high heat generation.
- Stand installations that contain sufficient electrical equipment, luminaires or lamps that are liable to generate excessive heat shall not be installed, unless adequate ventilation provisions are made (such as. a well-ventilated ceiling constructed of incombustible material).

 Note: See BS 7671:2018 Chapter 711 for more detailed requirements relating to exhibitions, shows and stands.

7.10.7.6 Protection against fire caused by electrical equipment

According to Government statistics, during 2015/16 54.4% of fires in England were caused by misuse of electrical appliances and products or electrical distribution! It is, therefore, essential that householders and businesses are aware of the potential dangers of using electrical equipment and take adequate precautions to prevent accidents and fire form happening.

The current version of the Wiring Regulations emphasises this point by stating that:

 BS 7671:2018 Requirement

"Persons, livestock and property shall be protected against harmful effects of heat or fire which may be generated or propagated in electrical installations."

It recommends that, wherever possible, every termination of a live conductor or connection or joint between live conductors is contained within an enclosure.

Further requirements are sprinkled throughout this British Standard and the following are a selection of the most important:

- Fixed electrical equipment shall be selected and erected such that its temperature in normal operation will not cause a fire.
- Heat generated by electrical equipment shall not cause danger or harmful effects to adjacent fixed material.
- Where fixed equipment may attain surface temperatures which could cause a fire hazard to adjacent materials, equipment shall:

    o   be mounted on a support which has low thermal conductance; or
    o   be within an enclosure which will withstand the surface temperatures being generated; or
    o   be screened by low-thermal-conductance materials that are capable of withstanding the heat emitted by the electrical equipment; or
    o   be mounted in a manner that allows safe dissipation of heat and at a sufficient distance from adjacent material on which such temperatures could have deleterious effects.

 Note: Any means of support shall be of low thermal conductance.

- Where high-temperature arcs, sparks or particles could be emitted by fixed equipment during their normal service, the equipment shall be:

    o   totally enclosed in arc-resistant material; or
    o   screened by arc-resistant material from materials upon which the emissions could have harmful effects; or
    o   mounted so as to allow safe extinction of emissions sufficiently far away from materials where these emissions could have harmful effects.

 Note: Arc-resistant material used for this protective measure shall be non-ignitable, of low thermal conductivity and of adequate thickness to provide mechanical stability.

- Fixed equipment that could cause a concentration and focus of heat shall be at a sufficient distance from any fixed object or building element.
- Precautions shall be taken to prevent the spread of liquid, flame and other products of combustion to electrical equipment containing a significantly high amount of flammable liquid.
- Materials used for the construction of enclosures for electrical equipment shall be capable of resisting heat and fire in accordance with an appropriate product standard.

7.10.7.7 Safety services

Safety services, such as:

- carbon monoxide (CO) detection and alarm systems;
- emergency lighting;
- essential medical systems;
- fire detection and alarm systems;
- fire evacuation systems;
- fire pumps;
- fire rescue service lifts;
- fire services communication systems;
- industrial safety systems;
- smoke ventilation systems;

by their very nature, need to be frequently regulated by statutory authorities – whose requirements are mandatory and have to be followed. For example:

> Devices designed to protect a conductor against overload should be installed so as to reduce to a minimum the risk of danger to persons.
>
> Circuits for safety services shall not pass through locations exposed to fire risk (BE2) unless they are fire resistant.
>
> Safety services may be required to operate at all material times that people or livestock are at risk, including during mains and local supply failure and through fire conditions. To meet this requirement, it is necessary to ensure that specific sources, equipment, circuits and wiring are selected, installed and used.

In order that safety services can operate in fire conditions:

- equipment shall be provided (either by construction or by erection) with protection ensuring fire-resistance of adequate duration;
- a safety source is generally additional to the normal source of supply and needs to be capable of maintaining a supply for an adequate length of time.

7.10.7.8 Protection against fault current

A device for protection against fault current need not necessarily be provided if the wiring has been installed in such a manner as to reduce to a minimum the risk of fire or danger to persons.

Where an overload protection device in a circuit supplying current-using equipment could cause danger or damage to persons, overload protection devices may be omitted. Examples of such circuits are:

- a circuit supplying a fire-extinguishing device;
- a circuit supplying a safety service, such as a fire alarm or a gas alarm.

In such situations consideration should be given to the provision of an over-load alarm.

7.10.7.9 Wiring systems

Other BS 7671:2018 requirements include:

> **BS 7671:2018 Requirement**
>
> *"The risk of spread of fire shall be minimized by the selection of appropriate materials and erection."*

- A wiring system shall be installed so that the general building structural performance and fire safety are not reduced.
- Cables complying with the requirements BS EN 60332-1-2 may be installed without any additional or special precautions.
- Cables connecting appliances to the permanent wiring system and not complying with the flame propagation requirements of BS EN 60332-1-2 will need to be limited to short lengths and shall not pass from one fire-segregated compartment to another.
- Cables shall satisfy the requirements of the Construction Projects Regulation (CPR) with respect of their reaction to fire.
- Devices that provide luminaires with protection against the risk of fire shall:

  o continuously monitor the power demand of the luminaires;
  o automatically disconnect the supply circuit within 0.3 s of a short-circuit or failure occurring which causes a power increase of more than 60 W;
  o provide automatic disconnection upon connection of the supply circuit if there is a failure which causes a power increase of more than 60 W;
  o be fail-safe devices.

- If a heating cable passes through, or is in close proximity to, material classified as a fire hazard, the cable:

  o shall be enclosed in material having the ignitability characteristic 'P' as specified in BS 476-12; and
  o shall be protected from any mechanical damage caused during installation and use;
  o if both live circuit conductors are uninsulated, they shall either:

    - be provided with a protective device; or
    - the system shall comply with BS EN 60598-2-23.

- Where the risk of flame propagation is high the cable shall meet the requirements of the appropriate part of the BS EN 60332-3 series.

7.10.7.10 Sealing of wiring system penetrations

A wiring system (such as a conduit system, cable ducting system, cable trunking system, busbar or busbar trunking system) which:

- crosses (or is near to) underground telecommunication or power cables shall have a minimum clearance of 100 mm and be provided with a fire-retardant partition between the cables;
- passes through elements of building construction (such as floors, walls, roofs, ceilings, partitions or cavity barriers) shall ensure that any openings remaining after passage of the wiring system are sealed according to the degree of fire-resistance of that particular element;
- penetrates elements of building construction that have a specified fire resistance shall be internally sealed to the same degree of fire resistance of the respective element before and after penetration;
- requires temporary sealing arrangements will ensure that this is completed during the erection of the wiring system.

It is to be ensured that any sealing damaged during alteration work is repaired as soon as is practicable.

All sealing work should resist external influences to the same degree as the wiring system with which it is used and, in addition, it shall:

- be compatible with the material of the wiring system with which it is in contact;
- be of adequate mechanical stability to withstand the stresses which may arise through damage to the support of the wiring system due to fire;
- be resistant to the products of combustion to the same extent as the elements of building construction which have been penetrated;
- permit thermal movement of the wiring system without reduction of the sealing quality;
- provide the same degree of protection from water penetration as that required for the building construction element in which it has been installed.

7.10.7.11 Firefighter's switches

A firefighter's switch shall be provided in the low-voltage circuit supplying:

- exterior electrical installations operating at a voltage exceeding low voltage; and
- interior discharge lighting installations operating at a voltage exceeding low voltage.

For single premises:

- wherever practicable, outdoor installations are to be controlled by a single firefighter's switch;
- similarly, every internal installation shall be controlled by a single firefighter's switch that is independent of the switch for any exterior installation.

For other buildings:

- the switch shall be outside the building and adjacent to the equipment for an exterior installation;
- the switch shall be in the main entrance to the building for an interior installation;
- in all cases, every firefighter's switch should:
  - o be placed in a conspicuous position that is reasonably accessible to firefighters, and be not more than 2.75 m from the ground;
  - o be easily accessible and clearly marked to indicate the installation or part of the installation which it controls.

A firefighter's switch shall:

- be coloured **red** and have fixed on (or near) it a permanent name-plate marked with the words **'FIREFIGHTER'S SWITCH'** or **'FIRE SWITCH'** in lettering not less than 10 mm high;
- be seen clearly by a person standing on the ground at the intended site, without opening the enclosure;
- have its ON and OFF positions clearly indicated by lettering not less than 10 mm high, with the OFF position at the top.

7.10.7.12 Medical locations

In the event of mains power failure in medical locations, some fire equipment will require a minimum amount of luminance. Examples of such equipment include:

- locations where central fire alarm and monitoring systems are installed;
- lifts for firefighters;
- fire detection and fire alarms;
- fire-extinguishing systems.

7.10.7.13 Inspection

Note: Inspection shall precede testing and shall normally be done with that part of the installation under inspection disconnected from the supply.

The inspection shall be made not only to verify that the installed electrical equipment is in compliance, correctly selected and erected and not visibly damaged or defective, but shall also include:

- checking (during erection) of presence of fire barriers, suitable seals and protection against thermal effects;
- verifying that, where RCDs are used for protection against fire, the conditions for protection by automatic disconnection of the supply are correct.

7.10.7.13.1 Periodic inspection and testing

Periodic inspection comprising of a detailed examination of the installation shall be carried out without dismantling.

> *Although having said that, the Regulations do permit a certain amount of 'partial dismantling' if required.*

These measures are supplemented by appropriate tests to show that the requirements for disconnection times for protective devices are complied with, to provide for:

- confirmation that the installation is not damaged or deteriorated so as to impair safety;
- protection against damage to property by fire and heat arising from an installation defect;
- the identification of installation defects and departures from the requirements of these Regulations that may give rise to danger;
- the protection of persons and livestock against the effects of electric shock and burns.

7.10.7.14 Testing

In locations exposed to fire hazard, a measurement of the insulation resistance between the live conductors should be applied.

Author's end note

Although the Wiring Regulations are primarily aimed at electrical installations in and around commercial and residential buildings, BS 7671:2018 also covers the often-overlooked need for electrical safety in a number of special installations and locations that are subject to additional requirements due to the extra dangers they pose – for example agricultural premises and mobile homes.

Chapter 8 now looks at these alternative locations and installations, and covers their additional wiring and equipment requirements.

8

Special installations and locations

Author's start note

Whilst the Regulations apply to electrical installations in all build-ings, there are also some indoor and out-of-doors special installa-tions and locations that are subject to additional requirements due to the extra dangers they pose. This chapter considers the require-ments for these special locations and installations, and whilst per-haps not being a complete list, it represents the most important requirements.

*The majority of these particular Regulations are **additional** to all of the other requirements contained in BS 7671:2018 – and are not meant as alternatives to them.*

8.1 General requirements

The following are intended to act as a reminder of the general requirements that are applicable to special installations and locations conductors (Figure 8.1).

8.1.1 Outdoor lighting

An outdoor lighting installation comprises one or more luminaires, a wiring system and accessories and includes lighting installations for:

- roads, parks, car parks, gardens, places open to the public, sporting areas, illumination of monuments and floodlighting;
- places such as telephone kiosks, bus shelters, advertising panels and town plans;
- road signs;

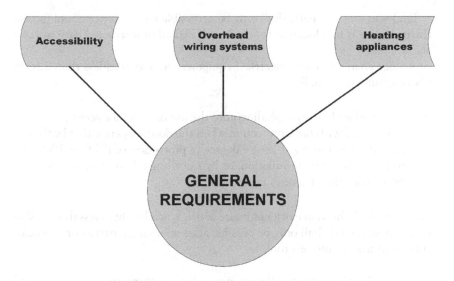

Figure 8.1 Special installations and locations – general requirements.

The following are excluded:

- temporary festoon lighting;
- owner-operated equipment;
- luminaires fitted to the outside of a building and supplied directly from the internal wiring of the building;
- road traffic signal systems.

8.1.1.1 Requirements for outdoor lighting, highway power supplies and street furniture

The protective measures of placing out of reach and obstacles shall **not** be used, except where:

- the maintenance of equipment is restricted to skilled persons who are specially trained;
- the items of street furniture are within 1.5 m of a low-voltage overhead line.

The protective measures of non-conducting location and Earth-free local equi-potential bonding shall **not** be used.

Where the protective measure automatic disconnection of supply is used, all live parts of electrical equipment shall be protected by insulation or by barriers or enclosures providing basic protection.

A door in street furniture, used for access to electrical lighting equipment, shall **not** be used as a barrier or an enclosure.

Enclosures for live parts shall only be accessible with a key or a tool (unless the enclosure is in a location to which only skilled or instructed persons have access).

A door giving access to electrical equipment and located less than 2.50 m above ground level shall:

- be locked with a key or shall require the use of a tool for access;
- be provided with basic protection when the door is open, either by the use of equipment having at least a degree of protection of IP2X or IPXXB by construction or by installation, or by installing a barrier or an enclosure giving the same degree of protection.

Access to the light source of a luminaire which is at a height of less than 2.80 m above ground level shall only be possible after removing a barrier or an enclosure requiring the use of a tool.

 For an outdoor lighting installation, a metallic structure (such as a fence, grid etc.), which is in the proximity of but not part of the outdoor lighting installation, need not be connected to the main earthing terminal.

Lighting arrangements in places such as telephone kiosks, bus shelters and town plans shall be provided with additional protection by an RCD.

A maximum disconnection time of 5 s shall apply to all circuits feeding fixed equipment used in highway power supplies.

The earthing conductor of a street electrical fixture shall have a minimum copper equivalent cross-sectional area not less than that of the supply neutral conductor at that point or not less than 6 mm², whichever is the smaller.

8.1.1.2 Double or reinforced insulation

For an outdoor lighting installation, where the protective measure for the whole installation is by double or reinforced insulation:

- no protective conductor shall be provided; and
- the conductive parts of the lighting column shall not be intentionally connected to the earthing system.

A device providing protection against the risk of fire shall meet the following requirements:

- The device shall continuously monitor the power demand of the luminaires.
- The device shall automatically disconnect the supply circuit within 0.3 s in the case of a short-circuit (or failure) which causes a power increase of more than 60 W.

- The device shall provide automatic disconnection while the supply circuit is operating with reduced power or if there is a failure which causes a power increase of more than 60 W.
- The device shall be fail-safe.

Suspension devices for extra-low-voltage luminaires, including supporting conductors, shall be capable of carrying five times the mass of the luminaires (including their lamps) intended to be supported. In all cases, the amount of support provided shall not be less than 5 kg.

8.2 Outdoor lighting installations

8.2.1 Cables

A cable passing through a joist within a floor or ceiling construction or through a ceiling support (e.g. under floorboards) shall:

- include in an earthed metallic covering; or
- be enclosed in an earthed conduit; or
- be enclosed in earthed trunking or ducting; or
- be mechanically protected against damage sufficient to prevent penetration of the cable by nails, screws etc.; or
- be at least 50 mm measured vertically from the top, or bottom as appropriate, of the joist or batten.

A cable concealed in a wall or partition at a depth of less than 50 mm from a surface of the wall or partition shall (in addition to the above requirements):

- be installed in an area within 150 mm from the top of the wall or partition or within 150 mm of an angle formed by two adjoining walls or partitions.

If the cables of an installation that is not intended to be under the supervision of a skilled or instructed person are concealed in a wall or partition (the internal construction of which includes metallic parts, other than metallic fixings such as nails, screws and the like) then it shall:

- incorporate an earthed metallic covering; or
- be enclosed in earthed conduit: or
- be enclosed in earthed trunking or ducting; or
- be mechanically protected sufficiently to avoid damage to the cable during construction of the wall or partition and during installation of the cable, or
- be provided with additional protection by means of an RCD.

Note: Where the installation is not intended to be under the supervision of a skilled or instructed person, consideration shall be given to providing additional protection by means of an RCD.

If a cable is not put in in a conduit or duct but is directly buried in the ground, then it must include a suitable protective conductor such as earthed armour or metal sheath, or both.

All overhead conductors shall be insulated.

Poles and other supports for overhead wiring shall be located or protected so that they are unlikely to be damaged by any foreseeable vehicle movement.

Overhead conductors shall be at a height above ground of not less than 6 m in all areas subjected to vehicle movement and 3.5 m in all other areas.

8.2.1.1 Underground cables

Underground distribution cables shall, unless provided with additional mechanical protection, be buried at a sufficient depth (i.e. a minimum of 0.5 m) to avoid being damaged by, for example, heavy vehicle movement.

8.2.1.2 Supplies

The nominal supply voltage to pleasure craft or houseboats is 230 V a.c. single-phase, or 400 V a.c. three-phase.

In the UK the Electricity Safety, Quality and Continuity Regulations (ESQCR) prohibit the connection of a PME earthing device to any metalwork of a boat.

8.2.1.3 Wiring systems

The following wiring systems are suitable for distribution circuits of marinas:

- underground cables;
- overhead cables or overhead insulated conductors, such as:

  o cables with copper conductors and thermoplastic or elastomeric insulation and sheath (to guard against external influences such as movement, impact, corrosion and ambient temperature); or
  o cables with armouring and thermoplastic or elastomeric insulation material;
  o mineral-insulated cables with a PVC protective covering; or
  o other cables and materials that are no less suitable than those listed above.

The following wiring systems shall **not** be used on or above a jetty, wharf, pier or pontoon:

- cables in free air suspended from or incorporating a support wire;
- non-sheathed cables in cable management systems;
- cables with aluminium conductors;
- mineral-insulated cables.

Cables need to be selected and installed so that mechanical damage due to tidal and other movement of floating structures is prevented.

Cable management systems shall be installed to allow the drainage of water by drainage holes and/or installation of the equipment on an incline.

8.2.2 Luminaires in fountains

A luminaire installed in zones 0 or 1 shall be fixed and shall comply with BS EN 60598-2-18.

Electrical equipment in zones 0 or 1 shall be provided with mechanical protection to medium severity (AG2), for example by use of mesh glass or by grids which can only be removed by the use of a tool.

An electric pump shall comply with the requirements of BS EN 60335-2-41.

8.2.2.1 Electrical equipment of fountains

Electrical equipment in zones 0 or 1 shall be provided with mechanical protection by using a mesh glass or grids which can only be removed by the use of a tool.

A luminaire installed in zones 0 or 1 shall be fixed and shall comply with BS EN 60598-2-18.

An electric pump shall comply with the requirements of BS EN 60335-2-41.

8.2.2.2 Limitation of wiring systems according to the zones

In zones 0 and 1, a wiring system shall be limited to that necessary to supply equipment situated in these zones.

8.2.3 Accessibility

In accordance with the Wiring Regulations (BS 7671:2018 Section 528.3.4):

Where an electrical service is located in close proximity to one or more non-electrical services, it shall meet the following conditions:

- the wiring system shall be suitably protected against the hazards likely to arise from the presence of the other services in normal use;
- fault protection shall be provided by means of automatic disconnection of supply.

Electrical equipment should be installed and arranged so that:

- following initial installation it is easily accessible for operation, inspection, testing, fault detection, maintenance and repair;

- connections and joints of heating appliances must be accessible for inspection, testing and maintenance, unless:

  o they are in a compound-filled or encapsulated joint;
  o the connection is between a cold tail and a heating element;
  o the joint is made by welding, soldering, brazing or a compression tool;

- live equipment is inaccessible to livestock.

8.2.4 Luminaires in temporary installations

Every luminaire and/or decorative lighting chain shall:

- have a suitable IP rating;
- be installed so as not to impair its ingress protection; and
- be securely attached to the structure or support intended to carry it.

Its weight shall not be carried by the supply cable, unless it has been selected and erected for this purpose.

Luminaires and decorative lighting chains mounted less than 2.5 m (arm's reach) above floor level or otherwise accessible to accidental contact shall be firmly fixed and so sited or guarded as to prevent risk of injury to persons or ignition of materials.

Access to the fixed light source shall only be possible after removing a barrier or an enclosure which shall require the use of a tool.

Lighting chains shall use HO5RN-F (BS 7919) cable or equivalent.

Insulation-piercing lampholders shall not be used unless the cables and lampholders are compatible and the lampholders are non-removable once fitted to the cable.

All lamps in shooting galleries and other sideshows where projectiles are used shall be suitably protected against accidental damage.

Where transportable floodlights are used, they shall be mounted so that the luminaire is inaccessible.

Supply cables shall be flexible and have adequate protection against mechanical damage.

Luminaires and floodlights shall be fixed and protected so that any focusing or concentration of heat is unlikely to cause ignition of any material.

8.2.4.1 Variables

The number and type of circuits required for lighting, heating, power, control, signalling, communication and information technology, etc. will generally depend on:

- the location and points of power demand;
- the loads to be expected on the various circuits;
- the daily and yearly variation of demand;

- any special conditions;
- requirements for control, signalling, communication and information technology, etc.

8.2.4.2 Protection against overvoltages

No additional protection against overvoltages of atmospheric origin is necessary for:

- installations that are supplied by low-voltage systems which do not contain overhead lines;
- installations that are supplied by low-voltage networks which contain overhead lines and their location is subject to less than 25 thunderstorm days per year;
- installations that contain overhead lines and their location is subject to less than 25 thunderstorm days per year;

provided that they meet the required minimum equipment impulse withstand voltage (in kV) for that particular installation (see Table 8.1).

Table 8.1 Required minimum impulse withstand voltage

Minimum impulse withstand voltage			
Category **I**	**II**	**III**	**IV**
Type of impulse voltage: Equipment with reduced impulse voltage	Equipment with normal impulse voltage	Equipment with high impulse voltage	Equipment with very high impulse voltage
Typical equipment: Sensitive home electronic devices such as home CCTV and computers etc.	Domestic appliances and tools	Distribution boards, switches and socket outlets	Energy meters and telecontrol systems
Nominal voltage of the installation			
230/240 V 1.5	2.5	4	6
400/690 V 2.5	4	6	8
1000 V 4	6	8	12

Note: Suspended cables having insulated conductors with earthed metallic coverings are considered to be *'underground cables'*.

Installations that are supplied by (or include) low-voltage overhead lines must incorporate protection against overvoltages of atmospheric origin if the location is subject to more than 25 thunderstorm days per year.

This protection must be provided either by:

- a surge protective device with a protection level not exceeding Category II; or
- other means providing at least an equivalent attenuation of overvoltages.

8.2.5 Overhead wiring systems

Whilst the only economic method of transmitting power from a grid station is by means of lines suspended from pylons, at lower voltages there is a choice between running them overhead or underground (Figure 8.2a). The supply to most domestic buildings (particularly in towns) is predominantly underground, but for electrical installations such as in agricultural buildings the most cost-effective way is via overhead cables.

The downside of this, of course, is the potential for the overhead cable to become a safety hazard, and protective methods must be used to guard against this possibility (Figure 8.2b).

8.2.5.1 Overhead power lines

Bare or insulated overhead distribution lines between buildings and structures need to be installed in accordance with the Electricity Safety, Quality and Continuity Regulations.

All overhead conductors, particularly in campsites, caravan sites and construction sites must be:

- insulated;
- protected against mechanical damage (taking into account the environment and activities of the site);
- at least 2 m away from the boundary of any caravan pitch;
- at a height not less than 6m in vehicle movement areas and 3.5 m in all other areas.

Poles and other overhead wiring supports shall be protected against any reasonably foreseeable vehicle movement.

Mains-operated electric fence controllers must:

- take account of the effects of induction when in the vicinity of overhead power lines;
- not be fixed to any supporting pole of an overhead power or telecommunication line.

With regards to safety management (and in addition to the above):

- overhead cables shall not be used over waterways;
- self-supporting suspension cables shall be installed at a height of at least 6 m.

Figure 8.2 Overhead wiring systems: (a) Power lines; (b) Urban wiring installation. (Courtesy Stingray.)

8.2.5.2 *Protection against indirect contact*

Protective measures against indirect contact may only be dispensed with if:

- overhead line insulator brackets (and metal parts connected to them) are not within arm's reach;
- the steel reinforcements of concrete poles are not accessible;
- exposed-conductive-parts (including small isolated metal parts such as bolts, rivets, nameplates and cable clips) cannot be gripped or cannot be contacted by a major surface of the human body:
  - o there is no risk of fixing screws used for non-metallic accessories coming into contact with live parts;
  - o inaccessible lengths of metal conduit do not exceed 150 mm;
  - o metal enclosures mechanically protecting equipment comply with requirements for Class II protection;
- unearthed street furniture that is supplied from an overhead line is inaccessible whilst in normal use.

8.2.6 Heating appliances

BS 7671:2018 Requirement

The equipment, system design, installation and testing of an electric surface heating system must meet the requirements of BS 6351.

Heating appliances must be fixed so as to minimise the risks of burns to livestock and of fire from combustible material.

Building Regulations Requirement

Any location that contains a forced air heating system or space heating appliance must comply with the appropriate parts of the Building Regulations.

Heating appliances **must** always be fixed.

In both sets of Regulations, the general requirements for heating appliances (e.g. water heaters, boilers, heating units, heating conductors and cables, surface and underfloor heating systems etc.) are very important for special installations and locations and the following, whilst perhaps not being a complete list, represents the most important requirements:

- Electrical heating appliances used for the breeding and rearing of livestock shall comply with BS EN 60335-2-71.

- In agricultural and horticultural premises, only electrical heating appliances with visual indication of the operating position shall be used.
- Electrical separation shall not be used for wall heating systems.
- Any circuit that supplies a heating unit needs to use an RCD so as to provide additional protection.
- Precautions shall be taken to prevent heating elements creating high temperatures to adjacent material.
- Heating units shall be connected to the electrical installation via cold tails or suitable terminals.
- Heating units shall be inseparably connected to cold tails (e.g. by a crimped connection).
- Heating units shall not cross any expansion joints of the building or structure.

Electric floor heating systems shall **not** use the protective measure '*protection by electrical separation*'.

Electric heating units that are embedded in the floor (and intended for heating the location) may be installed, provided that:

- they are protected by SELV;
- they are covered by an earthed metallic grid or sheath that is connected to local supplementary equipotential bonding.

If an electric heating unit is embedded in the floor in zone B or C of a swimming pool it must either:

- be connected to the local supplementary equipotential bonding by a metallic sheath; or
- be covered by an earthed metallic grid connected to the equipotential bonding.

Radiant heaters must be fixed not less than 0.5 m from livestock and from combustible material.

The designer (or installer) of an electrical heating system shall provide documentation for each heating system, which shall be fixed to (or next to) the distribution board of the heating system, and this documentation shall contain the following details:

- cables, earthed conductive shields etc.;
- description of the heating system and its function;
- description of the construction of the heating system, especially the installation depth of the heating units;
- layout of the heating units in the form of a sketch, drawing or picture;
- length/area of heating units;
- manufacturer and type of heating units;
- number of heating units installed;

- operation of the control equipment for the heating system in the dwelling area and the complementary heating zones, if any;
- operation of the heating installation;
- position and depth of the heating units;
- position of all junction boxes;
- product information, containing details of all approved materials in contact with the heating units (with necessary instructions for their installation);
- rated voltage;
- rated current of overcurrent protective device;
- rated resistance (cold) of heating units;
- rated residual operating current of the RCD;
- surface power density;
- the insulation resistance of the heating installation and the test voltage used;
- the type of heating units used and their maximum operating temperature.

 See BS EN 60335-2-96 for more detailed requirements relating to electric surface heating systems.

8.2.6.1 Electrode water heaters and boilers

An electrode water heater or boiler is a device which heats the water it contains, or raises steam, and can only be used on a.c. supplies; otherwise, electrolysis will occur which will break down the water into its components of hydrogen and oxygen.

When an electrode water heater or boiler:

- is directly connected to a supply exceeding low voltage, the installation shall include an RCD;
- is directly connected to a three-phase low-voltage supply, it shall be connected to the neutral of the supply as well as to the earthing conductor;
- is not piped to a water supply or is in physical contact with any earthed metal, a fuse in the line conductor may be substituted for the circuit-breaker and the shell of the electrode water heater or electrode boiler need not be connected to the neutral of the supply.

Other installation and usage requirements contained in BS 7671:2018 and the Building Regulations include:

- Electric appliances producing hot water or steam must be protected against overheating.
- If located in a bathroom, electrical appliances must be fixed and permanently wired into the wall and controlled by a pull-cord inside the bathroom or by a switch located outside.

- Heaters used for liquid or other substances must incorporate or be provided with an automatic device to prevent a dangerous rise in temperature.
- If the supply to a boiler/heater is single-phase and one electrode is connected to a neutral conductor earthed by the distributor, then the shell of that boiler/heater shall be connected to the neutral of the supply as well as to the earthing conductor.
- Metal parts (other than the current-carrying parts of single-phase water heaters and boilers) that are in contact with the water shall be solidly and metallically connected to the metal water pipe supplying that heater/boiler.
- The protective conductor has to be connected to the shell of the water heater or electrode boiler.
- The heater/boiler must be permanently connected to the electricity supply via a double-pole linked switch that is either:

  o separate from and within easy reach of the heater/boiler; or
  o part of the boiler/heater (provided that the wiring from the heater or boiler is directly connected to the switch without use of a plug and socket outlet).

- The shell of the heater/boiler must be bonded to the metallic sheath and armour (if any) of the incoming supply cable.

8.2.6.2 Forced air heating systems

Building Regulations Requirement

Locations that contain forced air heating systems and appliances that produce hot water or steam, and space heating appliances, must all comply with the appropriate parts of the Building Regulations.

There are also various requirements for forced air heating systems to be found in BS 7671:2018, but the main ones concern:

- that two, independent, temperature-limiting devices must be made available;
- electric heating elements of forced air heating systems (other than those of central-storage heaters) should:

  o not be capable of being activated until the prescribed airflow has been established;
  o deactivate when the airflow is reduced or stopped;

- frames and enclosures of electric heating elements must be of non-ignitable material.

8.2.6.3 *Heating conductors and cables*

Heating cables passing through (or in close proximity to) a fire hazard:

- must be enclosed in material with an ignitability characteristic '13' as specified in BS 476 Part 12;
- must be protected from any mechanical damage.

Heating cables that are going to be laid (directly) in soil, concrete, cement screed, or other material used for road and building construction must be:

- capable of withstanding mechanical damage;
- resistant to damage from dampness or corrosion.

A heating cables that is going to be laid (directly) in soil, a roadway, or the structure of a building must be installed so that it is:

- completely embedded in the substance it is intended to heat;
- not damaged by movement, either by it, or the substance in which it is embedded;
- compliant (in all respects) with the maker's instructions and recommendations.

8.2.6.4 *Locations with risks of fire due to heat storage equipment enclosures (such as heaters)*

These must not attain higher surface temperatures than:

- 90°C under normal conditions; and
- 115°C under fault conditions.

Heat storage appliances must not ignite combustible dust or fibres.

Heating appliances mounted close to combustible materials must be protected by barriers.

Where heating and ventilation systems containing heating elements are installed:

- the dust or fibre content and the temperature of the air must not present a fire hazard;
- temperature-limiting devices must have a manual reset.

8.3 Special installations and locations

In addition to the normal safety protection methods against direct and indirect contact listed in other parts of the Regulations, special installations and locations such as:

- agricultural and horticultural premises;
- conducting locations with restricted movement;
- construction and demolition sites;
- electrical installations in caravans, caravan parks and camping parks;
- electric vehicle charging installations;
- exhibitions, shows and stands;
- floor and ceiling heating systems;
- locations containing a bath or shower;
- marinas and similar locations;
- medical locations;
- mobile and transportable units;
- operating and maintenance gangways;
- rooms and cabins containing saunas;
- solar, photovoltaic (PV) power supply systems;
- swimming pools and other basins;
- temporary electrical installations;

must also comply with the requirements for safety protection in respect of:

- electric shock;
- thermal effects;
- overcurrent;
- undervoltage;
- isolation and switching.

These requirements are described in more detail below.

8.3.1 Agricultural and horticultural premises

In contrast to normal domestic installations, an agricultural installation is usually prone to damp conditions and so contact with Earth will be better and people and animals are more liable to electric shock. For animals (whose body resistance is much lower than humans) this situation is worsened as their contact with Earth will be greater and so even a small voltage could prove lethal to them. Animals can also cause a lot of physical damage to electrical installations and animal effluents present a greater risk of corrosion.

Horticultural installations are also subject to the same wet/high Earth contact conditions, and for these reasons special requirements have been introduced for all agricultural or horticultural installation (outdoors and indoors) and for locations where livestock is kept (e.g. cow sheds, stables, chicken houses, piggeries etc.) plus food processing stations and storage areas for hay, straw and fertilisers.

8.3.1.1 *General*

As the possibility of animals unintentionally coming in direct contact with a live installation is greater than for humans (they obviously cannot read the warning notices!), the following protective methods have to be used.

 In agricultural and horticultural premises, a TN-C system shall **not** be used.

8.3.1.1.1 Accessibility by livestock

It is a mandatory requirement of the Regulations that in locations intended for livestock, normal protective measures used in other locations (like placing obstacles out of reach) shall **not** be used. Instead, supplementary bonding needs to connect all exposed-conductive-parts and extraneous-conductive-parts that can be touched by livestock; and precautions need to be taken to ensure that:

- all electrical equipment shall, generally speaking, be inaccessible to livestock;
- equipment that may, unavoidably, be accessible to livestock (such as equipment for feeding and basins for watering) shall be designed, constructed and installed to minimise the risk of injury to livestock;
- all live parts are inside barriers or enclosures and suitable warning signs will be made available;
- barriers and/or enclosures are secured by a bolt or a key.

To ensure that these precautions are being obeyed, it is a mandatory requirement that all fixed agricultural and horticultural installations (outdoors and indoors), locations where livestock is kept (such as stables, chicken houses, piggeries), and feed-processing locations, lofts and storage areas for hay, straw and fertilizers shall be inspected on a regular basis, and that results of these inspections are fully recorded.

8.3.1.2 *Protective measures*

8.3.1.2.1   Identification

The following documentation shall be provided to the user of the installation:

- a plan indicating the location of all electrical equipment;
- a single-line distribution diagram;
- the routing of all concealed cables;
- an equipotential bonding diagram showing the locations of all bonding connections.

8.3.1.2.2 Electric fence controllers

Electric fencing systems have been developed to stop the free movement of animals across pastures. They are semi-permanent solutions that can be extended, altered, or removed to allow grazing to be divided up and/or protected from livestock access.

The system consists of plastic or wooden posts, insulators, conductive wire/rope/tape and an electrical energiser unit. This unit can be mains, battery or solar powered and it will send short electrical impulses along a conductive wire (tape, rope, etc.) so that when the conductive wire or fence is touched by an animal the current passes through it to the ground and causes the animal to feel a shock. The shock is sufficient to alarm an animal but not to harm it, and in time the animal will learn to stay away from the fence.

For specific requirements regarding electric fence installations see BS EN 60335-2-BS 7671:2008 and BS EN 6100-1.

8.3.1.2.3 Protection against external influences

In agricultural and/or horticultural premises:

- only waterproofed electrical equipment (to an IP44 standard) shall be used;
- socket outlets shall be installed so that they do not to come into contact with combustible material;

Note: This requirement does not apply to residential locations, offices, shops etc. belonging to agricultural and horticultural premises and where socket outlets need to comply with BS 1363-2 or BS 546.

- electrical equipment shall be protected against corrosive substances (e.g. in dairies or cattle sheds).

8.3.1.2.4 Protection against electric shock

The Regulations specifically state that the protective measures of:

- obstacles and placing out of reach; and
- non-conducting location and Earth-free local equipotential bonding

are **not** permitted.

8.3.1.2.5 Protection against fire

Fire is a particular hazard in agricultural premises as there are normally large quantities of straw and other flammable material stored in these locations so strict precautions must be taken. For example:

- electrical heating appliances used for the breeding and rearing of livestock shall be installed (in accordance with BS EN 60335-2-71) in order to minimise any risks of burns to livestock and of fire;

Note: For radiant heaters the clearance shall be not less than 0.5 m.
For fire protection purposes, RCDs:

- shall be installed with a rated residual operating current not exceeding 300 mA;
- shall disconnect all live conductors; and
- in locations where a fire risk exists conductors of circuits supplied at extra-low voltage shall be protected:

  o either by barriers or enclosures; or
  o in addition to their basic insulation, by an enclosure of insulating material.

8.3.1.2.6 Supplementary equipotential bonding

In locations intended for livestock, supplementary bonding shall connect all exposed-conductive-parts and extraneous-conductive-parts that can be touched by livestock.

Nonessential conductive-parts in, or on, the floor (such as concrete reinforcement in cellars used to hold liquid manure) shall be connected to the supplementary equipotential bonding.

Protective bonding conductors shall be protected against mechanical damage and corrosion, and electrolytic effects.

 Note: It is recommended that:

- spaced floors made of prefabricated concrete elements are part of the supplementary equipotential bonding;
- the supplementary equipotential bonding and the metal grid (if any) shall be erected so that it is protected against mechanical stresses and corrosion.

8.3.1.3 Selection and erection of electrical equipment

The selection of equipment is usually the responsibility of the designer, but most of the time it is the installer who actually selects the equipment and so it is important that this is done in accordance with the latest edition of the appropriate wiring rules and regulations. In addition, the installer must ensure that:

- the electrical equipment fully complies with the relevant product standards;
- on completion of all electrical installations, an initial inspection of the equipment and installation is carried out;
- periodic checks are completed to an agreed work instruction and time schedule.

BS 7671:2018 Requirement

Only electrical heating appliances with visual indication of the operating position shall be used.

8.3.1.4 Supplies

In special installations and locations, whenever SELV or PELV is used (regardless of values of the nominal voltage) protection **must** be provided by either:

- basic insulation; or
- barriers.

 A TN-C system shall **not** be used.

8.3.1.4.1 Automatic disconnection of supply

Automatic disconnection of supply (ADS, previously known as earthed equipotential bonding and automatic disconnection of supply (EEBADS)) is the most widely used method for achieving protection against electric shock in an installation. It works by limiting the size and duration of any voltages between exposed-conductive-parts and extraneous-conductive-parts or Earth.

In circuits, whatever the type of earthing system, the following disconnection device shall be provided:

- in final circuits supplying socket outlets with rated current not exceeding 32 A, an RCD with a rated residual operating current not exceeding 30 mA;
- In final circuits supplying socket outlets with rated current more than 32 A, an RCD with a rated residual operating current not exceeding 100 mA;
- in all other circuits, RCDs with a rated residual operating current not exceeding 300 mA.

8.3.1.4.2 Isolation and switching

Isolator switches are designed to isolate an electrical circuit from its power source. Such a switch does not have any interruption capacity, nor does it have any making or breaking capacity and is only intended to be operated on open circuit. That means isolator switches must be closed on open circuit and then other means or devices (e.g. a circuit-breaker (CB)) to energise line or isolator switches must be opened after opening upstream a device like a CB to de-energise line.

 Isolator switches are purely mechanical devices which can conduct normal operational currents as well as under short-circuit current for a defined period, and the Regulations specifically state that:

- a single isolation device shall be provided for each building or, if necessary, part of a building;
- circuits used occasionally (e.g. during harvest time) shall have isolation means for all live conductors, including the neutral conductor;
- all isolation devices shall be clearly marked according to the part of the installation to which they belong;
- isolation and switching and devices for emergency stopping or emergency switching shall not be erected where they are accessible to livestock or in any position where access may be impeded by livestock.

8.3.1.4.3 Safety services

Farmers and farm workers can easily be injured by livestock such as cattle, pigs, horses, sheep, dogs and other farm animals as they can be highly disordered. Therefore, it is recommended that farmers follow the following guidelines:

- Animals, being unpredictable, especially during the joining (mating) season, should be treated with caution at all times.
- Make sure yards, sheds and equipment are in good repair.
- Keep the walkways and laneways dry and non-slip wherever possible.
- Ensure that workers are appropriately trained and familiar with the temperament of the animals on your farm.
- Always wear suitable protective clothing (trousers, boots) and use appropriate animal-handling facilities and aids such as cradles and crushes.
- Keep all equipment in good repair, gates moving and hung, latches working and hinges greased.
- Take care when visiting other people's farms or saleyards.

With regard to electrical safety:

- Where the supply of food, water, air and/or lighting to livestock is not guaranteed in the event of power supply failure:

  o a secure source of supply should always be provided (such as an alternative or back-up supply); and
  o separate final circuits for ventilation and lighting units shall be provided.

- Where electrically powered ventilation is necessary in an installation, one of the following needs to be provided:

  o a standby electrical source ensuring sufficient supply for ventilation equipment; or
  o temperature and supply voltage monitoring.

 Note: A notice should be placed adjacent to the standby electrical source, indicating that it should be tested periodically according to the manufacturer's instructions.

8.3.1.5 Wiring systems

8.3.1.5.1   Selection and erection of wiring systems

All wiring systems must be erected so that they are inaccessible to livestock or suitably protected against mechanical damage caused by animal movement.

Overhead lines should always be insulated.

In areas of agricultural premises where vehicles and mobile agricultural machines are being used, the following methods of installation need to be employed:

* cables shall be buried in the ground at a depth of at least 0.6 m with added mechanical protection;
* cables in arable or cultivated ground shall be buried at a depth of at least 1 m;
* self-supporting suspension cables shall be installed at a height of at least 6 m.

 Remember when installing wiring systems that agricultural premises will always have many different kinds of fauna (e.g. rodents) present, in addition to your own livestock, and wiring systems need to suitably installed against damage from these!

8.3.1.5.2 Conduit systems, cable trunking systems and cable ducting systems

For locations where livestock is kept, all conduits shall have protection against corrosion of at least Class 2 (medium) for indoor use and Class 4 (high protection) outdoors according to BS EN 61386-21.

For locations where the wiring system may be exposed to impact and mechanical shock due to vehicles and mobile agricultural machines, etc.:

* conduits shall provide a degree of protection against impact of 5 J according to BS EN 61386-2;
* cable trunking and ducting systems shall provide a degree of protection against impact of 5 J according to BS EN 50085-2-1.

8.3.1.5.3 Socket outlets

Socket outlets used in agricultural and horticultural premises shall comply with:

* BS EN 60309-1; or
* BS EN 60309-2 (when interchangeability is required); or
* BS 1363, BS 546 (provided the rated current does not exceed 20 A).

 See BS 7671:2018 Section 705 for more detailed requirements relating to agricultural or horticultural installations.

8.3.2 Conducting locations with restricted movement

A restrictive conductive location is one in which the surroundings consist mainly of metallic or conductive parts, an example being a large metal container or boiler. People employed inside these locations (e.g. a person working inside the boiler whilst using an electric drill or grinder) would have their freedom of movement physically restrained and a large proportion of their body would be in contact with the sides of that location and, therefore, prone to shock hazards.

The particular requirements of this section apply to:

- fixed equipment in conducting locations where movement of persons is restricted by the location; and
- supplies for mobile equipment for use in such locations.

8.3.2.1 Protection against electric shock

 The protective measures of obstacles and placing out of reach are **not** permitted, as this sort of protection is reserved for locations to which only skilled or instructed persons have access.

In most installations, access to earthed material (e.g. metal) as well as a conductive surface is fairly limited and the contact area of the body is generally pretty small, and so the probability of receiving a fatal electric shock is relatively small.

In a conducting location with restricted movement, however, the situation is quite different because there will be little opportunity to move away from the conductive surface and, as the body's contact resistance is reduced as a consequence, the actual contact surface will increase.

The body's resistance to electrical contact is often lowered further due to perspiration caused by being in a confined location or being subjected to certain environmental conditions. This will enable a higher current through the body, and so the risk of ventricular fibrillation and the possibility of having a heart attack are increased.

As a result of the additional risk associated with agricultural and horticultural premises, they are considered to be a special location with additional risk of electric shock. So in a conducting location with restricted movement, the following protective measures apply to circuits supplying the following current-using equipment:

- For the supply to a hand-held tool or an item of mobile equipment:

  o electrical separation;
  o SELV.

- For the supply to handlamps:

  o SELV.

- For the supply to fixed equipment:

    o   automatic disconnection of the supply with supplementary equipo-
         tential bonding; or
    o   electrical separation; or
    o   SELV; or
    o   PELV.

If a functional Earth is required for certain equipment (e.g. measuring and control equipment):

- live parts shall be completely covered with insulation which can only be removed by destruction;
- supplementary equipotential bonding shall be provided between all exposed-conductive- and extraneous-conductive-parts inside a conducting location (particularly if it has restricted movement) and the functional Earth;
- the unearthed source shall have simple separation and shall be positioned **outside** any conducting location with restricted movement (unless the source is part of the fixed installation within that particular conducting location);
- the protective measures of obstacles, placing out of reach, non-conducting location, Earth-free local equipotential bonding (or electrical separation for the supply to more than one item of current-using equipment) are not permitted and shall **not** be used;
- where more than one voltage is in use, plugs and sockets must be non-interchangeable to prevent misconnection.

These requirements do **not** apply to:

- construction site offices, cloakrooms, meeting rooms, canteens, restaurants, dormitories, toilets etc.;
- installations covered by BS 6907.

For any circuit supplying one or more socket outlets with a rated current exceeding 32 A, an RCD having a rated residual operating current not exceeding 500 mA shall be provided.

8.3.2.2 Supplies

As a protective measure in agricultural and horticultural premises, the unearthed source shall have simple separation and shall be situated outside the conducting location.

Whatever the nominal voltage, where SELV or PELV is used:

- live parts shall be completely covered with insulation; or
- basic protection shall be provided by barriers or enclosures.

8.3.2.2.1 Automatic disconnection of supply

If a functional Earth is required for certain equipment (e.g. measuring and control equipment), equipotential bonding shall be provided between all exposed-conductive-parts (and any extraneous-conductive-parts that are inside the conducting location) and the functional Earth.

 See BS 7671:2018 Section 706 for more detailed requirements relating to extra-construction and demolition sites.

 Electrical installations at construction sites are primarily there so as to provide lighting and power for enabling work to proceed. As workmen will probably be working ankle-deep in wet, muddy conditions and using a selection of portable tools such as drills and grinders, they will be particularly susceptible to electric shock. To avoid this happening, a PME earthing facility shall **not** be used for earthing an installation, and all extraneous-conductive-parts must be reliably connected to the main earthing terminal.

8.3.2.3 Selection and erection of equipment

All equipment used for the distribution of electricity on construction and demolition sites shall meet the requirements of BS EN 61439-4.

A plug or socket outlet with a rated current equal to or greater than 16 A shall comply with the requirements of BS EN 60309-2.

8.3.2.4 Supplies

Supplies will normally be obtained from the local electrical supply company but remote sites could need an IT supply (such as a generator). If this is the case, then care must be taken in complying with the safety requirements for this particular source.

Equipment shall be identified by (and be compatible with) the particular supply from which it is energized, and shall contain only components connected to one and the same installation, except for control or signalling circuits and inputs from standby supplies.

8.3.2.4.1 Automatic disconnection of supply

Circuits supplying one or more socket outlets with a rated current exceeding 32 A shall be provided with an RCD having a rated residual operating current not exceeding 500 mA.

A PME earthing facility should **not** be used for the supply to a construction site, unless all extraneous parts are reliably connected to the main earthing terminal.

8.3.2.4.2 Isolation devices

Each assembly for construction sites (ACS) must include suitable devices for the switching and isolating of the incoming supply. These devices shall be capable of being permanently secured in the OFF position (e.g. by padlock or locating the device inside a lockable enclosure).

Current-using equipment shall be supplied by ACSs comprising:

- overcurrent protective devices; and
- devices affording fault protection; and
- socket outlets, if required.

8.3.2.5 Wiring systems

'Construction wiring' refers to wiring systems installed in order to provide an electrical supply for construction and demolition work and is **not** intended to form part of the permanent electrical installation. The term includes:

- consumer mains and sub-mains supplying site switchboards;
- sub-mains to site facilities in which electricity is used (e.g. sheds, amenities or transportable structures);
- final sub-circuits connected at circuit-breakers on a site switchboard, supplying plant, construction equipment (e.g. temporary construction lighting, auxiliary socket-outlet panels, hoists, cranes, and personnel lifts).

Construction wiring does not include flexible extension cords or flexible cables used to connect portable plug-in electrical equipment or luminaries to a socket outlet.

The Regulation's requirements include that:

- cables shall not be installed across a site road or a walkway unless they are adequately protected against mechanical damage;
- for reduced low-voltage systems, low-temperature 300/500 V thermoplastic or equivalent flexible cables shall be used;
- for applications exceeding reduced low voltage, flexible cable shall be HO7RN-F (BS 7919) type or equivalent having 450/750 V rating and resistant to abrasion and water;
- a wiring system buried in a floor shall be sufficiently protected to prevent damage caused by the intended use of the floor.

 See BS 7671:2018 Section 704 for more detailed requirements relating to construction and demolition site installations.

8.3.3 Electrical installations in caravans, caravan parks and camping parks

In order not to mix regulations on different subjects, such as those for the electrical installation of caravan and camping parks with those for the actual electrical installation inside a caravan or campervan, two sections have been created within BS 7671:2018, namely:

- Section 708 – electrical installations within caravan parks, camping parks and similar locations;
- Section 721 – electrical installations inside caravans and motor caravans.

8.3.3.1 Electrical installations in caravan parks, camping parks and similar locations

In a caravan park or camping park, special consideration is given to the:

- protection of people – due to the fact that the human body may be in contact with Earth potential;
- protection of wiring – due to tent pegs or ground anchors; and
- movement of heavy or high vehicles.

The actual Requirements for supplying electricity to leisure accommodation vehicles in caravan/camping parks and similar locations are as follows.

8.3.3.1.1 General

Cables that are run below caravan pitches must be provided with additional protection, as shown in Figure 8.3.

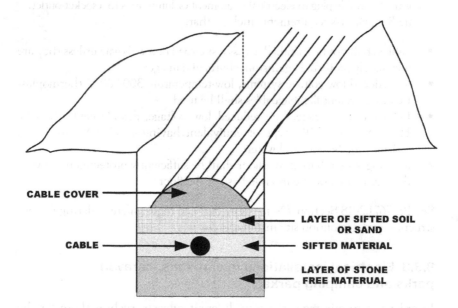

Figure 8.3 Cable covers.

8.3.3.1.2 Protective measures

A campsite or caravan pitch can be a dangerous place unless certain actions are taken, such as:

- all socket outlets (and final circuits providing a fixed connection of a supply to a mobile home or a residential park home) are protected by an RCD complying with BS 4293, BS EN 61008-1 or BS EN 61009-1, with a 30 mA rating, either individually or in groups not exceeding three sockets;
- equipment installed in a caravan pitch or campsite is protected against mechanical damage (see BS EN 62262 for more detailed information);
- relying on protective measures such as obstacles, placing out of reach, non-conducting location, and Earth-free local equipotential bonding are (in accordance with the Regulations) **not** allowed.

8.3.3.1.3 Selection and erection of equipment

The Regulations clearly state that:

- equipment installed in a campsite has to be protected against mechanical damage (see BS EN 62262 for more detailed information);
- electrical equipment that is going to be installed in an outside area must ensure that adequate precautions must be taken against:
    - o   mechanical stress;
    - o   the presence of water;
    - o   the presence of foreign solid bodies;
- the preferred method of supply to caravan pitch or tent pitch electrical equipment is via underground distribution circuits;
- an underground distribution circuit, unless provided with additional mechanical protection, must be buried at a depth of at least 0.6 m in order to avoid being damaged by tent pegs, ground anchors or by the movement of vehicles.

 For conduit systems buried underground, see BS EN 61386-24.

8.3.3.1.4 Supplies

The nominal supply voltage to leisure accommodation vehicles shall not exceed 230 V a.c. single-phase, 400 V a.c. three-phase or 48 V d.c.

The requirements of the Electricity Supply Regulations do not allow the supply neutral to be connected to any metalwork from (or in) a caravan (which means that only TT or TN-S systems may be used), and in general, the supply of electrical energy to caravan and tent sites must ensure that:

- wherever possible the supply is via underground cables;
- overhead supplies use insulated (as opposed to bare) cables;

- cables that are installed outside the area of the caravan pitch are at least 3.5 m above ground level (increased to 6 m where vehicle movements are possible);
- each socket has its own individual overcurrent protection in the form of a fuse or circuit-breaker.

The equipment supplying a caravan pitch needs to be located:

- adjacent to that pitch; and
- not more than 20 m from the connection facility to a leisure accommodation vehicle or tent when on its pitch.

In order to avoid the supply cable crossing a pitch other than the one intended to be supplied, no more than four socket outlets should be grouped in one location.

8.3.3.1.5 Wiring systems

A wiring system such as a conduit system, cable ducting system, cable trunking system, busbar or busbar trunking system which penetrates elements of building construction such as a leisure home which has specified fire resistance shall be internally sealed in accordance with the fire resistance of the building before penetration, in addition to being externally sealed.

8.3.3.1.5.1 Busbar trunking systems and powertrack systems Where a busbar trunking system is used as a protective conductor, it shall satisfy the following requirements:

- its electrical continuity shall be assured;
- it shall be protected against mechanical, chemical or electrochemical deterioration either by construction or by suitable connection;
- it shall permit the connection of other protective conductors at every predetermined tap-off point.

Other requirements include that:

- a busbar trunking systems shall comply with BS EN 60439-6; and
- a powertrack system shall comply with the appropriate part of the BS EN 61534 series.

8.3.3.1.5.2 Overhead conductors All overhead conductors shall:

- be insulated;
- be not less than 6 m above ground level in all areas subject to vehicle movement and 3.5 m in all other areas;

- ensure that all poles and other supports used to support overhead wiring are located and protected against any possible unforeseen vehicle movement.

8.3.3.1.5.3 Socket outlets   Socket outlets and their enclosures that are part of a caravan electricity supply equipment shall be protected against solid objects bigger than 1 mm, and against unavoidable water splashing.
  Socket outlets shall:

- be individually protected individually by:

  o   an RCD;
  o   an overcurrent protective device;

- be placed at a height of 0.5–1.5 m from the ground, measured to the lowest part of the socket outlet.

 In special cases caused by environmental conditions (e.g. risk of flooding or heavy snowfall), the maximum height is permitted to exceed 1.5 m.

- have a current rating not less than 16 A; and
- at least one socket-outlet shall be provided for each caravan pitch.

8.3.3.1.5.4 Underground distribution circuits   Underground cables (unless provided with additional mechanical protection) shall be buried at a depth of at least 0.6 m to avoid being damaged by tent pegs, ground anchors or by the movement of vehicles.

8.3.3.2 *Caravans and motor caravans*

Caravans and motor caravans are designed as leisure accommodation vehicles which are either towed (e.g. by a car) or self-propelled to a caravan site. They will often contain a bath or a shower, and special requirements for such installations will apply. In addition to the normal dangers associated with fixed electrical installations, there is the potential hazard of totally unskilled people moving the caravan/motor caravan, connecting and disconnecting the mains supply, and not ensuring that it is correctly earthed.

 Note: Some of the information contained in this section is a repeat of what has been discussed in the previous section (i.e. 8.3.3.1 Electrical installations in caravan parks and camping parks). This has been done deliberately, so that you won't have to constantly flick back and forth through the book to find the correct information.

8.3.3.2.1 General

The Regulations include a number of special requirements for caravans and mobile homes (as listed below), but these do not deal with:

- electrical circuits and equipment covered by the Road Vehicles Lighting Regulations 1989;
- installations covered by BS EN 1648-1 and BS EN 1648-2.

A mobile home is defined as a *"'transportable leisure accommodation vehicle'* that does not meet the requirements for use as a road vehicle", and is usually a permanent fixture on a caravan park. They normally have recognised power supplies and earthing, and their internal electrical installation is outside the scope of the Regulations. Nevertheless, potential safety hazards still have to be considered.

 Note: Caravans that are used as mobile workshops will also be subject to the requirements of the Electricity at Work Regulations 1989, and locations containing baths or showers for medical treatment, or for disabled persons, may have additional special requirements.

8.3.3.2.2 Protective measures

 The protective measures of:

- electrical separation (with the exception of shaver socket outlets);
- obstacles and placing out of reach;
- non-conducting location and Earth-free local equipotential bonding;

are **not** permitted.

Also, the Electricity Safety, Quality and Continuity Regulations (ESQCR) prohibit the connection of a PME earthing facility to any metalwork in or on a leisure accommodation vehicle (including a caravan).

8.3.3.2.2.1 Protection against overcurrent – final circuits   Each final circuit shall be protected by an overcurrent protective device which disconnects all live conductors of that circuit.

8.3.3.2.2.2 Protective conductors   The continuity of conductors and connections to unprotected or superfluous conductive parts must be confirmed by measuring the resistance of:

- protective conductors;
- protective bonding conductors;
- live conductors (in the case of ring circuits).

All protective conductors shall be incorporated in a multi-core cable or in a conduit together with the live conductors.

8.3.3.2.2.3 Protective equipotential bonding Structural metallic parts which are accessible from within the caravan shall be connected through main protective bonding conductors to the main earthing terminal within the caravan.

8.3.3.2.2.4 Protective measure: Electrical separation The protective measure of electrical separation shall not be used, except for a shaver socket outlet.

8.3.3.2.2.5 Protective measure – automatic disconnection of supply Where protection by automatic disconnection of supply is used, an RCD shall be provided and the wiring system shall include a circuit protective conductor which shall be connected to:

- the protective contact of the inlet; and
- the exposed-conductive-parts of the electrical equipment; and
- the protective contacts of the socket outlets.

8.3.3.2.3 Selection and erection of equipment

The means of connection to the caravan pitch socket outlet shall be provided from the caravan and shall comprise the following:

- a plug complying with BS EN 60309-2; and
- a flexible cord or cable of length 25 m (±2 m), incorporating a protective conductor.

Accessories that are exposed to the effects of moisture it shall be constructed (or enclosed) so as to provide protection from water splashing coming from all directions.

8.3.3.2.4 Supplies

8.3.3.2.4.1 Electrical inlets The nominal supply system voltage is chosen from BS EN 60038 which states that:

- the a.c. supply voltage of the installation of the caravan shall not exceed 230 V single-phase, or 400 V three-phase;
- the d.c. supply voltage of the installation of the caravan shall not exceed 48 V; and
- if there is more than one electrically independent installation, installations must be segregated and have a separate connecting device.

The a.c. electrical inlet on a caravan must comply with BS EN 60309-1 and, if interchangeability is required, the inlet needs to comply with BS EN 60309-2.
    The inlet shall:

- not be more than 1.8 m above ground level; and
- be located in a readily accessible position; and

- be protected against solid objects that are bigger than 1 mm, and from water splashing from all directions (with or without a connector engaged); and
- not protrude significantly beyond the body of the caravan.

8.3.3.2.4.2 Isolation   Each installation shall be provided with a main disconnector that will disconnect all live conductors.

If an installation consists of just one final circuit, then the isolating switch may be the overcurrent protective device.

A notice shall be permanently fixed near the main isolating switch inside the caravan showing what checks have to be completed before connecting and disconnecting the caravan installation from the mains supply (see Figure 8.721 on page 308 of BS 7671:2018).

Annex A721 to BS 7671:2018 provides guidance for extra-low-voltage d.c. installations.

8.3.3.2.4.3 Socket outlets   Socket outlets supplied at extra-low voltage shall have their voltage visibly marked.

8.3.3.2.5 Wiring systems

8.3.3.2.5.1 Electrical installations in caravans and motor caravans   All wiring shall be protected against mechanical damage (e.g. from vibration) either by location or by enhanced mechanical protection.

Wiring passing through metalwork shall be protected by bushes or grommets, and securely fixed in position.

Precautions shall be taken to avoid mechanical damage due to sharp edges or abrasive parts.

All cables (unless enclosed in rigid conduit) and all flexible conduit shall be supported at intervals not exceeding 0.4 m for vertical runs and 0.25 m for horizontal runs.

The nominal supply voltages for caravans shall (in accordance with BS EN 60038) be as follows:

- a.c. supplies: 230 V single-phase or 400 V three-phase;
- d.c. supplies: 48 V.

BS 7671:2018 also states that:

- structural metal parts within the caravan shall be connected via bonding conductors to the main earthing terminal within the caravan;
- electrical separation shall not be used except for the shaver socket.

The use of a PME earthing facility for earthing a caravan is prohibited by The Electricity Safety, Quality and Continuity Regulations (ESQCR).

It is recommended that at least one socket outlet is provided for each caravan pitch, on the proviso that each socket outlet and its enclosure complies with BS EN 60309-2 and meets the degree of protection of at least IP44 in accordance with BS EN 60529.

Each socket outlet shall:

- be provided with individual overcurrent protection;
- be protected individually by an RCD;
- have a current rating not less than 16 A.

The socket outlets shall be placed at a height of 0.5–1.5 m from the ground, measured to the lowest part of the socket outlet. In special cases (due to environmental conditions such as risk of flooding or heavy snowfall) the maximum height is permitted to exceed 1.5 m.

Socket-outlet protective conductors shall **not** be connected to any PEN conductor of the electricity supply.

8.3.3.2.5.2 *Types of wiring system*   The wiring systems shall be installed using one or more of the following:

- insulated single-core cables, with flexible class 5 conductors, in non-metallic conduit;
- insulated single-core cables, with stranded class 2 conductors (minimum of seven strands), in non-metallic conduit;
- sheathed flexible cables.

All cables shall meet the requirements of BS EN 60332-1-2.
   Non-metallic conduits shall comply with BS EN 61386-21.
   Cable management systems shall comply with BS EN 61386.

8.3.3.2.5.3 *Identification*   Instructions shall be provided with the caravan for its safe use. These instructions shall comprise (as a minimum):

- the importance of safety within the caravan;
- instructions on the types of appliances that can be used and from what source of supply;
- a description of the RCD(s) and their function;
- a description of the main isolating switch and its function;
- the use of the test button(s);
- instructions on the maintenance and recharging of an auxiliary battery if fitted.

 Note: The text for these instructions can be found in Figure 8.721 on page 308 of BS 7671:2018.

8.3.3.2.5.4 Proximity of wiring systems to other services   Cables of low-voltage systems shall be run separately from the cables of extra-low-voltage systems, so that there is no risk of physical contact between the two wiring systems.

ELV cables and electrical equipment may only be installed within the LPG cylinder compartment **if** the installation indicates that the gas cylinder is for use within the compartment (e.g. indication of empty gas cylinders).

Such electrical installations and components shall be constructed and installed so that they are not a potential source of ignition.

Where cables have to run through such a compartment, they shall be protected against mechanical damage by installation within a conduit system or within a duct passing through the compartment.

8.3.3.2.5.5 Luminaires in caravans and motor caravans   Each luminaire in a caravan shall preferably be fixed directly to the structure or lining of the caravan.

Where a pendant luminaire is installed in a caravan, it shall be made secure so as to prevent damage when the caravan is in motion.

A luminaire intended for dual voltage operation shall comply with the appropriate standard.

Accessories for the suspension of pendant luminaires shall be chosen with respect to the mass suspended and the forces associated with vehicle movement.

See the following Sections of BS 7671:2018:

- Section 708 for more detailed requirements relating to electrical installations within caravan parks, camping parks and similar locations; and
- Section 721 for more detailed requirements relating to electrical installations inside caravans and motor caravans.

8.3.4 Electric vehicle charging installations

With the increased number of all-electric or hybrid-electric cars currently being manufactured, the 2018 edition of BS 7671 now contains a new Section (722) to cover the requirements for circuits intended to supply electric vehicles for charging purposes.

This current Section 8.2.5 of my book is intended to provide the reader with a brief background overview of what, where and how electric vehicle charging installations are being provided in the UK. The Requirements of this section apply to circuits intended to supply electric vehicles for charging purposes, which can either come from an EV (electric vehicle) charging point or a dedicated home charging point.

Plug-in hybrid electric vehicles have both an electric motor and a conventional petrol or diesel engine. Compared to a battery-driven electric vehicle, this extends the total driving range but obviously lowers the all-electric range.

In 2019, nearly 38,000 pure electric cars were registered in the UK, a rise of 144 per cent on 2018's 15,500. It is, however, still only 1.6 per cent of a market of 2.3 million cars – but what about the future of electric cars?

8.3.4.1 General

The UK Government (similar to its French counterpart) has announced in recent months that all sales of petrol and diesel cars will cease by 2040, and a number of car manufacturers have already stated that they will produce only electric or hybrid vehicles from 2019.

Although many EV charge points are free to use, the majority of fast and rapid chargers require payment. Charging tariffs tend to comprise a flat connection fee, a cost per charging time (pence per hour) and/or a cost per energy consumed (pence per kWh).

You can also charge an electric car at home by using a dedicated home charging point (or a standard three-pin plug with an EVSE cable as a last resort), and most electric car drivers choose a home charging point in order to benefit from faster charging speeds and built-in safety features. In fact, charging an electric car is almost like charging a mobile phone – you plug it in overnight and top up during the day.

8.3.4.1.1 What happens if your electric car runs out of battery?

It's inadvisable to run out of electricity in your electric car and manufacturers warn against this, as it can cause damage to the battery. Running out of battery, or deep discharging as it's otherwise known, will deteriorate the battery cells and reduce their performance in the long run.

8.3.4.2 Protective measures

8.3.4.2.1 Safety protection

The protective measures of obstacles and placing out of reach, non-conducting location and Earth-free local equipotential bonding shall **not** be used; and where emergency switching off is required, such devices shall be capable of breaking the full load current of the relevant parts of the installation and disconnect all live conductors, including the neutral conductor.

Protective measures against d.c. fault current shall be taken, aside from when being provided by the EV charging equipment where, except for circuits using the protective measure of electrical separation, each charging point is protected by its own RCD of at least Type A, having a rated residual operating current not exceeding 30 mA.

 RCDs must comply with one of the following standards: BS EN 61008-1, BS EN 61009-1, BS EN 60947-2 or BS EN 62423.

8.3.4.2.2 Electrical separation

This protective measure is limited to one electric vehicle being supplied from one unearthed source. The circuit shall be supplied through a fixed isolating transformer complying with BS EN 61558-2-4.

8.3.4.2.3 Devices for protection against overcurrent

Each EV charging point shall be supplied individually by a final circuit protected by an overcurrent protective device complying with BS EN 60947-2, BS EN 60947-6-2 or BS EN 61009-1 or with the relevant parts of the BS EN 60898 series or the BS EN 60269 series.

Where an EV charging point is built into a low-voltage switchgear or controlgear assembly, it must comply with the requirements of the relevant part of BS EN 61439.

8.3.4.3 Selection and erection of equipment

A dedicated final circuit must be provided for the connection to electric vehicles. At EV charging points, this is normally restricted to a single charging point used at its rated current, but others are equipped with a dedicated distribution circuit capable of supplying multiple electric vehicle charging points if load control is available.

When installed outdoors, the equipment shall be selected with a degree of protection against:

- the presence of water;
- the presence of solid foreign bodies;
- mechanical damage (impact of medium severity from other vehicles or foreign bodies).

Each a.c. charging point shall incorporate one socket outlet complying with BS 1363-2 marked 'EV' on its rear with a label on the front face or adjacent to the socket outlet or its enclosure stating: *'suitable for electric vehicle charging'*, or:

- one socket outlet or connector complying with BS EN 60309-2 (which is interlocked);
- procedures to prevent the socket contacts being live when accessible; or
- one Type 1 vehicle connector complying with BS EN 62196-2 for use with mode 3 charging only; or
- one Type 2 socket outlet or vehicle connector complying with BS EN 62196-2 for use with mode 3 charging only; or
- one Type 3 socket outlet or vehicle connector complying with BS EN 62196-2 for use with mode 3 charging only.

Each socket outlet shall be mounted in a fixed position. (Portable socket outlets shall not be used, but tethered vehicle connectors are allowed.)

8.3.4.4 Supplies

With a conventional car, you don't have to worry about paying extra for the engine, but when it comes to an electric or hybrid car, you either have to buy or lease the battery and a potential hybrid or electric vehicle buyer may well ponder questions like, 'Will those batteries die like my cell phone?', and, more importantly, 'How much will it cost to replace one?'.

The battery on an electric car is a proven technology that will last for many years. In fact, the majority of EV manufacturers guarantee that their electric car batteries will last a minimum of eight years or 100,000 hours. Currently (i.e. at the time of publication) a basic replacement third-generation battery pack would cost in the region £1,000, but like all modern products, over the years the price will undoubtedly fall.

8.3.4.4.1 Types of EV chargers

There are three main types of EV charging – rapid, fast, and slow. These represent the power outputs (measured in kilowatts (kW)), and therefore charging speeds available to charge an EV will vary.

Rapid chargers are obviously the fastest way to charge an EV, and predominantly cover d.c. charging. This can be split into two categories; ultra-rapid and rapid. Ultra-rapid points can charge at 100+ kW – often 150 kW and up to 350 kW – and are d.c. only. Conventional rapid points make up the majority of the UK's rapid-charging infrastructure and charge at 50 kW d.c., with 43 kW a.c. rapid charging also often available.

Fast chargers include those which provide power from 7 kW to 22 kW, which typically fully charge an EV in 3–4 hours.

 Note: The most common public charge point found in the UK is a 7 kW untethered Type 2 inlet, though tethered connectors are available too for both Type 1 and Type 2.

Slow units (up to 3 kW) are best used for overnight charging and usually take between 6 and 12 hours for a purely electric EV, or 2–4 hours for a PHEV (plug-in hybrid electric vehicle). EVs charge on slow devices using a cable which connects the vehicle to a three-pin or Type 2 socket.

In EV charging modes 3 and 4, an electrical or mechanical system shall be provided to prevent the plugging/unplugging of the plug unless the socket outlet or the vehicle connector has been switched off from the supply.

8.3.4.4.1.1 EV charge points   As previously mentioned, although many EV charge points are free to use, the majority of fast and rapid chargers require payment. Charging tariffs tend to comprise a flat connection fee, a cost per charging time (pence per hour) and/or a cost per energy consumed (pence per kWh).

8.3.4.4.1.2 Home charging point   Again, as previously touched on, charging an electric car from a dedicated home charging point is like charging a mobile phone (i.e. you plug in overnight and top up during the day), and the majority of electric car drivers choose to home charge as it provides faster charging speeds and built-in safety features.

 Note: A home charging point is a compact weatherproof unit that mounts to a wall with a connected charging cable or a socket for plugging in a portable charging cable.

It's useful to have a standard three-pin plug with an EVSE (electric vehicle supply equipment) as a backup charging option, but they are not designed to withstand these loads and should not be used long term.

Charging power for electric cars is measured in kilowatts (kW) and a home charging point can charge your car at 3.7 kW or 7 kW giving about 15–30 miles of range per hour of charge (compared to 2.3 kW from a three-pin plug which provides up to 8 miles of range per hour).

8.3.4.5 Wiring systems

The main earthing terminal of the installation is connected to an installation Earth electrode by a protective conductor. A PME earthing facility shall not be used as the means of earthing for the protective conductor contact of a charging point located outdoors unless one of the following methods is used:

- the charging point forms part of a three-phase installation that also supplies loads other than for electric vehicle charging;
- protection against electric shock is provided by a device which disconnects the charging point from the live conductors of the supply and from protective Earth within 5 s in the event of the voltage between the circuit protective conductor and Earth exceeding 70 V rms.

For a TN system, the final circuit supplying a charging point for electric vehicles shall not include a PEN conductor and the lowest part of any socket outlet shall be placed at a height of 0.5 to 1.5 m from the ground.

 See BS 7671:2018 Section 722 for more detailed requirements relating to electric vehicle charging installations.

8.3.5 Exhibitions, shows and stands

Whilst Chapter 51 of BS 7671:2018) still contains the requirements for external influences (see below), the 18th edition of the Wiring Regulations has now been further developed in accordance with new ISO standards and the BS EN 60721 and BS EN 61000 series on environmental conditions.

The particular requirements of this sub-section apply to the temporary electrical installations in exhibitions, shows and stands (including mobile and portable displays and equipment), to protect users.

8.3.5.1 General

8.3.5.1.1   External influences

Any external influences that could affect where a temporary electrical installation is to be erected (such as the presence of water or mechanical stresses)
needs to be taken into account when selecting a suitable location.

8.3.5.1.2 Isolation

Separate, temporary structures such as vehicles, stands or units that are
intended to supply outdoor installations shall:

- be provided with their own readily accessible and properly identifiable
  means of isolation that:
  - o   isolates them from each supply;
  - o   isolates them from all live conductors.

Means of isolation shall be designed and installed so as to prevent unintentional or inadvertent closure, using devices and precautions such as:

- locating them within a lockable space or lockable enclosure;
- padlocking them;
- locating them adjacent to the associated equipment.

8.3.5.2 Protective measures

The following requirements are intended for temporary electrical installations
at exhibitions, shows, stands and mobile (or portable) displays and equipment:

- any cable supplying these temporary structures shall be protected by an RCD;
- all accessible structural metallic parts shall be connected through the main
  protective bonding conductors to the main earthing terminal within the unit;
- all accessible socket-outlet circuits (other than those for emergency lighting) shall also be protected by and RCD;
- all live parts shall be covered by insulation, and located inside barriers
  and shelters;
- PME earthing shall **not** be used unless the installation is continuously monitored and a suitable means of Earth has been confirmed before the connection.

8.3.5.2.1 Protection against electric shock

The cable supplying temporary structures shall be protected at its origin by an
RCD (with a rated residual operating current not exceeding 300 mA).

Protective measures for a non-conducting location such as Earth-free local
equipotential bonding, obstacles and placing out of reach are **not** permitted.

8.3.5.2.2 Protection against fire and heat generation

Lighting equipment with high-temperature surfaces such as bright lamps, spot-lights, small projectors and other equipment or appliances shall be suitably guarded, installed and located.

Showcases and signs (especially those subject to nearby heat generation) need to be constructed of material which is heat-resistant, mechanically strong, electrically insulated and adequately ventilated.

If a stand installation contains a concentrated amount of electrical equipment, luminaires or lamps that are liable to generate excessive heat, then they shall not be installed unless sufficient ventilation is available.

8.3.5.2.3 Protection against thermal effects

If a luminaire is mounted below 2.5 m from floor level or within arm's reach and therefore subject to accidental contact, then it shall be securely fixed, sited and guarded in order to prevent the risk of injury to persons or ignition of materials.

8.3.5.2.4 Protective equipotential bonding

All structural, metallic parts which are accessible from within the stand, vehicle, wagon, caravan or container should be connected via the main protective bonding conductors to the main earthing terminal within the unit.

8.3.5.2.5 Selection and erection of equipment

All switchgear and controlgear must be placed in closed cabinets which can only be opened by the use of a key or a tool, except for those parts designed and intended to be operated by ordinary (i.e. unqualified) persons.

8.3.5.2.6 Electrical connections

Joints shall not be made in cables (except where absolutely necessary) as a connection into a circuit. If joints have to be made, then connectors or enclosures to guard against water ingress and damage shall be used.

Cable anchorages should be used if the cable is likely to be subjected to severe strain or possible water ingress which could be transmitted to the terminals.

8.3.5.2.7 Electric discharge lamp installations

All luminous tubes, signs or lamps that are used to illuminate a stand or exhibit shall comply with the following:

- Location – the sign or lamp shall either be installed out of arm's reach or be sufficiently protected so as to reduce the risk of injury to persons.

- Installation – the facia or stand fitting material behind luminous tubes, signs or lamps shall be non-ignitable.
- Emergency switching devices – a separate circuit (that is controlled by an emergency switching device) shall be used to supply signs, lamps or exhibits.

The switch shall be easily visible, accessible and clearly marked.

8.3.5.2.8 Luminaires and lighting installations

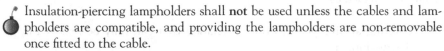

Insulation-piercing lampholders shall **not** be used unless the cables and lampholders are compatible, and providing the lampholders are non-removable once fitted to the cable.

Extra-low-voltage (ELV) lighting systems for filament lamps shall comply with BS EN 60598-2-23.

8.3.5.2.9 Socket outlets and plugs

An adequate number of socket outlets shall be installed to allow user requirements to be met safely.

Each socket outlet circuit not exceeding 32 A and all final circuits (other than for emergency lighting) shall be protected by an RCD.

Floor-mounted socket outlets shall be protected from accidental ingress of water and be strong enough to withstand expected traffic (human or mechanical) loads.

8.3.5.3 Supplies

The supply voltage for a temporary electrical installation in an exhibition, show or stand shall not exceed 230/400 V a.c. or 500 V d.c.

Except where an installation is within a building, a PME earthing facility shall **not** be used unless it is permanently under the supervision of a skilled or instructed person.

8.3.5.3.1 Electric motors – isolation

If the position of an electric motor could become dangerous then it will need to be given some form of isolation on all poles adjacent to the motor which it controls.

See BS EN 60204-1 for more detailed requirements.

8.3.5.3.2 Extra-low-voltage transformers and electronic converters

ELV transformers shall be mounted out of arm's reach of the public and shall have adequate ventilation.

A manual-reset protective device shall protect the secondary circuit of each transformer or electronic convertor.

Access by a competent person for the inspection, testing and maintenance of the device shall be provided.

Electronic converters shall conform to BS EN 61347-1.

8.3.5.3.3 Extra-low voltage provided by SELV or PELV

Where SELV or PELV is used, whatever the nominal voltage, basic protection shall be provided by:

- covering all live parts with an insulation that can only be removed by destruction; or
- protective barriers or enclosures.

8.3.5.4 Wiring systems

Flexible cables shall **not** be laid in areas accessible to the public unless they are protected against mechanical damage.

If there is a risk of mechanical damage to the wiring system, then armoured copper cables with a minimum cross-sectional area of 1.5 mm² shall be used.

8.3.5.4.1 Types of wiring system

If there isn't a fire alarm system installed in a building that is used for exhibitions etc., cable systems shall be either:

- flame retardant and low smoke; or
- unarmoured single- or multi-core cables enclosed in metallic or non-metallic conduit or trunking that provides sufficient fire protection.

8.3.5.4.2 Luminaires and lighting installations in exhibitions shows and stands

Exhibitions that contain a concentration of electrical equipment, luminaires and lamps that are liable to generate excessive heat shall not be installed unless adequate ventilation is available (e.g. a well-ventilated ceiling constructed of incombustible material).

Insulation-piercing lampholders shall **not** be used unless the cables and lampholders are compatible **and** provided the lampholders are non-removable once they have been fitted to the cable.

Luminaires mounted below 2.5 m (i.e. within arm's reach) from floor level, or otherwise accessible to accidental contact, shall be firmly fixed and located or guarded so as to prevent possible risk of injuring persons or igniting of materials.

8.3.5.4.3 Socket outlets

An adequate number of socket outlets shall be installed to allow user requirements to be met safely.

Floor-mounted socket outlets shall be sufficiently protected from accidental ingress of water and have sufficient strength to be able to withstand the expected traffic load.

 See BS 7671:2018 Section 711 for more detailed requirements relating to exhibitions, shows and stands.

8.3.6 Floor and ceiling heating systems

Section 753 of BS 7671 has been completely revised since it was last published in 2008 and now applies to the installation of embedded electric floor and ceiling heating systems. For example, to enable surface heating these can be erected as either a thermal storage heating system or a direct heating system.

The requirements of this Section also apply to electric heating systems for de-icing, frost prevention and similar applications, and cover both indoor and outdoor systems such as walls, ceilings, floors, roofs, drainpipes, gutters, pipes, stairs, roadways, and non-hardened compacted areas (e.g. football fields, lawns). The regulations do not, however, apply to the installation of wall heating systems nor heating systems for industrial and commercial applications.

8.3.6.1 General

8.3.6.1.1 Ambient temperature

Circuit wiring (i.e. cold tails and control leads that are installed in the zone of heated surfaces) must take into account the increase of ambient temperature.

8.3.6.1.2 Presence of solid foreign bodies

When installing heating units:

* care must be taken to ensure that no penetrating means, such as screws for door stoppers, are used in the area where floor or ceiling heating units are going to be fitted;
* precautions need to be taken to prevent the accumulation of dust or other substances in quantities which could affect heat dissipation from the wiring system;
* the completed wiring system shall meet the degree of protection relevant to the particular location. (For further details see BS EN 60529.)

8.3.6.2 Protective measures

 The following protective measures are **not** permitted:

- electrical separation for wall heating systems (with the exception of shaver socket outlets);
- non-conducting location and Earth-free local equipotential bonding;
- obstacles and placing out of reach;

8.3.6.2.1 Prevention of mutual detrimental influences

In order to prevent the possibility of mutual, detrimental influences occurring between electrical and non-electrical installations, the electrical installation must be arranged in such a way that the likelihood of electromagnetic interference occurring is minimised.

8.3.6.2.2 Protection by RCDs

A circuit supplying heating equipment of Class II construction or equivalent insulation shall be provided with additional protection by the use of an RCD.

8.3.6.3 Selection and erection of equipment

8.3.6.3.1 Heating units

Attachment of room fittings in heating-free areas shall be provided in such a way that the heat emission is not prevented by such fittings.

Heating units that are being installed in ceilings must be capable of resisting water that might hit the product at a 15° angle or less (i.e. IPX1).

Heating units that are intended to be installed in a floor of concrete or similar material shall be protected so that accidental submersion in up to 1 m of water for up to half an hour (for example, if a house were flooded during a gale) will not affect the heating unit, which will still be able to operate safely.

8.3.6.3.1.1 Protection against overheating   As heating units will cause higher temperatures or arcs under fault conditions, special measures should be taken when installing a heating unit close to easily ignitable building structures (e.g. placing on a metal sheet, in metal conduit or at a distance of at least 10 mm in air from the ignitable structure).

To avoid the overheating of floor or ceiling heating systems in buildings, one or more of the following measures need to be applied within the zone where heating units are installed to limit the temperature to a maximum of 80°C:

- proper installation of the heating system in accordance with the manufacturer's instructions;

- suitable design of the heating system;
- use of protective devices.

Heating units shall be:

- connected to the electrical installation via cold tails or suitable terminals;
- inseparably connected to cold tails (e.g. by a crimped connection);
- provided with a metal sheath or metal enclosure or fine-mesh metallic grid that is connected to the protective conductor of the supply circuit, when used as part of a wall heating system.

 Flexible sheet heating elements **must** comply with the requirements of BS EN 60335-2-96.

8.3.6.4 Identification and notices

The designer of the installation/heating system, or its installer, shall provide a plan for each heating system, which shall be fixed to, or located adjacent to, the distribution board of the heating system.

The installation plan shall contain the following details:

- cables, earthed conductive shields etc.;
- heated area;
- insulation resistance of the heating installation and the test voltage used;
- the layout of the heating units in the form of a sketch, drawing or picture;
- length/area of heating units;
- manufacturer and type of heating units;
- number of heating units installed;
- position/depth of heating units;
- position of junction boxes;
- rated power;
- surface power density;
- rated current of overcurrent protective device;
- rated residual operating current of RCD;
- rated resistance (cold) of heating units;
- the leakage capacitance;
- installation and maintenance instructions.

 See BS 7671:2018 Section 753 for more detailed requirements relating to heating cables and embedded heating systems.

8.3.7 Locations containing a bath or shower

When people use bathrooms and showers, most of the time they are naturally unclothed and wet and thus very vulnerable to electric shock due to their reduced body resistance (i.e. absence of shoes means less protection from shock, whilst water on their skin will tend to short-circuit its natural protection). Special measures are, therefore, required to ensure that the possibility of direct and/or indirect contact is reduced.

The following requirements apply to locations containing a fixed bath (bath tub or birthing pool), showers and their surrounding zones. They do **not** apply to emergency facilities in industrial areas and laboratories.

The requirements are based on three zones which take into account the limitations of walls, doors, fixed partitions, ceilings and floors. The three zones are as shown in Figure 8.4 and Table 8.2.

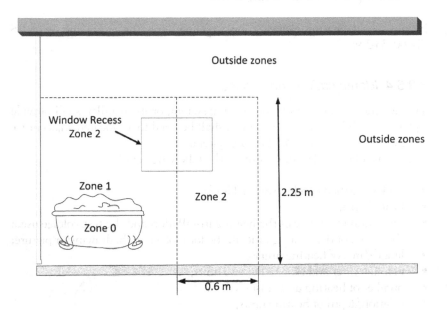

Figure 8.4 Zone limitations.

 Locations containing baths or showers for medical treatment, or for disabled persons, may have special requirements.

8.3.7.1 General

Horizontal or sloping ceilings, walls (with or without windows, doors, floors and fixed partitions) may be taken into account where these effectively limit the extent of locations containing a bath or shower as well as their zones.

8.3.7.2 Protective measures

 The protective measures of obstacles, placing out of reach, non-conducting location and Earth-free local equipotential bonding are **not** permitted, and protection by electrical separation shall **only** be used for:

Table 8.2 Limitations of zones

Zone 0   The interior of the bath tub or shower basin

 Note: for showers without a basin, the height of zone 0 is 0.10 m, with the same horizontal extent as zone 1

Zone 1   Zone 1 does not include zone 0 and is limited by:

- the finished floor level and the highest fixed shower head or water outlet or 2.25 m above the finished floor level (whichever is higher);
- the vertical surface

  o   around the bath tub or shower basin; and
  o   1.20 m from the centre point of the fixed water outlet on the wall (or ceiling for showers without a basin).

 Note: The space under the bath tub or shower basin is considered to be zone 1

Zone 2   Zone 2 is limited by:

- the finished floor level and the highest fixed shower head or water outlet or 2.25 m above the finished floor level (whichever is higher);
- the vertical surface at the boundary of zone 1 and 0.60 m from the border of zone 1.

 Note: For showers without a basin, there is no zone 2, but the horizontal dimension of zone 1 is increased to 1.20 m.

- circuits supplying one item of current-using equipment; or
- one single socket outlet;
- extra-low voltage provided by SELV or PELV.

Where SELV or PELV is used, basic protection for equipment in zones 0, 1 and 2 shall be provided by:

- basic insulation under the supervision of a suitably qualified and/or trained person; or
- barriers or enclosures affording a degree of protection for people and against access by water or solid foreign objects;
- protection by RCDs.

If existing basic protective measures or the fault protection fails (or the installation is prone to careless users), additional protection shall be provided for all low-voltage circuits serving the location, or passing through zones 1 and/or 2 not serving the location, by the use of one or more RCDs.

8.3.7.2.1 Electrical separation

Protection by electrical separation shall only be used for circuits supplying:

- one item of current-using equipment; or
- one single socket outlet.

8.3.7.2.2 Supplementary equipotential bonding

Local supplementary protective equipotential bonding shall include all simultaneously accessible exposed-conductive-parts of fixed equipment, and extraneous-conductive-parts including, where practical, the following:

- metallic pipes supplying services and metallic waste pipes (e.g. water, gas);
- metallic central heating pipes and air-conditioning systems;
- peripheral metallic structural parts of the building.

The equipotential bonding system shall be connected to the protective conductors of all equipment including those of socket outlets.

Where the location containing a bath or shower is in a building with a protective equipotential bonding system, supplementary equipotential bonding may be omitted provided that the following conditions are met:

- all final circuits of the location comply with the requirements for automatic disconnection;
- all final circuits of the location have additional protection by means of an RCD;
- all extraneous-conductive-parts of the location are effectively connected to the protective equipotential bonding.

 Special requirements exist for locations containing baths and showers for medical use or for disabled persons (see Building Regulations Part M) and Section 710 of BS 7671:2018.

8.3.7.3 Selection and erection of equipment

8.3.7.3.1   Erection of switchgear and controlgear

 The following requirements do not apply to switches and controls which are part of fixed current-using equipment that is used in that zone, or to insulating pull-cords of cord-operated switches.

In zone 0:

Switchgear or accessories shall **not** be installed.

In zone 1:

Only the switches of a SELV circuit may be installed, with the safety source being installed outside zones 0, 1 and 2.

In zone 2:
Switchgear, controlgear and accessories that include switches or socket outlets shall **not** be installed, with the exception of:

- switches and socket outlets of SELV circuits (provided that the safety source is installed outside zones 0, 1 and 2);
- shaver supply units complying with BS EN 61558-2-5.

 All other socket outlets are prohibited within a distance of 3 m horizontally from the boundary of zone 1.

8.3.7.3.2 Current-using equipment

Zone 0
Current-using equipment shall only be installed provided that all the following requirements are met:

- the equipment is fixed and permanently connected;
- the equipment is protected by SELV.

Zone 1
Only the following fixed and permanently connected current-using equipment shall be installed:

- electric showers;
- equipment protected by SELV or PELV (provided that the source is installed outside zones 0, 1 and 2);
- luminaires;
- shower pumps;
- towel rails;
- ventilation equipment;
- water heating appliances;
- whirlpool units.

8.3.7.3.3 External influences

Installed electrical equipment shall have at least the following degrees of protection:

- in zone 0: protection from accidental submersion in 1 m of water for up to 30 minutes;
- in zones 1 and 2: resistance to water splashes from any direction.

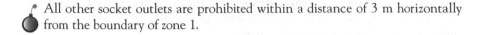 **Note:** This requirement does not apply to shaver supply units that comply with BS EN 61558-2-5 which are installed in zone 2 and located where direct

spray from showers is unlikely. However, electrical equipment exposed to low-pressure water jets (e.g. for cleaning purposes) shall be protected.

8.3.7.4 Electric floor heating systems

 The so-called protective measure, *'protection by electrical separation'* is **not** permitted for electric floor heating systems.

Only heating cables or thin-sheet flexible heating elements may be installed for electric floor heating systems, **provided** that they have either a metal sheath or a metal enclosure or a fine-mesh metallic grid which is connected to the protective conductor of the supply circuit.

Compliance with the latter requirement is not needed if the protective measure SELV is provided for the floor heating system.

See BS 7671:2018 Section 701 for more detailed requirements relating to locations containing a bathroom of shower.

8.3.8 Marinas and similar locations

The particular requirements of this section are applicable only to circuits intended to supply pleasure craft or houseboats in marinas and similar locations. They do **not** apply to the supply for houseboats (if they are supplied directly from the public network) **or** to the **internal** electrical installations of pleasure craft or houseboats.

8.3.8.1 General

8.3.8.1.1 External influences

For marinas, particular attention shall be given to the likelihood of corrosive elements, movement of structures, mechanical damage, presence of flammable fuel and the increased risk of electric shock owing to:

- contact of the body with Earth potential;
- presence of water;
- reduction in body resistance.

Equipment installed on or above a jetty, wharf, pier or pontoon must be protected against:

- atmospheric corrosive or polluting substances;
- ingress of small objects;
- mechanical impact damage of medium severity, by providing one or more of the following:

   o   installing equipment with a minimum degree of protection against external mechanical impact (see BS EN 62262);

o   position or location, selected to avoid being damaged by any reason-
      ably foreseeable impact;

o   using some form of local or general mechanical protection;

- water jets;
- water splashes;
- water waves.

8.3.8.2 Protective measures

 The protective measures of obstacles, placing out of reach, non-conducting
location and Earth-free local equipotential bonding are **not** permitted.

8.3.8.2.1 Protection against overcurrent

Each socket outlet shall be protected by an individual overcurrent protective
device.

   A fixed connection for the supply to each houseboat shall be protected by
an overcurrent protective device.

8.3.8.2.2 Protection by RCDs

Socket outlets shall be protected individually by an RCD.

   Final circuits intended for a fixed connection of supply to houseboats shall
be protected individually by an RCD which shall disconnect all poles, including the neutral.

8.3.8.3 Selection and erection of equipment

8.3.8.3.1   Isolation

At least one means of isolation (with a maximum of four outlet sockets) shall
be installed in each distribution cabinet.

 Note: This switching device shall disconnect all live conductors including the
neutral conductor.

8.3.8.3.2 Plugs and socket outlets

Socket outlets shall:

- be protected against solid objects that are bigger than 1mm and against
  water splashing, or be otherwise be protected by an enclosure;
- comply with BS EN 60309-1 (above 63 A) and BS EN 60309-2 (up to
  63 A).

Every socket outlet shall be:

- located as close as practicable to the berth to be supplied;
- installed in the distribution board or in separate enclosures that can hold a maximum of four socket outlets, grouped together;
- placed at a height of not less than 1 m above the highest water level.

One socket outlet shall supply one pleasure craft or houseboat – only!

In general, single-phase socket outlets (with rated voltage 200–250 V and rated current 16 A) shall be provided.

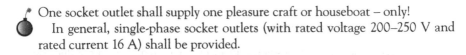

See BS 7671:2018 Section 709 for more detailed requirements relating to marinas and similar locations.

8.3.9 Medical locations

BS 7671:2018 Section 710 is a relatively new one, specifically aimed at electrical installations in medical locations, which was produced for the 2008 edition of this standard. It has now been brought up to date and in alignment with the Medical Devices Directive and other International, European and British standards; plus the Health Technical Memorandum (HTM) 06-01, published by the Department of Health. Unfortunately, for reasons best known to the authors, the numbering system of this particular sub-section seems to have a mind of its own!

BS 7671:2018 Section 710 is aimed at patient healthcare facilities and although the requirements mainly refer to hospitals, private clinics, medical and dental practices, healthcare centres and dedicated medical rooms in the workplace, this section also applies to electrical installations in locations designed for medical research and (where applicable) to veterinary clinics.

The requirements do **not**, however, apply to the actual medical electrical equipment itself, as this is fully covered in ISO 13485 (the Medical Devices Directive) and the BS EN 60601 series on medical electrical equipment and systems.

8.3.9.1 General

8.3.9.1.1   Official requirements for hospitals

HTM 06-01 (a Government guidance document entitled *Electrical services supply and distribution*) provides the legal requirements, design applications, operation and maintenance of the electrical infrastructure within healthcare premises.

The approved layout of a typical hospital operating theatre is shown in Figure 710.2 of BS 7671:2018, and a typical medical IT system shown in Figure 710.3.

But to save you having to hunt through this Standard, and other Government and official documents, for the meaning of some of the medical terms mentioned within this section, and elsewhere in my book, Annex 8.1 to Section 8.3.9 provides an *aide-memoire* to the most important ones.

 Within patient healthcare facilities, the risk to patients suffering through reduced body resistance is enhanced and it is extremely important that the following rules and regulations are strictly adhered to.

8.3.9.1.2 Protective measures

8.3.9.1.2.1 Exposed-conductive-parts of equipment   In Group 2 medical locations, where PELV is used, exposed-conductive-parts of equipment, (e.g. operating theatre luminaires) shall be connected to the circuit protective conductor.

8.3.9.1.2.2 Equipotential bonding busbar   Equipotential bonding busbars shall be located in or near the medical location using a protective conductor.

This is particularly relevant for fixed conductive non-electrical patient supports such as operating theatre tables, physiotherapy couches and dental chairs, which should be connected to the equipotential bonding conductor unless they are intended to be isolated from Earth.

8.3.9.1.2.3 Supplementary equipotential bonding   For Group 1 and Group 2 medical locations, supplementary equipotential bonding shall be installed for the following parts which are located (or that may be moved into) the *patient environment*:

* connections to conductive floor grids (if installed);
* extraneous-conductive-parts;
* metal screens of isolating transformers, via the shortest route to the earthing conductor;
* screening against electrical interference fields (if installed).

In addition, supplementary equipotential bonding connection points for medical electrical equipment (ME) shall be provided in each medical location, as follows:

* Group 1 medical location – one per patient location;
* Group 2 medical location – a minimum of four connection points, but, depending on the number of medical IT socket outlet per patient location, not less than 25% of the total number.

The equipotential bonding busbar shall be located in or near the medical location and all connections shall be accessible, labelled, clearly visible, and capable of being easily disconnected individually.

It is recommended that radial wiring patterns are used to avoid 'Earth loops', which may cause electromagnetic interference.

Unless they are intended to be isolated from Earth, fixed conductive non-electrical patient supports such as operating theatre tables, physiotherapy couches and dental chairs should be connected to the equipotential bonding conductor.

Note: The allocation of Group numbers for some of the safety services normally found within a particular location is shown at Annex 8.2 to Section 8.3.9.

8.3.9.1.2.4 Protection against electric shock   The protective measures of:

- obstacles and placing out of reach;
- non-conducting location;
- Earth-free local equipotential bonding; and
- electrical separation for the supply of more than one item of current-using equipment;

are **not** permitted in medical locations.

8.3.9.1.2.5 Functional extra-low voltage   Functional extra-low voltage (FELV) is not permitted as a method of protection against electric shock in medical locations.

8.3.9.1.2.6 Extra-low voltage provided by SELV or PELV   When using SELV and/or PELV circuits in Group 1 and Group 2 medical locations, protection by basic insulation of live parts (or by barriers or enclosures) must be provided.

Where PELV is used In Group 2 medical locations, all exposed-conductive-parts of equipment (such as operating theatre luminaires) shall be connected to the circuit protective conductor.

8.3.9.2 Supplies

In medical locations at least two different sources of supply shall be provided; one for the electrical supply system and one for safety services.

Automatic changeover facilities shall comply with BS EN 60947-6-1 and the distribution system shall be designed and installed so that in the case of a mains failure it facilitates the automatic changeover from the main distribution network to an electric safety source within:

- 0.5 s for luminaires in an operating theatre, light sources for essential medical electrical (ME) (e.g. endoscopes, monitors etc.) and life-support ME equipment;
- 15 s for safety lighting and other services such as firefighters' lifts, vigilance systems for smoke extraction, paging communication systems and fire detection, etc.;

- more than 15 s for ME required for the maintenance of healthcare installations (e.g. sterilising equipment etc.).

8.3.9.2.1 Transformers

Transformers should always be installed in close proximity to a medical location and with the following additional requirements:

- They shall comply with the requirements of BS-EN-61558-2-15 (*'Safety of transformers, reactors, power supply units and combinations thereof'*) and BS EN 61439 (*'Low-voltage switchgear and controlgear assemblies'*).
- The leakage current of the output winding to Earth and the leakage current of the enclosure shall not exceed 0.5 mA.
- At least one single-phase transformer per room (or functional group of rooms) shall be used to form the IT systems for mobile and fixed equipment and the rated output shall be not less than 0.5 kVA and shall not exceed 10 kVA.
- If several transformers are required to supply equipment in one room, they shall **not** be connected in parallel.
- If the supply of three-phase loads via an IT system is also required, a separate three-phase transformer **must** be provided for this purpose.

 Capacitors **shall not** be used in transformers for medical IT systems.

8.3.9.3 Selection and erection of equipment

 Distribution boards shall meet the requirements of the BS EN 61439 series.

8.3.9.3.1 Risk of explosion

The gases used as anaesthetics in operating theatres are extremely flammable if present in high concentrations. However, provided that there is adequate ventilation (e.g. 20 air changes per hour), no special precautions are necessary for the electrical installation in these medical locations.

All electrical devices (such as socket outlets and/or switches) should be installed at least 0.2 m below any medical-gas outlet, so as to minimise the risk of ignition of flammable gases.

8.3.9.4 Wiring systems

8.3.9.4.1   Arc fault detection devices

In Group 1 and Group 2 medical locations, arc fault detection devices (AFDDs) are not required to be installed. In medical locations of Group 0, AFDDs shall be used subject to a risk assessment.

8.3.9.4.2 Electromagnetic disturbances

Owing to the growing amount of medical implants (for example a cardioverter defibrillator to guard against heart failure), special considerations have to be made concerning electromagnetic interference (EMI) and electromagnetic compatibility (EMC) as not all medical equipment is housed inside a Faraday cage.

8.3.9.4.3 RCDs

Care shall be taken to ensure that simultaneous use of many items of equipment connected to the same circuit cannot cause unwanted tripping of the associated RCD.

In Group 1 and Group 2 medical locations, the following shall apply:

- Where RCDs are required, only type A (according to BS EN 61008 and BS EN 61009) or type B (according to IEC 62423) shall be selected, depending on the possible fault current arising.

 Type a.c. RCDs **shall** not be used.

- The voltage offered between simultaneously accessible exposed-conductive-parts and/or extraneous-conductive-parts shall not exceed 25 V a.c. or 60 V d.c..
- Care shall be taken to ensure that simultaneous use of large numbers of equipment items connected to the same circuit cannot cause unwanted tripping of the RCD.
- Where a medical IT system is used, additional protection by means of an RCD shall **not** be used.

8.3.9.4.4 TN systems

In final circuits of Group 1 rated up to 63 A, RCDs shall be used.

In final circuits of Group 2 medical locations (except for a medical IT system), protection by automatic disconnection of supply by means of RCDs shall only be used on the following circuits:

- circuits for the supply of movements of fixed operating tables; or
- circuits for large equipment with a rated power greater than 5 kVA; or
- circuits for X-ray units.

 Note: This requirement is mainly applicable to mobile X-ray units that are brought **into** Group 2 locations.

8.3.9.4.5 IT systems

Each IT system shall be equipped with an insulation monitoring device (IMD) in accordance with Annex A and Annex B of BS EN 61557-8.

In Group 2 ME locations, an IT system shall always be used for:

- final circuits supplying medical electrical equipment and systems intended for life support;
- surgical applications; and
- other electrical equipment located (or which may be moved into) the *patient environment*.

For each group of rooms serving the same function, at least one IT system is necessary and this shall be equipped with an IMD that has an acoustic and visual alarm system that is located in a suitable place so that it can be permanently monitored (via audible and visual signals) by the medical staff and the technical staff.

This alarm system shall have:

- a green signal lamp to indicate normal operation;
- a yellow signal lamp which lights when the minimum value set for the insulation resistance is reached;

 It shall **not** be possible for this light to be cancelled or disconnected – but the **yellow** signal should automatically go out when the normal condition is restored.

- an audible alarm which sounds when the minimum value set for the insulation resistance is reached. This audible alarm may (in certain circumstances) be capable of being be silenced.

documentation of all faults occurring in a medical location shall be maintained and shall include:

- the meaning of each type of signal; and
- the procedure to be followed in case of an alarm at first fault.

Consideration should also be given to the installation of fault location systems which localise insulation faults in any part of the medical IT system.

Note: In an IT system, the omission of devices for protection against overload is permitted for circuits supplying current-using equipment where unexpected disconnection of the circuit could cause danger or damage (e.g. a circuit supplying medical equipment used for life support) in specific medical locations.

8.3.9.4.5.1 Monitoring devices in an IT system   In IT systems, continuous insulation monitoring devices shall be provided which give an auditable and visual indication in the event of a first fault, which shall continue as long as the fault persists.

The following monitoring devices and protective devices may be used:

- insulation monitoring devices (IMDs);
- insulation fault-location systems;
- overcurrent protective devices.

Consideration should be given to the possibility that, if a line conductor of an IT system is earthed accidentally, then the insulation (or components) that are normally rated for the voltage between line and neutral conductors can be temporarily stressed with the line-to-line voltage.

The object of this particular regulation is to provide requirements for the safety of a low-voltage installation in the event of:

- a fault between the high-voltage system and Earth in the transformer sub-station that supplies the low-voltage installation;
- loss of the supply neutral in the low-voltage system;
- short-circuit between a line conductor and neutral in the low-voltage installation;
- accidental earthing of a line conductor of a low-voltage IT system.

In locations where there is a strong risk of fire due to the nature of pro-cessed or stored materials, the wiring system of an IT system (except for mineral-insulated cables, busbar trunking systems or powertrack systems) shall be protected against insulation faults, by an IMD with audible and visual signals.

8.3.9.4.5.2 IT system ME socket outlets   It is a mandatory requirement that socket-outlet circuits in the medical IT system for Group 2 medical locations:

- intended to supply ME equipment shall be unswitched;
- used on medical IT systems shall be coloured **blue** and clearly and perma-nently marked '*Medical equipment only*'.

At each patient's place of treatment (e.g. bedheads):

- each socket outlet shall be supplied by an individually protected circuit; or
- several socket outlets shall be separately supplied by a minimum of two circuits.

8.3.9.5 Safety services

Medical locations shall have safety standby power supply systems available which will energise installations that are required for continuous operation in case of failure of the general power system.

The safety power supply system shall be capable of automatically taking over if the main power supply drops for more than 0.5 s and by more than 10%.

Safety services which have been provided for a number of locations with different classifications should meet that classification which gives the highest security of supply.

A list of examples with suggested reinstatement times is given in Table A710 of Annex A710 to BS 7671:2018.

8.3.9.5.1 General requirements for safety power supply sources of Group 1 and Group 2

- Primary cells are not allowed as safety power sources.
- An additional main incoming power supply (from the general power supply) is not regarded as a source of a safety power supply.

The availability (readiness for service) of safety power sources shall be monitored, and where socket outlets are supplied from a safety power supply source they shall be readily identifiable according to their safety services classification.

8.3.9.5.2 Power supply sources with a changeover period greater than 15 s

Hospital maintenance and healthcare services equipment such as:

- catering equipment;
- cooling equipment;
- sterilisation equipment;
- storage battery chargers;
- technical building installations (e.g. air-conditioning, heating and ventilation systems, building services and waste-disposal systems);

shall be connected (either automatically or manually) to a safety power supply source capable of maintaining it for a minimum period of 24 h.

8.3.9.5.3 Emergency lighting systems

In the event of mains power failure, the changeover period to the standby the safety services sources such as:

- emergency lighting and exit signs;
- switchgear and controlgear for emergency generating sets;
- normal power supply and the standby safety services power supply distribution boards;
- essential service rooms;
- locations of central fire alarm and monitoring systems;

shall not exceed 15 s.

- Group 1 medical location rooms shall be supplied (from the standby for safety services) with at least one luminaire in case of emergency;

- Group 2 medical locations rooms shall have at least of 90% of their normal lighting requirements supplied from the standby safety service.

 Note: Escape route luminaires shall be arranged on alternate circuits.

8.3.9.5.4 Other services

Other services such as:

- electrical equipment of medical gas supply ((e.g. compressed air, vacuum supply and narcosis (anesthetics)) – exhaustion as well as their monitoring devices;
- fire detection and fire alarms;
- fire extinguisher systems;
- firefighter lifts;
- paging systems;
- smoke extraction ventilation systems;
- vitally important ME equipment used in Group 2 medical locations;

shall be provided with a changeover system to a standby safety service within 15 s.

8.3.9.6 Earthing

 In accordance with the Electricity Safety, Quality and Continuity Regulations (ESQCR), PEN conductors **shall not** be used in medical locations and medical buildings downstream of the main distribution board.

It is recommended that radial wiring patterns are used to avoid 'Earth loops' that may cause electromagnetic interference.

8.3.9.6.1 Medical facility Earth faults

 In the event of a first fault to Earth, a total loss of supply in medical Group 2 locations shall be prevented.

Cables intended to supply temporary structures shall be protected at their origin by an RCD.

Earth-free local equipotential bonding for the supply to more than one item of current-using equipment shall not be used.

Supplementary protective equipotential bonding shall be installed (using metal screens of isolating transformers) via the shortest route to the earthing conductor.

RCDs **should** be used for:

- additional protection for lighting in places such as telephone kiosks, bus shelters, advertising panels;

- socket-outlet circuits not exceeding 32 A and all final circuits other than for emergency lighting;
- surgical applications;
- electrical installations including a PV power supply system without at least simple separation between the a.c. side and the d.c. side;
- *'other'* electrical equipment located or that may be moved into the *patient environment*.

RCDs should **not** be used:

- for final circuits supplying medical electrical equipment and systems intended for life support;
- where a medical IT system is functioning.

 Note: For each circuit that is protected by an RCD, the possibility of the RCD's unwanted tripping due to excessive protective conductor currents produced by equipment in normal operation needs to be considered.

8.3.9.7 Inspection and testing

The International Electrotechnical Commission (IEC) and the British Standards Institution (BSI) manufacturers' standards for medical electrical equipment consist of two types of testing: type testing and routine testing.

8.3.9.7.1 Type testing

Type testing is carried out by an approved test house (under closely specified and closely controlled environmental conditions) on a single representative sample equipment item for which certification of compliance with a standard is being sought. These tests are not intended for routine use – indeed, it has been documented that repetition of many of the tests would certainly cause deterioration in performance and safety of the equipment under test.

8.3.9.7.2 Routine testing

Routine testing, on the other hand, is intended to provide an indication of the inherent safety of the equipment without subjecting it to undue stress that would be liable to cause deterioration.

8.3.9.7.3 Documentation related to the installation

Once the installation is completed, an overall plan of the electrical installation together with records, drawings, wiring diagrams and modifications relating to the medical location shall be provided. These should consist of, but are not limited to:

- single line overview diagrams of the distribution system of the normal power supply and the power supply for safety services;
- distribution board block diagrams showing switchgear, controlgear and distribution boards;
- schematic diagrams of controls;
- functional description for the operation of the safety power supply system;
- verification of compliance with the requirements of standards.

8.3.9.7.4 Initial verification

In addition to the requirements of BS 7671:2018 Chapter 64 and HTM 06-01 (Part A), the following tests **shall** be carried out, both prior to commissioning and after alteration or repairs, and before recommissioning:

- complete functional tests of all insulation monitoring devices (IMDs) associated with the medical IT system, including insulation failure, transformer high temperature, overload, discontinuity and the acoustic/visual alarms linked to them;
- measurements of leakage current from the IT transformers of the output circuit and enclosure in no-load condition;
- measurements to verify that the resistance of the supplementary equipotential bonding is within stipulated limits.

8.3.9.7.5 Periodic inspection and testing

Periodic inspection and testing should be carried out in accordance with Health Technical Memorandum (HTM) 06-01 (Part B) and local health authority requirements, as follows and at the given intervals (the tests are the same as those above for the designated stages):

- **Annually** – complete functional tests of all IMDs associated with the medical IT system, including insulation failure, transformer high temperature, overload, discontinuity and the acoustic/visual alarms linked to them;
- **Annually** – measurements to verify that the resistance of the supplementary equipotential bonding is within the stipulated limits;
- **Every three years** – measurements of leakage current of the output circuit and of the enclosure of the medical IT transformers in no-load condition.

The dates and results of each verification **shall** be recorded.

 See BS 7671:2018 Section 710 for more detailed requirements relating to Medical locations.

Annex 8.1 to Section 8.3.9

This Annex is intended as an *aide-memoire* to the meaning of some of the most important medical terms used within this section, and elsewhere in the book.

Applied part refers to that part of the medical electrical equipment that, whilst in normal use, comes into physical contact with the patient.

Intracardiac procedure is a procedure whereby an electrical conductor is placed within the heart of a patient or is likely to come into contact with the heart, such conductor being accessible outside the patient's body. In this context, an electrical conductor includes insulated wires such as cardiac pacing electrodes or intracardiac ECG electrodes, or insulated tubes filled with conducting fluids.

Medical electrical equipment (ME equipment). Electrical equipment having an applied part for transferring energy to or from the patient or detecting such energy transfer to or from the patient, and which is:

- provided with not more than one connection to a particular supply mains;
- intended by the manufacturer to be used in the diagnosis, treatment or monitoring of a patient, or for compensation or alleviation of disease, injury or disability.

ME equipment also includes the accessories that have been defined by the manufacturer as being necessary for enabling normal use of the ME equipment.

Medical electrical system (ME system). Combination, as specified by the manufacturer, of items of equipment (and their accessories), at least one of which is ME equipment to be interconnected by functional connection or by use of a multiple socket outlet.

Medical IT system. IT electrical system fulfilling specific requirements for medical applications.

 Note: These supplies are also known as *'isolated power supply systems'*.

Medical location. A location intended for purposes of diagnosis, treatment including cosmetic treatment, monitoring and care of patients.

Patient. Living being (person or animal) undergoing a medical, surgical or dental procedure.

 Note: A person under treatment for cosmetic purposes may also be considered as a patient.

Patient environment. Any volume in which intentional or unintentional contact can occur between a patient and parts of the ME equipment or ME system, or between a patient and other persons touching parts of the ME equipment or system.

Safety

Electrical installations in medical locations (and also electrical installations in locations designed for medical research and (where applicable) veterinary clinics) must be capable of ensuring the continued safety of patients and medical staff. For convenience, these medical locations are classified into groups depending on the level of safety required. For example:

- the type of contact between applied parts and the patient;
- the threat to the safety of the patient that represents a discontinuity (failure) of the electrical supply; and
- the purpose for which the location is used.

To ensure the protection of patients from possible electrical hazards, additional protective measures need to be applied in medical locations.

Care should also be taken to ensure that other installations do not compromise the level of safety provided by installations meeting the requirements of this section.

Medical locations

Medical locations are split into groups as follows:

- **Group 0** – Medical location where no applied parts are intended to be used and where discontinuity (failure) of the supply cannot cause danger to life.
- **Group 1** – Medical location where discontinuity of the electrical supply does not represent a threat to the safety of the patient.

Classes and types of medical electrical equipment

All medical electrical equipment is categorised into classes according to the type of protection against electric shock that it uses.

Class I equipment   Class I equipment has a protective Earth where the basic means of protection is the insulation between live parts and exposed conductive parts such as the metal enclosure. In the event of a fault (that would otherwise cause an exposed-conductive-part to become live), the supplementary protection (i.e. the protective Earth) comes into effect.

 Note: Large fault current flows from the mains part to Earth via the protective Earth conductor which will cause a protective device (usually a fuse) in the mains circuit to disconnect the equipment from the supply.

Class II equipment   Class 2 equipment uses either double insulation or reinforced insulation. In double-insulated equipment, the basic protection is provided by the first layer of insulation. If the basic protection fails then supplementary protection is provided by a second layer of insulation, thus preventing contact with live parts.

Class III equipment   Class III equipment is either battery operated or supplied by a SELV transformer and, as the voltages do not exceed 25 V a.c or 60 V d.c., protection against electric shock is a minimal requirement.

Patient safety

Since the hazard to people will depend on the treatment being administered, hospital locations are divided into groups as follows:

- **Group Zero** – where no treatment or diagnosis using medical electrical equipment is administered (e.g. consulting rooms).
- **Group One** – where medical electrical equipment is in use, but not for treatment of heart (intracardiac) conditions.
- **Group Two** – where medical electrical equipment is in use for heart (intracardiac) conditions.

Annex 8.2 To section 8.3.9

This Annex contains examples for some of the group numbers that are allocated for the classification of medical location safety services (Table 8.3).

Table 8.3 Medical location services

Medical location	Group		
	0	1	2
Anaesthetic area			X
Angiographic examination room			X
Bedrooms		X	
Delivery room		X	
ECG, EEG, EHG room		X	
Endoscopic room		X	
Examination or treatment room		X	
Haemodialysis room		X	
Heart catheterisation room			X
Hydrotherapy room		X	
Intensive care room			X
Intermediate Care Unit (IMCU)			X
Magnetic resonance imaging (MRI) room		X	X
Massage room	X	X	
Nuclear medicine		X	
Operating plaster room			X
Operating preparation room			X
Operating recovery room			X
Operating theatre			X
Premature baby room			X
Physiotherapy room		X	
Radiological diagnostic and therapy room		X	X
Urology room		X	

8.3.10 Mobile and transportable units

For the purposes of this section, the term '*unit*' is intended to mean a vehicle and/or mobile (self-propelled or towed) or transportable structure (e.g. a container or cabin) in which all or part of an electrical installation is contained and which is provided with a temporary supply by means of, for example, a plug and socket outlet.

8.3.10.1 General

Where mobile equipment is likely to be used, the equipment can be fed from an adjacent and conveniently accessible socket outlet, taking into account the length of flexible cord normally fitted to portable appliances and luminaires.

The socket outlet, plugs and connecting devices used to connect the unit to the supply shall comply with BS EN 60309-2 and meet the following requirements:

- plugs shall be within an enclosure of insulating material;
- connecting devices located outside the unit shall be protected against solid objects and water splashes when in use;
- enclosures containing these connectors shall not only be protected against solid objects and water splashes when in use, but also the ingress of dust.

8.3.10.2 Protective measures

 The protective measures of obstacles, placing out of reach, non-conducting location and Earth-free local equipotential bonding are **not** permitted.

 Automatic disconnection of the supply shall be provided by means of an RCD.

8.3.10.2.1 Protective equipotential bonding

Accessible conductive parts of the unit, such as the chassis, shall be connected through the main protective bonding conductors to the main earthing terminal within the unit.

The main protective bonding conductors shall be finely stranded.

8.3.10.2.2 Socket outlets

Additional protection by an RCD shall be provided for every socket outlet intended to supply current-using equipment outside the unit, with the exception of socket outlets which are supplied from circuits with protection by:

- SELV; or
- PELV; or
- electrical separation.

8.3.10.2.3 PME earthing systems

A PME earthing facility shall not be used as a means of earthing, except:

- where the installation is continuously under the supervision of a skilled or instructed person; and
- the suitability and effectiveness of the means of earthing has been confirmed before the connection is made.

8.3.10.2.4 IT systems

An IT system, every socket outlet intended to supply current-using equipment outside the unit, shall be protected by an RCD, with the exception of socket outlets that are being supplied from circuits already protection by:

- SELV; or
- PELV; or
- electrical separation.

8.3.10.2.5 TN systems

A PME earthing facility shall **not** be used with a TN system as a means of earthing, except:

- where the installation is continuously under the supervision of a skilled or instructed person; and
- the suitability and effectiveness of the means of earthing has been confirmed before the connection is made.

8.3.10.2.6 Additional RCD protection

In a.c. systems, additional protection by means of an RCD shall be provided for mobile equipment with a current rating not exceeding 32 A when used outdoors.

8.3.10.3 Selection and erection of equipment

8.3.10.3.1 Identification of equipment

A permanent notice shall be fixed to the unit in a prominent position stating:

- the type of supplies which may be connected to the unit;
- the voltage rating of the unit;
- the number of supplies, phases and their configuration;
- the on-board earthing arrangement;
- the maximum power requirement of the unit.

8.3.10.3.2 Plugs and connectors

Connecting devices used to connect the unit to the supply shall:

- be within an enclosure of insulating material (with at least a degree of IP55); and
- provide a degree of protection not less than IP44.

8.3.10.3.3 Proximity to non-electrical services

No electrical equipment (including wiring systems), except extra-low-voltage (ELV) equipment for gas supply control, shall be installed in any gas cylinder storage compartment.

Where cables have to run through such a compartment, they shall be protected against mechanical damage by installation within either:

- a conduit complying with the BS EN 61386 series; or
- a ducting system complying with the appropriate part of the BS EN 50085 series.

The conduit or ducting system shall be able to withstand an impact equivalent to AG3 without visible physical damage.

ELV cables and electrical equipment may only be installed within the LPG cylinder compartment **if** the installation indicates the operation of the gas cylinder (e.g. indication of empty gas cylinders) or is for use within the compartment.

 Such electrical installations and components shall be constructed and installed so that they are **not** a potential source of ignition.

8.3.10.3.4 Socket outlets

Socket outlets located outside the unit shall be provided with an enclosure affording a degree of protection not less than 1P44.

8.3.10.4 Supplies

 Generating sets that are able to produce voltages other than SELV or PELV (and which are mounted in a mobile unit) **shall be** automatically switched off in case of an accident to the unit.

8.3.10.4.1 IT systems

An IT system can be provided by:

- an isolating transformer or a low voltage generating set, with an insulation monitoring device or an insulation fault-location system installed; or

- a transformer providing simple separation (in accordance with BS EN 61558-1) with an RCD and an Earth electrode installed to provide automatic disconnection in case the transformer fails.

8.3.10.5 Wiring systems

Flexible cables (for connecting the unit to the supply), or cables of equivalent design, having a minimum cross-sectional area of 2.5 mm² copper shall be used.

The flexible cable shall enter the unit by an insulating inlet in such a way as to minimise the possibility of any insulation damage or fault which might energise the exposed-conductive-parts of the unit.

The wiring system shall be installed using one or more of the following:

- unsheathed flexible cable with thermoplastic or thermosetting insulation installed in either a conduit, trunking or ducting;
- sheathed flexible cable thermoplastic or thermosetting insulation.

Where cables have to run through such a compartment, they shall be protected against mechanical damage by installation within a conduit system (complying with the BS EN 61386 series) or within a ducting system (complying the BS EN 50085 series).

Where installed, this conduit or duct shall be able to withstand an impact equivalent to AG3 without visible physical damage.

 See BS 7671:2018 Section 711 for more detailed requirements relating to mobile and transportable units

8.3.11 Operating and maintenance gangways

This (comparatively small) Section of BS 7671:2018 centres on the operation and safe maintenance of switchgear and controlgear within areas that include gangways and where access is restricted to skilled or instructed persons.

8.3.11.1 Restricted access areas

Restricted access areas shall:

- be clearly and visibly marked by appropriate signs;
- not provide access to unauthorised persons; and
- provide doors that, although normally in the closed position, allow easy evacuation in case of danger, without the use of a key, tool or other device that is not part of the opening mechanism.

8.3.11.2 Requirements for operating and maintenance gangways

The width of gangways and access areas shall be adequate for work, operational access, emergency access, emergency evacuation and for transport of equipment.

Gangways shall permit at least a 90° opening of equipment doors or hinged panels. (For further details, please refer to Annex A 729 of BS 7671:2018.)

Gangways longer than 10 m shall be accessible from both ends.

 See Figure 729.3 of BS 7671:2018 for examples of positioning of doors in closed restricted access.

8.3.11.3 Restricted access areas where basic protection is provided by barriers or enclosures

Where basic protection is provided by barriers or enclosures, the following minimum dimensions apply after barriers and enclosures are in a fixed position and circuit-breakers and switch handles are in the most difficult position, including 'isolation' (Table 8.4). (Also see Figure 729.1 of BS 7671:2018.)

Table 8.4 Minimum gangway dimensions

Gangway	Dimensions
Gangway width including between:	700 mm
• obstacles and switch handles or circuit-breakers in the most onerous position, and obstacles or switch handles or circuit-breakers in the most onerous position and the wall.	
Gangway width between barriers or enclosures or other barriers or enclosures and the wall	700 mm
Height of gangway to barrier or enclosure above floor	2000 m
Live parts that have been placed out of reach	2500 mm

In some cases, additional workspace may be needed for extra-large, specialised switchgear and controlgear assemblies.

 See BS 7671:2018 Section 729 for more detailed requirements relating to operating and maintenance gangways.

8.3.12 Rooms and cabins containing saunas

Saunas, similar to swimming pools, bathrooms and showers etc., are primarily used by people who are unclothed and wet and thus very vulnerable to electric shock due to their reduced body resistance (i.e. absence of shoes means less protection from shock, whilst water on their skin will tend to short-circuit its natural protection). Special measures are, therefore, needed to ensure that the possibility of direct and/or indirect contact is reduced.

8.3.12.1 General

The Regulations requirements for hot air saunas are based on three zones which take into account the limitations of walls, doors, fixed partitions, ceilings and floors and the electric heater itself. The three zones are shown in Figure 8.5.

Zone	Insulation requirements and limitations
1	In zone 1, only the sauna heater and equipment belonging to the sauna heater shall be installed. The zone containing the sauna heater is limited by the floor, the cold side of the thermal insulation of the ceiling and a vertical surface around the sauna heater, 0.5 m from its surface.
2	Zone 2 is the volume outside zone 1, limited by the floor, the cold side of the thermal insulation of the walls and a horizontal surface located 1.0 m above the floor. In zone 2 there is no special requirement concerning heat resistance of equipment.
3	Zone 3 is the volume outside zone 1, limited by the cold side of the thermal insulation of the ceiling and walls and a horizontal surface located 1.0 m above the floor. Equipment shall withstand a minimum temperature of 125°C. Insulation and sheaths of cables shall withstand a minimum temperature of 170°C.

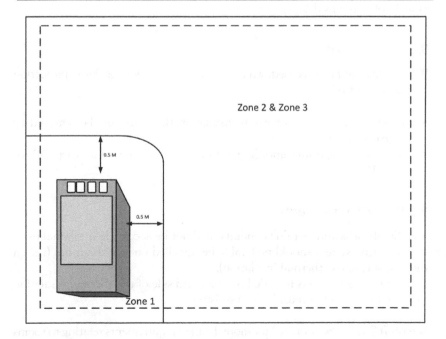

Zone 2 & Zone 3

0.5 M

0.5 M

Zone 1

Figure 8.5 Zones of ambient temperature.

8.3.12.2 *Protective measures*

The protective measures of obstacles, placing out of reach, non-conducting location and Earth-free local equipotential bonding are **not** permitted.

8.3.12.2.1 External influences

The equipment shall have a degree of protection against water splashes and a low-pressure water jet spray.

8.3.12.2.2 Protection by RCDs

Additional protection shall be provided for all circuits of the sauna, by the use of one or more RCDs.

8.3.12.3 *Selection and erection of equipment*

8.3.12.3.1 Isolation, switching, control and accessories

Socket outlets shall **not** be installed within the location containing the sauna heater.

Switchgear and controlgear forming part of the sauna heater equipment or of other fixed equipment installed in zone 2 may be installed within the sauna room or cabin, whilst all other switchgear and controlgear (e.g. for lighting) must be placed outside.

8.3.12.4 *Supplies*

Where SELV or PELV is used, whatever the nominal voltage, basic protection shall be provided by:

- covering all live parts with an insulation that can only be removed by destruction; or
- barriers or enclosures affording a degree of protection of at least IP4X or IPXXD.

8.3.12.5 *Wiring systems*

Metallic sheaths and metallic conduits shall **not** be accessible in normal use.

The wiring system should preferably be installed outside the zones (i.e. on the cold side of the thermal insulation).

If the wiring system is installed on the warm side of the thermal insulation in zones 1 or 3, then it must be heat-resisting.

See BS 7671:2018 Section 703 for more detailed requirements relating to rooms and cabins containing sauna heaters.

8.3.13 Solar, photovoltaic (PV) power supply systems

Photovoltaics (PV) is a technology that converts light directly into electricity by using photons from sunlight to knock electrons into a higher state of energy, thereby creating electricity. Solar cells are packaged in photovoltaic modules (often electrically connected in multiples as solar photovoltaic arrays) and convert energy from the sun into electricity.

8.3.13.1 General

PV modules:

- may be connected in series up to the maximum allowed operating voltage of the modules and the PV convertor;
- shall be installed in such a way that there is adequate heat dissipation during the greatest solar radiation for the site.

8.3.13.2 Protective measures

Where protective bonding conductors are installed, they shall be parallel (and in close contact) to the d.c. cables, a.c. cables and accessories. The protective measures of non-conducting location, however, and Earth-free local equipotential bonding shall **not** be used on the d.c. side.

8.3.13.2.1 Double or reinforced insulation

Protection by the use of Class II or equivalent insulation is the preferred option on the d.c. side.

8.3.13.2.2 Extra-low voltage provided by SELV or PELV

For SELV and PELV systems, the uninterrupted operating control must not exceed 120 V d.c.

8.3.13.3 Selection and erection of equipment

 The selection and erection of equipment shall enable safe maintenance.

8.3.13.3.1 Protection against electromagnetic interference in buildings

Wiring loops shall be as small as possible, so as to minimise the effects of lightning causing electromagnetic interference (EMI).

8.3.13.4 Supplies

Electrical equipment on the d.c. side must all be suitable for direct voltage and direct current.

If blocking diodes are used, they shall be connected in series with the PV strings.

PV equipment on the d.c. side shall be considered to be energised, even when the system is disconnected from the a.c. side.

On the a.c. side, the PV supply cable shall be connected to the supply side of the overcurrent protective device for automatic disconnection of circuits supplying current-using equipment.

To allow maintenance of the PV convertor, the equipment should be capable of being isolated both from the d.c. side and the a.c. side.

Where an electrical installation includes a PV power supply system that does not possess a simple separation between the a.c. side and the d.c. side, an RCD (type B) shall be installed to provide fault protection by automatically disconnecting the supply.

 Note: If the PV convertor is not constructed to be able to feed d.c. fault currents into the electrical installation, an RCD of type B is not required.

 The protective measures of non-conducting location and Earth-free local equipotential bonding are **not** permitted on the d.c. side.

8.3.13.5 Devices for isolation

In the selection and erection of devices for isolation and switching being installed between the PV installation and the public supply, the public supply shall be considered the source, and the PV installation shall be considered the load.

A switch-disconnector shall be provided on the d.c. side of the PV convertor.

All junction boxes (PV generator and PV array boxes) shall carry a warning label indicating that parts inside the boxes may still be live after isolation from the PV convertor.

8.3.13.5.1 Earthing arrangements

Earthing of one of the live conductors of the d.c. side is permitted, provided that there is some form of simple separation between the a.c. side and the d.c. side.

 Note: Any connections with Earth on the d.c. side should be electrically connected so as to avoid corrosion (see BS EN 13636 and BS EN 15112).

8.3.13.5.2 Protection against fault current

The PV supply cable on the a.c. side shall be protected against fault current by an overcurrent protective device installed at the connection point to the a.c. mains.

8.3.13.5.3 Protection against overload on the d.c. side

Overload protection may be omitted:

- for PV string and PV array cables when the continuous current-carrying capacity of the cable is equal to, or greater than, 1.25 I_{sc} STC at any location;
- to the PV main cable if the continuous current-carrying capacity is equal to or greater than 1.25 I_{sc} STC of the PV generator.

8.3.13.6 Wiring systems

8.3.13.6.1   Selection and erection of wiring systems

To minimise the risk of Earth faults and short-circuits, PV string cables, PV array cables and PV d.c. main cables shall be selected and erected.

Wiring systems shall withstand the expected external influences such as wind, ice formation, temperature and solar radiation.

See BS 7671:2018 Section 712 for more detailed requirements relating to solar, photovoltaic (PV) power supply systems.

8.3.14 Swimming pools and other basins

The particular requirements of this section apply to the basins of swimming pools, the basins of fountains and the basins of paddling pools, and also to the surrounding zones of these basins. Swimming pools are, by design, 'wet areas', as people using them are normally wet which will increase their vulnerability to electric shock. Special measures are, therefore, needed to ensure that all possibility of direct and/or indirect contact is reduced.

Note: Except for areas especially designed as swimming pools, the requirements of this section do not apply to natural waters, lakes in gravel pits, coastal areas and the like. However, special requirements may be necessary in swimming pools designed for medical purposes.

8.3.14.1 General

8.3.14.1.1   Classification of external influences

As shown in Figure 8.6, the following requirements concern the dimensions of three zones.

Zones 1 and 2 may be limited by fixed partitions having a minimum height of 2.5 m.

Zone 0   This zone is the interior of the basin of the swimming pool or fountain, including any recesses in its walls or floors, basins for foot cleaning, and waterjets or waterfalls and spaces below them.

Zone 1   This zone is limited by:

- zone 0;
- a vertical plane 2 m from the rim of the basin;
- the floor or surface expected to be occupied by persons;
- the horizontal plane 2.5 m above the floor or the surface expected to be occupied by persons.

Where' the swimming pool or fountain contains diving boards, springboards, starting blocks, chutes or other components expected to be occupied by persons, zone 1 comprises the zone limited by:

- a vertical plane situated 1.5 m from the periphery of the diving boards, springboards, starting blocks, chutes and other components such as accessible sculptures, viewing bays and decorative basins;
- the horizontal plane 2.5 m above the highest surface expected to be occupied by persons.

Zone 2   This zone is limited by:

- the vertical plane external to zone 1 and a parallel plane 1.5 m from the former;
- the floor or surface expected to be occupied by persons;
- the horizontal plane 2.5 m above the floor or surface expected to be occupied by persons.

 Note: There is no zone 2 for fountains.

 Note: For a swimming pool where it is not possible to locate a socket outlet or switch outside zone 1, a socket outlet or switch (preferably with a non-conductive cover or coverplate) is permitted in zone 1 **if:**

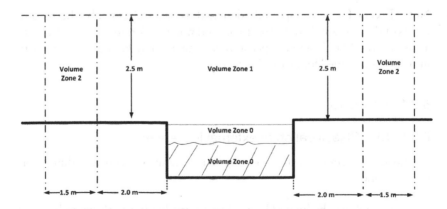

Figure 8.6 Swimming and paddling pool zone dimensions.

- it is installed at least 1.25 m horizontally from the border of zone 0;
- is placed at least 0.3 m above the floor, and is protected by:

    o   SELV; or
    o   automatic disconnection of supply; or
    o   electrical separation.

8.3.14.2 Protective measures

In swimming pools and other basins, the protective measures of obstacles, placing out of reach, non-conducting location; and Earth-free local equipotential bonding are **not** permitted.

8.3.14.2.1 Zones 0 and 1 (swimming pools and other basins)

Except for fountains, in zone 0 and zone 1, **only** protection by SELV is permitted with the SELV source installed outside of zones 0, 1 and 2.

Extraneous-conductive-parts in zones 0, 1 and 2 must always be connected via supplementary protective bonding conductors to the protective conductors of exposed-conductive-parts of equipment situated in these zones.

Any equipment that is located inside the basin that is **only** intended to be in operation when people are **not** inside zone 0 shall be supplied by a circuit protected by:

- SELV;
- automatic disconnection of the supply (using an RCD); or
- electrical separation.

The socket outlet of any circuit supplying such equipment (and similarly the control device of such equipment) shall have a notice warning users that this equipment may be used **only** when the swimming pool is **not** occupied by persons.

8.3.14.2.2 Zones 0 and 1 (fountains)

In zone 0, a cable for electrical equipment shall be installed as far outside the basin rim as is reasonably practicable.

In zone 1, a cable shall be selected, installed and provided with mechanical protection to medium severity (AG2) and be suitable for submersion in water up to 10 m in depth.

In zones 0 and 1 of fountains, one or more of the following protective measures shall be employed:

- SELV;
- automatic disconnection of supply (using an RCD);
- electrical separation.

Note: There is no zone 2 for fountains.

8.3.14.2.3 Requirements for SELV circuits

Where SELV is used, whatever the nominal voltage, basic protection shall be provided by:

- covering all live parts with an insulation that can only be removed by deliberate action or destruction; or
- barriers or enclosures.

8.3.14.3 Selection and erection of equipment

8.3.14.3.1 Current-using equipment of swimming pools

In zones 0 and 1, it is **only** permitted to install fixed current-using equipment that has been specifically designed for use in a swimming pool.

Equipment that is intended to be in operation **only** when people are outside zone 0 may be used in all zones, provided that it is supplied via a protected circuit.

It is, nevertheless, permitted to install an electric heating unit that has been embedded in the floor, provided that it:

- is protected by SELV; or
- includes an earthed metallic sheath connected to the supplementary equipotential bonding and its supply circuit is additionally protected by an RCD; or
- is covered by an embedded earthed metallic grid connected to the supplementary equipotential bonding with its supply circuit protected by an RCD.

8.3.14.3.2 Erection according to the zones

In zones 0, 1 and 2, any metallic sheath (or metallic covering) of a wiring system shall be connected to the supplementary equipotential bonding.

Note: Cables should preferably be installed in conduits made of insulating material.

8.3.14.3.3 External influences

Electrical equipment shall have at least the following degree of waterproof protection, in accordance BS EN 60529:

- zone 0: IPX8;
- zone 1: IPX4, IPX5 (where waterjets are likely to occur for cleaning purposes);

- zone 2:

    o   IPX2 for indoor locations;
    o   IPX4 for outdoor locations;
    o   IPX5 where waterjets are likely to occur for cleaning purposes.

8.3.14.3.4 Junction boxes

A junction box shall **not**:

- be installed in zone 0;
- be installed in zone 1 unless it is a SELV circuit.

8.3.14.3.5 Special requirements for the installation of electrical equipment in zone 1 of swimming pools and other basins

Fixed equipment (such as filtration systems, jet stream pumps etc.) that is designed for use in swimming pools and other basins and which is supplied at low voltage, is permitted in zone 1, subject to all the following requirements:

- The equipment shall be located inside an insulating enclosure providing at least Class II or equivalent insulation and protection against mechanical impact of medium severity (AG2).

Note: This regulation applies irrespective of the classification of the equipment.

- The equipment shall only be accessible via a hatch (or a door) by means of a key or a tool which shall disconnect all live conductors and the supply cable.
- The equipment's supply circuit shall be protected by:

    o   SELV; or
    o   an RCD; or
    o   electrical separation.

Switchgear, controlgear and socket outlets shall **not** be installed in zone 0.
    In zone 2, a socket outlet or a switch is permitted **only** if the supply circuit is protected by one of the following protective measures:

- SELV;
- automatic disconnection of supply using an RCD;
- electrical separation.

For swimming pools without a zone 2, lighting equipment supplied by other than a SELV source at 12 V, an a.c. rms or 30 V ripple-free d.c. supply a socket outlet or switch (preferably with a non-conductive cover or coverplate) is permitted in zone 1 **if** it is installed in a wall or ceiling provided that:

- it is at least 2 m above the lower limit of zone 1;
- the circuit is protected by automatic disconnection of the supply;
- additional protection is provided by an RCD;
- every luminaire is protected from mechanical impact of medium severity.

8.3.14.3.6 Underwater luminaires for swimming pools

A luminaire for use in the water or in contact with the water shall be fixed and shall comply with BS EN 60598-2-18.

Underwater lighting located behind watertight portholes, and serviced from behind, shall comply with BS EN 60598 so as to ensure that all such lighting is installed in such a way so that no intentional or unintentional conductive connection between any exposed-conductive-part of the underwater luminaires and any conductive parts of the portholes can occur.

8.3.14.4 Wiring systems

8.3.14.4.1 Additional requirements for the wiring of fountains

For a fountain, the following additional requirements shall be met:

- Electrical equipment in zones 0 or 1 shall be provided with mechanical protection (such as using mesh glass or grids which can only be removed by a specialised tool).
- A luminaire installed in zones 0 or 1 shall be fixed and shall comply with BS EN 60598-2-18.
- An electric pump shall comply with the requirements of BS EN 60335-2-41.

8.3.14.4.2 Protection, isolation, switching, control and monitoring

In zones 0 or 1, switchgear, controlgear and socket outlets shall **not** be installed.

In zone 2, a socket outlet or a switch is permitted, but **only** if the supply circuit is protected by an RCD or SELV, or by electrical separation.

See BS 7671:2018 Section 702 for more detailed requirements relating to swimming pools and other basins.

8.3.15 Temporary electrical installations

This section specifies the minimum electrical installation requirements for the safe design, installation and operation of temporarily erected mobile or transportable electrical machines and structures which incorporate electrical equipment. These machines and structures are typically installed repeatedly and temporarily, at fairgrounds, amusement parks, circuses and/or similar places – ideally without loss of safety.

One of the main requirements is that an adequate number of socket outlets shall be installed to allow the user's requirements to be safely met.

 Note: In booths, stands and for fixed installations, one socket outlet for each square metre or linear metre of wall is generally considered adequate.

8.3.15.1 General

The design and installation of temporary electrical installations should always take into account that:

- socket outlets dedicated to lighting circuits that are placed out of arm's reach should be encoded or marked according to their purpose;
- when used outdoors, plugs, socket outlets and couplers must comply with BS EN 60309;
- all plug and socket outlets (except for SELV) must be of the non-reversible type and capable of being connected to a protective conductor;
- all socket outlets deigned for household and similar use should be of the shuttered type and, for an a.c. installation, preferably of a type complying with BS 1363.

 In certain instances, additional requirements may be necessary for isolation and switching and BS 7671:2018 Table 537.4 summarises the functions provided by the devices for isolation and switching, together with indication of the relevant product standards.

All low-voltage plug and socket outlets shall conform to BS EN 60309.

A plug and socket outlet must **not** be used as a device for connecting a water heater and/or boiler to the supply.

If a final circuit has a number of socket outlets (or connection units intended to supply two or more items of equipment), and where it is known that the total protective conductor current in normal service will exceed 10 mA the circuit must be provided with:

- a high-integrity protective conductor connection; or
- a radial final circuit with a ring protective conductor; or
- a radial final circuit with a single protective conductor.

Plugs and socket outlets in a SELV system must **not** have a protective conductor contact.

8.3.15.2 Protective measures

Switchgear and controlgear needs to be placed in cabinets which can be opened only by the use of a key or a tool, except for those parts designed and intended to be operated by ordinary persons.

The protective measures of obstacles, placing out of reach, non-conducting location and Earth-free local equipotential bonding are **not** permitted.

In addition, a PME earthing facility shall also not be used.

 One of the other requirements in the Regulations is that all electrical equipment shall be protected against solid objects and water splashes.

8.3.15.2.1 Protection against electric shock

Automatic disconnection of supply to the temporary electrical installation has to be provided at the origin of the installation by one or more RCDs with a rated residual operating current not exceeding 300 mA.

8.3.15.2.2 Protection against thermal effects

A motor which is automatically or remotely controlled (and which is not continuously supervised) must be fitted with a manually reset protective device against excess temperature.

8.3.15.2.3 RCDs

Owing to the increased risk of damage to cables in temporary installations, all final circuits for:

- lighting;
- socket outlets; and
- mobile equipment connected by means of a flexible cable or cord;

must be protected by RCDs.

The supply for a battery-operated emergency lighting circuit should be connected to the same RCD protecting the lighting circuit, unless:

- the circuits are protected by SELV or PELV; or
- the circuits protected by electrical separation; or
- the lighting circuits are placed out of arm's reach.

8.3.15.2.4 Supplementary equipotential bonding

- Extraneous-conductive-parts in, or on, the floor (such as concrete reinforcement in general or in reinforcement of cellars for containing liquid manure) shall be connected to the supplementary equipotential bonding.
- In locations intended for livestock, supplementary bonding shall connect all exposed-conductive-parts and extraneous-conductive-parts that can be touched by livestock.
- The supplementary equipotential bonding and the metal grid, if any, shall be erected so that it is permanently protected against mechanical stresses and corrosion.
- Where a metal grid is laid in the floor, it shall be included within the supplementary bonding of the location.

8.3.15.2.5 Selection and erection of equipment

Switchgear and controlgear should be placed in cabinets which can only be opened by the use of a key or a tool, except for those parts designed and intended to be operated by ordinary persons.

Electrical equipment needs to be protected against solid objects and water splashing.

8.3.15.2.6 Electrical connections

Joints shall **not** be made in cables except where necessary as a connection into a circuit.

If joints are required, they shall either use connectors in accordance with the relevant British or Harmonised Standard or made in an enclosure that is dust resistant.

If there is a likelihood of the electrical terminals suffering from the connection, then the connector can be made to include cable anchorage(s).

 The electrical installation between its origin and the electrical equipment must be inspected and tested after it has been assembled on site.

8.3.15.2.7 Electric discharge lamp installations

All luminous tubes, signs or lamps must be installed out of arm's reach or adequately protected to reduce the risk of injury to persons.

8.3.15.2.8 Emergency switching device

A separate circuit (controlled by an emergency switch) **shall** be used to supply luminous tubes, signs or lamps.

The switch **shall** be easily visible, accessible and marked in accordance with the requirements of the local authority.

8.3.15.2.9 Floodlights

Where transportable floodlights are used, they should be mounted so that the luminaire is inaccessible.

Supply cables must be flexible and have adequate protection against mechanical damage.

Luminaires and floodlights need to be fixed and protected against a possible concentration of heat.

8.3.15.2.10 Lampholders

Insulation-piercing lampholders shall **not** be used unless the cables and lampholders are compatible and the lampholders are non-removable once fitted to the cable.

8.3.15.2.10.1 Lamps in shooting galleries   It is essential that all lamps in shooting galleries and other sideshows where projectiles are used are suitably protected against accidental damage.

8.3.15.2.11 Luminaires

Every luminaire and decorative lighting chain shall:

* have a suitable IP rating;
* be installed so as not to spoil its ingress protection; and
* be securely attached to the structure, or support, intended to carry it.

Its weight shall not be carried by the supply cable, unless it has been specifically selected for this purpose.

Luminaires and decorative lighting chains mounted lower than 2.5 m above floor level shall be firmly fixed, sited or guarded so to prevent risk of injury to persons or ignition of materials.

Access to the fixed light source should only be possible after removing a barrier or an enclosure, which shall require the use of a tool.

Lighting chains shall use HO5RN-F (BS 7919) cable or equivalent.

8.3.15.2.12 Socket outlets and plugs

To allow the user's requirements to be met safely, sufficient socket outlets need to be installed.

 Note: In booths, stands and for installations, one socket outlet for each square metre (or linear metre) of wall is generally considered adequate.

8.3.15.2.13 Switchgear and controlgear – isolation

Electrical installations in booths, stands and amusement devices must have their own means of isolation, switching and overcurrent protection (in other words, be readily accessible).

Isolation devices must disconnect all live conductors (i.e. line and neutral conductors).

Temporary electrical installations designed for amusement devices and for supplying outdoor installations, must have their own, readily accessible, properly identified, means of isolation.

8.3.15.3 Supplies

8.3.15.3.1 Automatic disconnection of supply

For supplies to a.c. motors, RCDs (where used) should be of the time-delayed type (BS EN 60947-2) or of the S-type (BS EN 61008-1 or BS EN 61009-1) to prevent unwanted tripping.

8.3.15.3.2 Electric dodgems

Electric dodgems must be electrically separated from the supply mains by means of a transformer or a motor-generator set and operate at voltages not exceeding 50 V a.c. or 120 V d.c.

 Note: Placing out of arm's reach is also acceptable for electric dodgems.

8.3.15.3.3 Electrical supply

Each amusement device must have a connection point that is readily accessible and permanently marked to indicate its rated voltage, current and frequency.

8.3.15.3.4 Low voltage generating sets – generators

All generators should be located or protected so as to prevent any danger and injury to people by unintended contact with hot surfaces and dangerous parts.

Electrical equipment associated with the generator shall be mounted securely and, preferably on anti-vibration mountings.

Where a generator supplies a temporary installation that is part of a TN, TT or IT system, care shall be taken to ensure that the earthing arrangements are in accordance with the Regulations.

The neutral conductor of the star-point of the generator shall (except for an IT system) be connected to the exposed-conductive-parts of the generator.

8.3.15.3.5 Safety isolating transformers and electronic convertors

- A manually reset protective device shall protect the secondary circuit of each transformer or electronic convertor.

- Access for competent persons for inspection, testing and maintenance testing shall be provided.
- Enclosures containing rectifiers and transformers shall be adequately ventilated and the vents shall not be obstructed when in use.
- Electronic converters shall conform to BS EN 61347-2-2.
- Safety isolating transformers shall comply with BS EN 61558-2-6 or provide an equivalent degree of safety.
- Safety isolating transformers shall be mounted out of arm's reach or be mounted in a location that provides equal protection, and shall have adequate ventilation.

8.3.15.3.6 Supply from the public network

The line and neutral conductors from different sources of supply shall **not** be interconnected downstream from the origin of the temporary electrical installation.

8.3.15.3.7 TN system

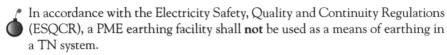

In accordance with the Electricity Safety, Quality and Continuity Regulations (ESQCR), a PME earthing facility shall **not** be used as a means of earthing in a TN system.

8.3.15.4 *Wiring systems*

8.3.15.4.1   Cables and cable management systems

- All cables shall meet the requirements of BS EN 60332-1-2.
- Armoured cables or cables that are protected against mechanical damage shall be used wherever there is a risk of mechanical damage due to external influence.
- Buried cables shall be protected against mechanical damage and (if buried in the ground) be marked by electrical tape (Figure 8.7).
- Cables shall have a minimum rated voltage of 450/750 V.
- Cable trunking systems and cable ducting systems shall comply with BS EN 50085 regarding protection against impact.
- Conduit systems shall comply with the relevant part of the BS EN 61386 series.
- Where flexible conduit systems are provided, they shall comply with BS EN 61386-23.
- Mechanical protection shall be used in public areas and in areas where wiring systems are crossing roads or walkways.
- Tray and ladder systems shall comply with BS EN 61537.
- Where subjected to movement, wiring systems shall be of flexible.

Figure 8.7 Warning tape – cable below.

 See BS 7671:2018 Section 740 for more detailed requirements relating to temporary electrical installations for structures, amusement devices and booths at fairgrounds, amusement parks and circuses.

> ### Author's end note
>
> *Having thoroughly reviewed the various basic requirements of the Wiring Regulations; its relationship with the Building Regulations; the need for pay special attention to electrical safety; earthing; external influences; and some of the special installations and locations in which electric and electromagnetic equipment have to be controlled, in the next chapter we look at the methods to install, maintain and (when required) repair these electrical installations.*

9

Installation, maintenance and repair

Author's start note

Although this is a small chapter, it is nevertheless an extremely important one as it provides some guidance on the requirements for installation, maintenance and repair. It also lists the Regulations' requirements for maintenance etc. with respect to electrical installations – but as per the previous chapters which contain similar lists, please remember that this is only the author's impression of the most important aspects of the Wiring Regulations, and electricians should always consult the latest edition of BS 7671 to satisfy compliance. Finally, in Annex 9A, there is a complete set of check lists for the quality, safety and environmental control of electrical equipment and electrical installations.

According to studies recently completed by CEN/CENELEC, the installation and maintenance engineer is the primary cause of reliability degradations during the in-service stage of most electrical installations. The problems associated with poorly trained, poorly supported and/or poorly motivated personnel with respect to reliability and dependability, therefore, requires careful assessment and quantification.

The most important factor, however, that affects the overall reliability of a modern product (system or installation) is the increased number of individual components that are required in that product. Since most system failures are actually caused by the failure of a single component (equipment item or sub-assembly), the reliability of each individual component must be considerably better than the overall system reliability.

Because of this requirement, quality standards for the installation, maintenance, repair and inspection of in-service products have had to be laid down in engineering standards, handbooks and local operating manuals (written for specific items and equipment). These publications are used by maintenance engineers and should always include the most recent amendments. It is essential, therefore, that the quality-, safety- and environ-mental-assurance personnel also use the same procedures for their inspec-tions as were used for the installation.

 Note: Schedules for installing wiring systems (together with guidance for selecting the appropriate size of cable, current ratings etc.) are shown in Appendix 4, Tables 9.4A1 and 9.4A2 of BS 7671:2018.

One part of the Regulations is devoted entirely to inspection and testing (i.e. Part 6) and emphasises the need for continual improvement by stating:

> *"Every installation (or alteration to an existing installation) shall, during erection and on completion before being put into service, be inspected and tested to verify, so far as is reasonably practicable, that the requirements of the Regulations have been met.*
>
> *The verification shall be made by a competent person and on com-pletion of the verification, a certificate shall be prepared."*

9.1 General

As previously shown, the requirements of BS 7671:2018 apply to the design, erection and verification of electrical installations (see Fig 9.1).

This is particularly relevant with installations such as those of:

- agricultural and horticultural premises;
- caravans, caravan parks and similar sites;
- commercial premises;
- construction sites, exhibitions, shows, fairgrounds and other installa-tions for temporary purposes (including professional stage and broadcast applications);
- domestic buildings;
- external lighting and similar installations;
- industrial premises;
- highway equipment and street furniture;
- low voltage generating sets;
- marinas;
- medical locations;
- mobile or transportable units;
- photovoltaic systems;
- prefabricated buildings;
- public premises;
- residential premises.

 Note: *'Premises'* covers the land **and** all facilities including buildings belonging to it.

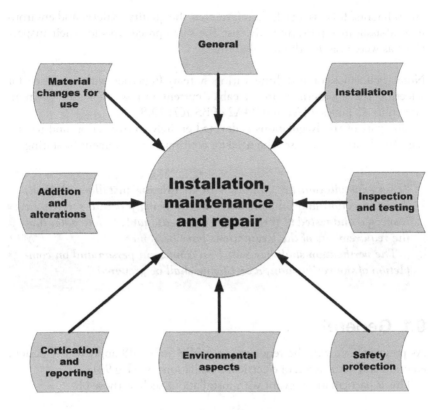

Figure 9.1 Installation, repair and maintenance.

The Regulations do **not** apply to any of the following installations:

- aircraft equipment;
- electrical equipment of machines covered by BS EN 60204;
- equipment for mobile and fixed offshore installations;
- equipment for motor vehicles;
- equipment onboard ships covered by BS 8450;
- lightning protection systems for buildings and structures covered by BS EN 62305;
- systems for the distribution of electricity to the public;
- railway traction equipment, rolling stock and signalling equipment;
- those aspects of lift installations covered by relevant parts of BS 5655 and BS EN 81-1;
- those aspects of mines and quarries specifically covered by Statutory Governmental Regulations;
- radio interference suppression equipment (except so far as it affects the safety of the electrical installation).

9.2 Installation

Many requirements and recommendations for the installation of electrical equipment are to be found in BS 7671:2018. Most of these have already been mentioned in other parts of this book and so the intention of this particular chapter is to bring to your attention some of the most important ones as shown in Figure 9.2. It should **not**, however, be viewed as a complete list!

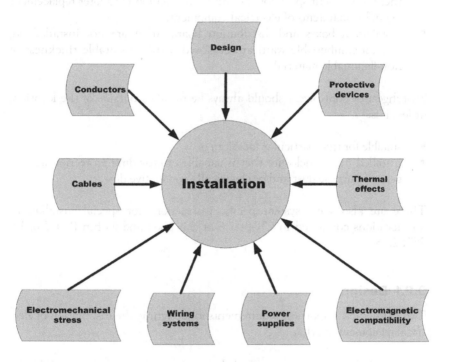

Figure 9.2 Installation.

 Note: Also see Chapter 8 of this book for specific requirements concerning special installations and locations, and Chapter 10 for inspections and tests.

Electrical equipment must be installed so that:

- all live parts are installed inside enclosures or behind barriers and a suitable warning label has been provided – particularly if piece of equipment in question could retain a dangerous electrical charge after it has been switched off;
- electrical joints and connections are properly constructed with regard to conductance, insulation, mechanical strength and protection;
- precautions have been taken to prevent persons or livestock from unintentionally touching live parts;

- the characteristics of the electrical equipment are not damaged or weakened during the installation process;
- the design temperatures are not exceeded;
- the risk of ignition of flammable materials due to high temperature or electric arc is minimised;
- the equipment is fully accessible for operation, inspection, testing, fault detection, maintenance and repair;
- there is sufficient space for the initial installation (and later replacement of) individual items of electrical equipment;
- installation boxes and distribution boards (that are not installed on, or in, a combustible wall) are enclosed within a suitable thickness of non-flammable material.

Switchgear or controlgear should always be installed **outside** of the location, unless it is:

- suitable for that particular location; or
- installed in an enclosure that is capable of providing protection against general purpose dust and/or electrically conductive dust.

There are also some system-specific instructions for special installations and locations contained in Chapter 8 of this book and within Part 7 of BS 7671:2018.

9.2.1 Design

Two of the most important requirements for designing the installation of electrical equipment are that:

- installed equipment items, and their connections, must be accessible for operational, inspection and maintenance purposes; and
- equipment must be arranged to allow easy access for periodic inspection, testing and maintenance.

Every installation needs to be divided into circuits in order to:

- avoid danger and minimise inconvenience in the event of a fault;
- ensure safe inspection, testing and maintenance;
- prevent the indirect energising of a circuit that is otherwise intended to be isolated;
- reduce the possibility of unwanted tripping of RCDs due to excessive protective conductor (PE) currents that are not due to a fault;
- reduce the effects of electromagnetic interferences (EMI);
- take account of any danger that may arise from the failure of a single circuit (e.g. a lighting circuit).

Separate circuits shall be provided for parts of the installation which need to be individually controlled.

The electrical installation must be designed to ensure:

- the correct functioning of an electrical installation is sufficient for its intended use;
- the installation of equipment (e.g. fixing, connection of conductors, etc.) does not affect the protection afforded by an enclosure;
- the maximum demand of an installation is achievable;
- the protection of persons, livestock and property is guaranteed.

The design needs to take into consideration the possibility of common, detrimental influences occurring between the electrical installation and other non-electrical installations. Other considerations include:

- equipment being selected and installed to provide for the safety and proper functioning for the intended use of the installation;
- installed equipment being appropriate for all predicted external influences;
- that a switching device has **not** been inserted in the neutral conductor, alone.

 If the use of a new material or invention is different from the Regulations, then it is essential that resulting degree of safety of the installation is not less than that obtained by compliance with the Regulations.

9.2.2 Protective devices

Overload fault protective devices (other than those locations where there is a possibility of fire risk or explosion occurring or where a particular installation specifically requires different conditions) should be installed at any point where a reduction in the current-carrying capacity of an installation's conductors is likely to occur.

An overload device protecting a conductor may be installed along the run of that conductor provided that that part of the run (i.e. between the point where a change occurs and the position of the protective device) does not include any branch circuits or outlets for connection of current-using equipment and is:

- protected against fault current;
- shorter than 3 m;
- installed so as to reduce the risk of a fault occurring to an absolute minimum; and
- installed so as to reduce to a minimum the risk of fire or danger to persons.

A device (such as a circuit-breaker with a short-circuit release, or a fuse) that is designed to provide protection against fault current shall **only** be installed where overload protection is achieved by other means.

Unless an RCD or other protective device is installed to interrupt the supply in the event of the first Earth fault, an RCM (or an insulation fault location system) should be provided so as to indicate that there is a fault from a live part to an exposed-conductive-part, or to Earth.

9.2.3 Thermal effects

Electrical installation must be installed so that:

- the risk of igniting any flammable materials due to a high temperature or an electric arc is minimised;
- during normal operation of the electrical equipment, there will be minimal risk of burns to persons or livestock.

9.2.4 Electromagnetic comparability

All electrical installations and equipment shall be in accordance with the current electromagnetic compatibility (EMC) regulations and the relevant EMC standard for that particular installation (e.g. IEC 60601-1-2 for medical products).

The design of an electrical installation should aim at reducing the effect of induced voltage disturbances and electromagnetic interferences (EMI), and the installation design shall take into consideration the anticipated electromagnetic disturbances generated either by the installation itself or the installed equipment.

9.2.5 Electromechanical stresses

Every conductor or cable shall have adequate strength and be installed so that it can withstand electromechanical forces that may be caused by any current, including fault current, it may have to carry in service.

9.2.6 Power supplies

The type of earthing system to be used for an electrical installation needs to take into consideration the characteristics of the source of energy and (in particular) any current earthing facilities.

In a TN, TT or IT system, an RCD with a rated residual operating current of not more than 30 mA will need to be installed to protect every circuit.

The safety sources for safety services shall be installed as fixed equipment that cannot be affected by failure of the normal source. The following electrical sources for safety services are recognised:

- storage batteries;
- primary cells;
- generator sets (independent of the normal supply);
- independent feeders from the supply network.

Note: Stationary batteries need to be installed in a secure location so that they are accessible **only** to skilled or instructed persons.

Suitable precautions shall be taken to prevent the installation or installed equipment suffering from danger or damage owing to an interruption of supply.

A disconnection device must be installed so that it can easily and effectively allow switching and/or isolation of the electrical installation, circuits or individual items of equipment as and when required for operation, inspection, testing, fault detection, maintenance and repair.

9.2.7 Wiring systems

Electrical installations shall be so arranged that:

- the risk of ignition of flammable materials due to high temperature or an electric arc is minimised;
- during normal operation of the electrical equipment, there shall be minimal risk of burns to persons or livestock.

The type of wiring system and the method of installation will depend on:

- the nature of the location;
- the type of structure supporting the wiring;
- accessibility of wiring to persons and livestock;
- the voltage;
- the electromechanical stresses likely to occur due to short-circuit and Earth fault currents;
- electromagnetic interference;
- other external influences (e.g. mechanical, thermal or anything associated with fire) to which the wiring is likely to be exposed during the erection of the electrical installation or in service.

The installation of wiring systems will meet requirements if:

- the rated voltage of the cable(s) is not less than the nominal voltage of the system; and
- adequate mechanical protection of the basic insulation is provided by (one or more) of the following:

  o the cable's non-metallic sheath;
  o non-metallic trunking or ducting (complying with BS EN 50085);
  o non-metallic conduit (complying with BS EN 61386).

Where an electrical service is located near to one or more non-electrical service, it shall meet the following conditions:

- The wiring system has to be suitably protected against the hazards likely to arise from the presence of these other services whilst in normal use.

- Fault protection shall be provided by means of automatic disconnection of supply.
- The wiring system has to have been installed so that the general building structural performance and fire safety are not reduced.

A wiring system which passes through the location but is not intended to supply electrical equipment within that location shall:

- have no connection or joint actually in the location (unless the connection or joint is installed in an enclosure); and
- be protected against overcurrent; and
- not use bare live conductors;
- fully meet the tests on electric cables described in BS EN 60332-1-2.

A wiring system shall **not** be installed near to any services that produce heat, smoke or fumes which could damage the wiring, **unless** it is protected from these harmful effects by some form of shielding that dissipates the heat from the wiring.

9.2.8 Cables

The installation of cables should be undertaken in accordance with the following precautions, recommendations and requirements:

- Cables should preferably be installed in conduits made of an insulating material.
- Buried cables, conduits and ducts shall be at a sufficient depth to avoid being damaged by any reasonably foreseeable disturbance of the ground and (unless they are installed in a conduit or duct) use an earthed armour or metal sheath as a protective conductor.
- The location of buried cables shall be marked by cable covers or a suitable marker tape.
- Surface-run and overhead cables shall be protected against mechanical damage.
- Cables installed under a floor or above a ceiling shall not be at risk of being damaged by contact with the floor, or the ceiling, or their fixings.
- Cables should not be installed across a site road or a walkway unless they are protected against mechanical damage.
- Cable supports and enclosures should not have any sharp edges that could damage the wiring system.
- Cables should not be installed in a location where they are liable to be covered by thermal insulation.
- Cables need to be adequately supported against their premature collapse in the event of a fire throughout the installation.
- Cables shall not pass from one fire-segregated compartment to another.

- Where multi-core cables are installed in parallel, each cable shall contain one conductor of each line.
- Flexible cables used for fixed wiring shall be of the heavy-duty type.
- Cables complying with the requirements of BS EN 60332-1-2 may be installed without special precautions; l otherwise, they must be limited to short lengths for connection of appliances.
- For battery cables, special requirements may apply.

Cables shall **not** be run in a lift or hoist well unless it forms part of the lift installation.

9.2.9 Conductors

A bare live conductor **shall** always be installed on insulators.

A device protecting a conductor against overload may be installed along the run of that conductor provided that that part of the run (i.e. between the points where a change occurs and the position of the protective device) has neither branch circuits nor outlets for connection of current-using equipment, and:

- is protected against fault current;
- is less than 3 m in length;
- has been installed so as to reduce:

  o the risk of fault to a minimum; and
  o the risk of fire or danger to persons.

Note: The number and type of live conductors (e.g. single-phase, two-wire a.c., three-phase, four-wire a.c.) are established based on the source of energy and the type of circuit(s) used within the installation.

9.3 Inspection and testing

During erection, on completion of an installation, and addition or alteration to an installation (and before it is put into service) appropriate inspections and tests (as indicated in Figure 9.3) must be carried out by competent and skilled persons to verify that the requirements of BS 7671:2018 have been met.

In addition, an assessment should be made of the frequency and quality of maintenance the installation can reasonably be expected to receive during its intended life and to ensure that:

- any periodic inspection and testing, maintenance and repairs likely to be necessary during the intended life can be readily and safely carried out; and
- the effectiveness of the protective measures for safety during the intended life shall not diminish; and
- the reliability of equipment designed to enable proper functioning of the installation is appropriate to the intended life.

Figure 9.3 Inspection and testing.

 Note: Any defect or omission that will affect safety due to the addition or alteration that is revealed during inspection and testing must obviously be corrected before an Inspection Certificate is issued.

9.3.1 Frequency of inspection and testing

The amount and timing of inspections and maintenance of an installation will depend on the type of installation or equipment, its use and operation, the current frequency and quality of maintenance provided and any external influences to which it is subjected.

The results and recommendations of the previous report, if any, shall also be taken into account.

In the case of an installation that has an effective management system for preventive maintenance whilst in normal use, periodic inspection and testing may be replaced by an adequate regime of continuous monitoring and maintenance of the installation and all its constituent equipment by skilled persons, competent in such work.

 Full records of these inspections and tests **must**, however, be retained.

9.3.2 Periodic inspection

The aims of periodic inspection and testing are to:

*   confirm that the installation is in a satisfactory condition for continued service and has not been damaged;

- confirm that the safety of the installation has not deteriorated;
- ensure the continued safety of persons and livestock against the effects of electric shock and burns;
- identify installation defects and non-compliance with the requirements of the Regulations which may give rise to danger;
- prevent property from being damaged by fire and heat caused by a defective installation.

 The Regulations state that maintenance inspections **shall** consist of a careful scrutiny of the installation (dismantled or otherwise), using the appropriate tests described in Chapter 6 of BS 7671:2018.

As previously mentioned, the frequency of maintenance inspections will depend on:

- the type of installation, its use and operation;
- the frequency and quality of maintenance; and
- the external influences to which it is subjected.

Precautions need to be taken to ensure that maintenance inspections do not cause:

- danger to persons or livestock;
- damage to property and equipment (even if the circuit is defective).

The maintenance inspection is made to verify that the installed electrical equipment:

- complies with the requirements of the Regulations and the appropriate National Standards and European Harmonised Directives;
- is correctly selected and erected in accordance with the Regulations etc.;
- is not visibly damaged or defective so as to impair safety.

The inspection shall include the following items, where relevant:

- access to switchgear and equipment;
- cable routing in safe zones;
- choice and setting of protective and monitoring devices;
- connection of conductors;
- connection of single-pole devices for protection or switching in phase conductors only;
- correct connection of accessories and equipment;
- erection methods;
- identification of conductors;
- labelling of protective devices, switches and terminals;
- presence of danger notices and other warning signs;
- presence of diagrams, instructions and similar information;
- presence of fire barriers and suitable seals;

- presence of isolation and switching devices (and their correct location);
- prevention of mutual (i.e. detrimental) influences;
- presence of undervoltage protective devices;
- protection against electric shock (direct and indirect contact);
- protection against external influences;
- protection against mechanical damage;
- protection against overcurrent;
- protection against thermal effects;
- selection of conductors for current-carrying capacity and voltage drop.

There are various requirements for forced air heating systems, to be found in BS 7671.

Under workshop conditions it should be possible to gain access to all circuitry whilst it is operating (with a minimum of effort required), to partially dismantle the module concerned with the assumption that there will be only a small risk to the components, or to the testing maintenance staff.

Special connecting leads, printed wiring extension boards and any other special items required for maintenance purposes, together with the mating half of all necessary connectors, will probably have to be obtained from the manufacturer.

Equipment is normally expected to have been designed to have a useful life of not less than 20 years. 'Useful life' normally means "the period for which the equipment will continue to operate with the specified level of reliability".

No components, modules or equipment should, therefore, be used (so far as can be ascertained at the time of manufacture) for which spares cannot be fully guaranteed to be available **throughout** the life of the equipment.

Note: Generally speaking, the following CEN/CENELEC recommendations are relevant for all installed electrical and/or electronic equipment.

CENELEC Recommendations

For ease of maintenance, all equipment provided should have:

- *easily accessible test points to facilitate fault location;*
- *modules that have been constructed so as to facilitate the connection of test equipment (e.g. logic analysers, emulators and test ROMs etc.);*
- *fault location provision to allow functional areas within each module or equipment to be isolated.*

9.3.3 Accessibility of electrical equipment

It is essential that all parts of a wiring system which may require maintenance are made safe and that installations can guarantee acceptable access to all those qualified to complete these duties.

To achieve this requirement, electrical equipment must be arranged so that:

- its safe operation, inspection and maintenance is enabled;
- there is sufficient space for the initial installation and later replacement of individual items of electrical equipment;
- the equipment is fully accessible for operation, inspection, testing, fault detection, maintenance and repair, and access to each connection.

Every installation shall be divided into circuits, as necessary, to:

- avoid hazards and minimise inconvenience in the event of a fault;
- ensure safe inspection, testing and maintenance;
- prevent the indirect energising of a circuit that is intended to be isolated;
- reduce the possibility of unwanted tripping of RCDs due to excessive protective conductor currents produced by equipment;
- reduce the effects of electromagnetic interferences (EMI);
- take account of any danger that may arise from the failure of a single circuit (e.g. a lighting circuit).

 Note: The selection and erection of equipment for solar photovoltaic (PV) power supply systems must enable maintenance and/or service work to be carried out safely.

9.3.4 Restricted access areas

Section 729 of BS 7671:2018 is a fairly new and relatively small addition to the standard, which centres on the operation and safe maintenance of switchgear and controlgear within areas that included gangways and where access is restricted to skilled or instructed persons.

Restricted access areas shall:

- be clearly and visibly marked by appropriate signs;
- not provide access to unauthorised persons; and
- provide closed doors that, although normally shut, nevertheless will allow easy evacuation in case of danger without the use of a key, tool or any other device that is not part of the opening mechanism.

The width and access of gangways and access areas shall:

- be suitable for a normal working environment with operational access, emergency access, emergency evacuation and for transport of equipment;
- permit at least a 90° opening of equipment doors or hinged panels; and
- gangways longer than 10 m shall be accessible from both ends.

Where basic protection is provided by obstacles, barriers or enclosures, the following minimum dimensions will apply (Table 9.1):

Table 9.1 Minimum dimensions of gangways

Gangway	Dimensions
Gangway width between:	700 mm
• Barriers, enclosures, switch handles or circuit-breakers • obstacles and switch handles or circuit-breakers	
Gangway width between:	700 mm
• barriers or enclosures (and/or other barriers or enclosures) and the wall • obstacles (or other obstacles and enclosures) and the wall	
Height of gangway to barrier or enclosure above floor	2000 mm
Live parts placed out of reach	2500 mm

Note: Where additional workspace is needed (such as special switchgear and controlgear assemblies), larger dimensions may be required.

9.3.5 Connections

Every connection shall be accessible for inspection, testing and maintenance, except for:

- a connection between a cold tail and the heating element (e.g. in ceiling heating, floor heating or a trace heating system);
- a joint designed to be buried in the ground;
- a joint that is either compound filled or encapsulated;
- a joint made by welding, soldering, brazing or with a compression tool;
- a joint (or connection) made in the equipment by the manufacturer and not intended to be inspected or maintained;
- equipment complying with BS 5733 that is intended to be a maintenance-free accessory.

9.3.6 Switching off for mechanical maintenance

If mechanical maintenance could involve a risk of physical injury, then a device such as a:

- multipole switch;
- circuit-breaker;
- control and protective switching device (CPS);
- control switch operating a contactor;
- plug and socket outlet;

may be inserted in the main supply circuit:

- for switching off electrically powered installations and equipment; or
- to prevent electrically powered equipment from becoming unintentionally reactivated during mechanical maintenance;

provided that the device used for switching off the main power (or a control switch for such a service) will:

- require manual operation;
- be designed and/or installed so as to prevent inadvertent or unintentional switching on;
- be so placed and durably marked so as to be readily identifiable and convenient for the intended use;
- ensure that the open position of the contacts of the device shall be visible or clearly and reliably indicated by the use of the symbols O and I to indicate the open and closed positions, respectively.

 If a switch is used as a device for switching off for mechanical maintenance, it **shall** be capable of cutting off the full load current of the relevant part of the installation.

9.3.7 Disconnecting devices

Disconnecting devices shall be provided so as to allow electrical installations, circuits or individual items of equipment to be switched off or isolated for the purposes of operation, inspection, testing, fault detection, maintenance and repair.

9.3.8 Insulating enclosure

An insulating enclosure shall ensure that:

- any conductive part mounted inside that enclosure is not, and cannot be, connected to a protective conductor;
- insulating barriers or enclosures can only be opened with the use of a tool or key;
- the operation of the equipment protected in this way is not adversely affected;
- it is not criss-crossed by conductive parts which are likely to transmit a potential;
- it does not contain any screws or other fixing means which might need to be removed (e.g. during installation and maintenance) and which could be replaced by metallic screws or some other type of fixing that would then affect the enclosure's insulation.

9.3.9 Low-power supply sources

The power output of a low power supply system should be inspected so as to ensure that it is limited to 500 W for 3-hour duration or 1500 W for 1-hour duration.

Batteries shall be of heavy-duty industrial design (e.g. those complying with BS EN 60623 or BS EN 60896) and with a minimum design life of 5 years.

9.3.10 Multi-phase sequence

For multi-phase circuits, it shall be verified that the phase sequence is maintained at all relevant points throughout the installation, and that overcurrent detection has been provided for the neutral conductor.

9.4 Safety protection

Disconnecting devices shall be provided so as to allow electrical installations, circuits or individual items of equipment to be switched off (or isolated) for the purposes of operation, inspection, fault detection, testing, maintenance and repair.

Where it is necessary to remove a protective measure in order to carry out this sort of maintenance or repair etc., it shall be done so that the protective measure can be reinstated without any reduction of the degree of protection that it was originally designed for.

The protective measures of placing out of reach and obstacles shall **not** be used unless the maintenance of equipment is restricted to specially trained skilled persons.

Measuring and monitoring equipment (and its method of use) shall be chosen according to BS EN 61557. **If**, other measuring equipment is used, it shall provide no lesser degree of performance and safety.

 Note: Safety isolating transformers shall comply with BS EN 61558-2-6 or provide an equivalent degree of safety.

9.5 Environmental aspects

The design of the electrical installation shall take into account the environmental conditions to which it will be subjected, and all wiring systems shall be selected and erected so that:

- no damage is caused by condensation or ingress of water during installation, use and maintenance;
- all possible damage arising from mechanical stress (such as by impact, abrasion, penetration, tension or compression) during installation, use or maintenance is minimised;

- during installation, use or maintenance, damage to the sheath or insulation of cables and their terminations is minimised;
- the degree of protection of electrical equipment is fully maintained after the connection of the cables and conductors;
- a wiring system buried in a floor is sufficiently protected against damage caused by the intended use of the floor.

 The use of any lubricants that could have a detrimental effect on the cable or wiring system is **not** permitted.

9.6 Certification and reporting

As explained in Chapter 10 of this book (and as listed in Table 9.2), the following certificates shall be completed where appropriate:

 Please note that all Electrical Installation Certificates, Electrical Installation Condition Reports and Minor Electrical Installation Works Certificates **MUST** be compiled and signed (or otherwise authenticated) by a competent person or persons.

Table 9.2 Certificates and Report required

Type of Certificate	Content
Electrical Installation Certificate (EIC)	A safety certificate that is issued by a qualified electrician as confirmation that an electrical installation project complies with the requirements of BS 7671:2018 It is used only for the initial certification of a new installation or for an alteration or addition to an existing installation where new circuits have been introduced. It contains details of the installation together with a record of the inspections made and the test results It is **not** to be used for a periodic inspection, for which a Periodic Inspection Report form should be used
Electrical Installation Condition Report (EICR)	A periodic inspection report on a property's safety, relating to its fixed wiring. Its main purpose is to guarantee the safety of the residents and to ensure they are not susceptible to electrical shocks and/or fire It contains details of the condition of an existing installation – together with records of previous inspection, the results of testing and a recommendation for when the next periodic inspection should occur
Minor Electrical Installation Works Certificate (MEIC)	For all minor electrical installation work that does not include the provision of a new circuit

Although alterations and repairs are frequently completed during an installation, normal repairs that are carried out following a periodic inspection, and reports of these repairs shall be compiled and signed, or otherwise authenticated, by a competent person or persons.

Following the periodic inspection, an Electrical Installation Condition Report, together with a schedule of inspections and a schedule of test results, shall be given by the person carrying out the inspection, or a person authorised to act on their behalf, to the person ordering the inspection (normally the property owner).

Any damage, deterioration, defects, dangerous conditions and non-compliance with the requirements of the Regulations which may give rise to danger shall be recorded, together with any significant limitations of the inspection and testing, including their reasons.

Reports may be produced in any durable medium, including written and electronic media.

Regardless of the media used for original certificates and/or reports (or their copies), they shall have their authenticity and integrity verified by a reliable process or method. The process or method shall also verify that any copy is a true copy of the original.

 Records are an important part of the management of any electrical installation as they provide objective evidence of activities performed and/or results achieved. The following section provides some of the requirements that come from the Regulations.

9.7 Additions and alterations to an installation

The Regulations require every electrical installation to be inspected and tested during erection and on completion before being put into service to verify that the requirements of BS 7671:2018 have been met. By definition, this requirement also applies to alterations and/or additions to an existing installation, as well as to entirely new installations.

The following is a summary of the relevant requirements for additions and alterations to an installation.

No addition or alteration, temporary or permanent, shall be made to an existing installation unless:

- it has been ascertained that the rating and the condition of any existing equipment (including that of the distributor) will be adequate for the altered circumstances;
- the earthing and bonding arrangements used as a protective measure for the safety of the addition or alteration are adequate.

If wiring additions or alterations are made to an installation such that some of the wiring complies with the current 2018 edition of BS 7671 (i.e. the 18th edition of the Wiring Regulations), but there is also wiring to previous versions of these Regulations (e.g. the 17th Edition of the Wiring Regulations – BS 7671:2008), a warning notice shall be affixed at or near the appropriate distribution board with the following wording (Figure 9.4):

Figure 9.4 Warning notice – non-standard colours.

The contractor or other person responsible for all work competed (including additions and alterations), or a person authorised to act on their behalf, shall record on the Electrical Installation Certificate or the Minor Electrical Installation Works Certificate, any defects found, so far as is reasonably practicable, in the existing installation.

9.7.1 Alternative supplies

Where an installation includes alternative or additional sources of supply, warning notices shall be affixed at the following locations in the installation:

- at the origin of the installation;
- at the meter position (if remote from the origin);
- at the consumer unit or distribution board to which the alternative or additional sources are connected;
- at all points of isolation of all sources of supply.

The warning notice shall have the following wording (Figure 9.5):

Figure 9.5 Warning notice – alternative or additional sources of supply.

9.8 Material changes of use

Where there is a material change of use of a building, any work carried out shall ensure that the building complies with the applicable requirements of the following paragraphs of Schedule 1 of the Building Act 1984 (and its associated Approved Document):

   (a) In all cases:

- *means of warning and escape (B1);*
- *internal fire spread – linings (B2);*
- *internal fire spread – structure (B3);*
- *external fire spread (B4);*
- *access and facilities for the fire service B5);*
- *resistance to moisture (C1)(2);*
- *dwelling houses and flats formed by material change of use (E4);*
- *ventilation (F1);*
- *sanitary conveniences and washing facilities (G1);*
- *bathrooms (G2);*
- *foul water drainage (H1);*
- *solid waste storage (H6);*
- *combustion appliances (J1, J2 & J3);*
- *conservation of fuel and power – dwellings(L1);*
- *conservation of fuel and power – buildings other than dwellings (L2);*
- *electrical safety (P1);*

(b) In other cases [Table 9.3]:

Table 9.3 Building Act Requirements

Material Change of use	Requirement	Approved Document
The building is used as a dwelling, where previously it was not	Resistance to moisture	C2 E1, E2, E3
The public building consists of a new school	Acoustic conditions in schools	E4
The building contains a flat, where previously it did not	Resistance to the passage of sound	E1, E2, E3
The building is used as an hotel or a boarding house, where previously it was not	Structure	A1, A2, A3 E1, E2, E3
The building is used as an institution, where previously it was not	Structure	A1, A2, A3
The building is used as a public building, where previously it was not	Structure	A1, A2 A3 E1, E2, E3
The building is not a building described in Classes I to VI in Schedule 2, where previously it was	Structure	A1, A2, A3
The building, which contains at least one room for residential purposes, contains a greater or lesser number of dwellings than it did previously	Structure	A1, A2, A3 E1, E2, E3
The building, which contains at least one dwelling, contains a greater or lesser number of dwellings than it did previously	Resistance to the passage of sound	E1, E2, E3

 Note: In some circumstances (particularly when a historic building is undergoing a material change of use and where the special characteristics of the building needs to be recognised) it may **not** be practical to improve sound insulation to the standards set out in Part E1 or resistance to contaminants and water as set out in Part C. In these cases, the aim should be to improve the insulation and resistance as far as practically possible – always provided that the work does not prejudice the character of the historic building, or increase the risk of long-term deterioration to the building fabric and/or fittings.

Annex 9A – Example stage audit checks

Design stage

As shown in Table 9.4, the Design stage is mainly about the quality management of a stage audit check.

Table 9.4 Example stage audit checks

Item	Related item	Remark
1 Requirements	1.1 Information	Has the customer fully described his requirement?
		Has the customer any mandatory requirements?
		Are the customer's requirements fully understood by all members of the design team?
		Is there a need to have further discussions with the customer?
		Are other suppliers or subcontractors involved? If yes, who is the prime contractor?
	1.2 Standards	What international standards need to be observed? Are they available?
		What national standards need to be observed? Are they available?
		What other information and procedures are required? Are they available?
	1.3 Procedures	Are there any customer-supplied drawings, sketches or plans? Have they been registered?
2 Quality procedures	2.1 Procedures manual	Is one available?
		Does it contain detailed procedures and Instructions for the control of all drawings within the drawing office?
	2.2 Planning implementation and production	Is the project split into a number of work packages? If so:
		• Are the various work packages listed? • Have work package leaders been nominated? • Is their task clear? • Is their task achievable?
		Is a time plan available? Is it up to date? Regularly maintained? Relevant to the task?
3 Drawings	3.1 Identification	Are all drawings identified by a unique number?
		Is the numbering system strictly controlled?

Table 9.4 *(continued)*

Item	Related item	Remark
	3.2 Cataloguing	Is a catalogue of drawings maintained? Is this catalogue regularly reviewed and up to date?
	3.3 Amendments and modifications	Is there a procedure for authorising the issue of amendments, changes to drawings? Is there a method for withdrawing and disposing of obsolete drawings?
4 Components	4.1 Availability	Are complete lists of all the relevant components available?
	4.2 Adequacy	Are the selected components currently available and adequate for the task? If not, how long will they take to procure? Is this acceptable?
	4.3 Acceptability	If alternative components have to be used, are they acceptable to the task?
5 Records	5.1 Failure reports	Has the design office access to all records, failure reports and other relevant data?
	5.2 Reliability data	Is reliability data correctly stored, maintained and analysed?
	5.3 Graphs, diagrams, plans	In addition to drawings, is there a system for the control of all relevant graphs, tables, plans etc.? Are CAD facilities available? (If so, go to 6.1)
6 Reviews and audits	6.1 Computers	If a processor is being used: • Are all the design office personnel trained in its use? • Are regular back-ups taken? • Is there an anti-virus system in place?
	6.2 Manufacturing division	Is a close relationship being maintained between the design office and the manufacturing division?
	6.3	Is notice being taken of the manufacturing division's exact requirements, their problems and their choices of components etc.?

Installation stage

Table 9.5 lists the requirements of an installation and how they should be met.

Table 9.5 Installation stage

Item	Related item	Remark
1 Degree of quality	1.1 Quality control procedures	Are quality control procedures available?
		Are they relevant to the task?
		Are they understood by all members of the manufacturing team?
		Are they regularly reviewed and up to date?
		Are they subject to control procedures?
	1.2 Quality control checks	What quality checks are being observed? Are they relevant?
		Are there laid down procedures for carrying out these checks? Are they available? Are they regularly updated?
2 Reliability of product design	2.1 Statistical data	Is there a system for predicting the reliability of the product's design?
		Is sufficient statistical data available to be able to estimate the actual reliability of the design, before a product is manufactured?
		Is the appropriate engineering data available?
	2.2 Components and parts	Are the reliability ratings of recommended parts and components available?
		Are probability methods used to examine the reliability of a proposed design? If so, have these checks revealed design deficiencies such as:
		• assembly errors?
		• operator learning, motivational or fatigue factors?
		• latent defects?
		• improper part selection?
		(Note: If necessary, use additional sheets to list actions taken)

Acceptance stage

Table 9.6 provides a check list of all the final quality checks that have to be made before the product can be released.

Table 9.6 Acceptance stage

Item	Related item	Remark
1 Product performance		Does the product perform to the required function? If not what has been done about it?
2 Quality level	2.1 Workmanship	Does the workmanship of the product fully meet the level of quality required or stipulated by the user?
	2.2 Tests	Is the product subjected to environmental tests? If so, which ones?
		Is the product field tested as a complete system? If so, what were the results?
3 Reliability	3.1 Probability function	Are individual components and modules environmentally tested? If so, how?
	3.2 Failure rate	Is the product's reliability measured in terms of probability function? If so, what were the results?
		Is the product's reliability measured in terms of failure rate? If so, what were the results?
	3.3 Mean time between failures	Is the product's reliability measured in terms of mean time between failure (MTBF)? If so, what were the results?

In-service stage

Table 9.7 is a list of all the checks that need to be made once the product has been installed into a working environment.

Table 9.7 In-service stage

Item	Related item	Remark
1 System reliability	1.1 Product basic design	Are statistical methods being used to prove the product's basic design? If so, are they adequate? Are the results recorded and available?
		What other methods are used to prove the product's basic design? Are these methods appropriate?
2 Equipment reliability	2.1 Personnel	Are there sufficient trained personnel to carry out the task?
		Are they sufficiently motivated? If not, what is the problem?

Table 9.7 *(continued)*

Item	Related item	Remark
2.1.1	Operators	Have individual job descriptions been developed? Are they readily available?
2.1.1	Operators	Are all operators capable of completing their duties?
2.1.2	Training	Do all personnel receive appropriate training?
		Is a continuous on-the-job training (OJT) programme available to all personnel? If not, why not?
2.2	Product dependability	What proof is there that the product is dependable?
		How is the product dependability proved? Is this sufficient for the customer?
2.3	Component reliability	Has the reliability of individual components been considered?
		Does the reliability of individual components exceed the overall system reliability?
2.4	Faulty operating procedures	Are operating procedures available?
		Are they appropriate to the task? Are they regularly reviewed?
2.5	Operational abuses	Are there any obvious operational abuses?
		If so, what are they? How can they be overcome?
2.5.(1)	Extended duty cycle	Do the staff have to work shifts? If so, are they allowed regular breaks from their work?
		Is there a senior shift worker? If so, are his duties and responsibilities clearly defined?
		Are computers used? If so, are screen filters available? Do the operators have keyboard wrist rests?
2.5.(2)	Training	Do the operational staff receive regular on-the-job training?
		Is there any need for additional in-house or external training?
3 Design capability	3.1 Faulty operating procedures	Are there any obvious faulty operating procedures?
		Can the existing procedures be improved upon?

Author's end note

In accordance with BS 7671:2018, every installation (or alteration to an existing installation) shall, during erection and on completion before being put into service, be inspected and tested to verify, so far as is reasonably practicable, that the requirements of the Regulations have been met.

The next and final chapter of this book provides an extensive list of the types of inspections and tests that should be used, together with details of the sort of certificates and reports that will be required by the authorities.

10

Inspection and testing

Author's start note

In accordance with the Regulations, every installation (or alteration to an existing installation) during erection and on completion before being put into, or back into, service, has to be inspected and tested to verify, so far as is reasonably practicable, that the requirements of the Regulations have been met.

Part 6 of BS 7671:2018 has been completely restructured and now aligns with the CENELEC (European Committee for Electrotechnical Standardization) Standard in order to meet the latest requirements for initial and periodic inspection and testing.

This final chapter of the book provides some guidance on the requirements for installation, maintenance, inspection, certification and repair of electrical installations. It lists the Regulations' primary requirements for these activities with respect to electrical installations and (in Annex 10A) provides examples of test equipment used to test electrical installations.

Before being put into service, all installations need to be examined during erection and on completion in order to verify, so far as is reasonably practicable, that the requirements of the Regulations have been met. As shown in Figure 10.1, this will require a number of inspections and tests to be completed.

To meet these requirements, it is essential for any electrician engaged in inspection, testing and certification of electrical installations to have a **full** working knowledge of the Wiring Regulations contained in BS 7671:2018.

The electrician must also have above average experience and knowledge of the type of installation under test in order to carry out **any** inspection and testing. Without this prerequisite, it could be quite dangerous – particularly concerning installations such as that shown in Figure 10.2.

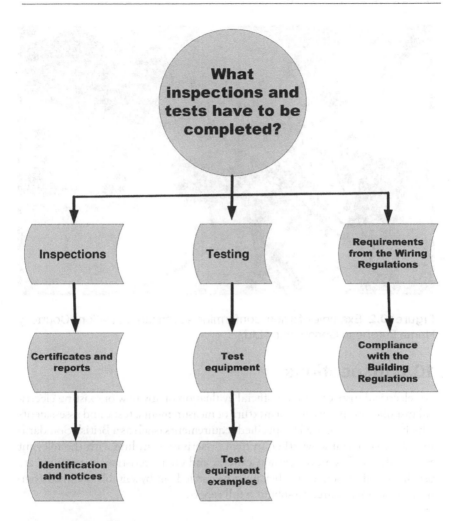

Figure 10.1 Inspections and tests.

10.1 What inspections and tests have to be completed and recorded?

Every installation must be inspected and tested during erection and on completion before being put into service:

- to verify that, so far as is reasonably practicable, that the requirements of the Wiring Regulations and the Building Regulations have been met;
- to verify that precautions have been taken to avoid danger to persons and damage to property and installed equipment during inspection and testing;
- to make an assessment of the frequency and type of maintenance (e.g. periodic inspection, testing, maintenance, repair etc.) that an installation can reasonably be expected to receive during its intended life.

Figure 10.2 Example of a non-conforming electrical installation. (Courtesy Herne European Consultancy Ltd.)

10.2 Inspections

An electrical inspection is an official evaluation of any new or existing electrical installation and involves a mixture of measurements, tests, and assessments which are then compared to specified requirements (such as a British Standard) in order to determine whether an item or activity is in line with the relevant Standard(s) and achieves certain criteria and characteristics. Inspections are usually non-destructive but they must be carried out by suitably qualified engineers, who are required to submit a full report.

10.2.1 General

An inspection should **always** precede testing and is normally done with that part of the installation under inspection disconnected from the supply.

Inspections need to be made to verify that the installed electrical equipment:

- complies with the requirements of the applicable British or Harmonised Standard appropriate to the intended use of the equipment;

Note: Equipment complying with a foreign national standard may be used only if it provides the same degree of safety afforded by a British or Harmonised Standard.

- is correctly selected and erected in accordance with the Regulations;
- is not visibly damaged or defective so as to impair safety.

10.2.2 Inspection check list

The Building Regulations specifically state that inspections **shall** include the design, construction, inspection and testing of any new electrical installation or new work associated with an alteration or addition to an existing installation.

Thus, in accordance with the requirements of BS 7671:2018 and for compliance with the Building Regulations 2010, the inspection shall include the following items:

- access to switchgear and equipment;
- cable routing;
- choice and setting of protective and monitoring devices;
- connection of accessories and equipment;
- connection of conductors;
- connection of single-pole devices for protection or switching in phase conductors;
- continuity of all protective conductors;
- continuity of all ring final circuit conductors;
- Earth electrode resistance;
- Earth fault loop impedance;
- erection methods;
- functional testing;
- identification of conductors;
- insulation of non-conducting floors and walls;
- insulation resistance;
- labelling of protective devices, switches and terminals;
- polarity;
- presence of danger notices and other warning signs;
- presence of diagrams, instructions and similar information;
- presence of fire barriers, suitable seals and protection against thermal effects;
- prevention of mutual (i.e. detrimental) influence;
- presence of undervoltage protective devices;
- prospective fault current;
- protection against electric shock:

  o exposed-conductive-parts;
  o insulating enclosures;
  o insulation of operational electrical equipment;
  o capability of equipment to withstand mechanical, chemical, electrical and thermal influences and stresses normally encountered during service;
  o verification of the quality of the insulation;

- protection against electric shock by direct and/or indirect contact:

  o SELV;
  o limitation of discharge of energy;

- protection against direct current;

  - barriers or an enclosure;
  - insulation of live parts;
  - obstacles;
  - PELV;
  - placing out of reach;

- protection against external influences;
- protection against indirect contact;

  - automatic disconnection of supply;
  - Earth-free local equipotential bonding;
  - earthed equipotential bonding;
  - earthing and protective conductors;
  - earthing arrangements for combined protective and functional purposes; non-conducting location (absence of protective conductors);
  - electrical separation;
  - main equipotential bonding conductors;
  - use of Class II equipment or equivalent insulation;
  - supplementary equipotential bonding conductors;

- selection of conductors for current-carrying capacity and voltage drop;
- selection of equipment appropriate to external influences;
- site-applied insulation;

  - protection against direct contact;
  - protection against indirect contact;
  - supplementary insulation.

In addition to the above list of mandatory inspections for compliance with the Wiring and the Building Regulations, the following are some of the additional inspections that electricians usually complete during initial and periodic inspections and tests of electrical installations:

- cables and conductors (current-carrying capacity, insulation and/or sheath);
- correct connection of accessories and equipment;
- electrical joints and connections (to ensure that they meet stipulated requirements concerning conductance, insulation, mechanical strength and protection);
- emergency switching;
- insulation;
- insulation monitoring devices (design, installation and security);
- inspection of another, associated, electrical installation;
- isolation and switching devices (and their correct location);
- locations with risks of fire due to the nature of processed and/or stored materials;

- plug and socket outlets;
- protection against electric shock – special installations or locations;
- protection against Earth insulation faults;
- protection against mechanical damage;
- protection against overcurrent;
- protection by extra-low-voltage systems (other than SELV);
- protection by non-conducting location;
- protection by residual current devices (RCDs);
- protection by separation of circuits;
- supplies;
- supplies for safety services;
- wiring systems (selection and erection, temperature variations).

10.2.3 Requirements from the building regulations

 Precautions shall be taken to avoid danger to persons and livestock, and to avoid damage to property and installed equipment, during inspection and testing.

Every electrical connection and joint shall be accessible for inspection, except for the following:

- a compound filled or encapsulated joint;
- a joint designed to be buried in the ground;
- a joint made by welding, soldering, brazing or appropriate compression tool;
- joints or connections made in the equipment by the manufacturer of the product and not intended to be inspected, tested or maintained;
- a connection between a cold tail and the heating element (e.g. in a ceiling heating, floor heating or trace heating system);
- equipment complying with BS 7671:2018 for a maintenance-free accessory and marked with the symbol.

Inspection shall precede testing, and ideally (i.e. from a safety point of view) should be completed with that part of the installation under inspection disconnected from the supply.

The inspection shall be made to verify that the installed electrical equipment is:

- correctly selected and erected in accordance with the Regulations; and
- in compliance with the relevant requirements of the applicable British or Harmonised Standard, appropriate to the intended use of the equipment; and
- not visibly damaged or defective so as to impair safety.

The inspection shall include at least the checking of the following items, where relevant to the installation and, where necessary, including any particular requirements for special installations or locations (see Part 7 of BS 7671:2018):

- presence of danger notices and other warning signs;
- presence of diagrams, instructions and similar information;
- adequacy of access to switchgear and equipment;
- connection of single-pole devices for protection;
- correct connection of accessories and equipment;
- identification and connection of conductors;
- erection methods;
- labelling of protective devices, switches and terminals;
- presence of fire barriers, suitable seals and protection against thermal effects;
- prevention of mutual detrimental influences;
- presence of appropriate devices for isolation and switching correctly located;
- presence of undervoltage protective devices;
- protection against electric shock;
- routing of cables in safe zones (or protection against mechanical damage);
- selection of conductors for current-carrying capacity and voltage drop, in accordance with the design;
- selection of equipment and protective measures appropriate to external influences.

10.3 Testing

Testing any electrical installation (even the simplest) can be very dangerous – not just to the electrician carrying out the testing, but also to bystanders and other people – unless it is carried out safely (see Figure 10.3).

As a minimum, the electrician must:

- ensure that the test equipment being used has recently been inspected, is correctly maintained and (where necessary) is calibrated either against a workshop standard or a national standard;
- have an above average experience and knowledge of the type of installation under test;
- have a thorough understanding of the correct application and use of the relevant test instruments (and their associated leads, probes and accessories);
- observe the safety measures and procedures set out in the 4th edition of HSE Guidance Note GS38 concerning the safe use of instruments and their accessories.

The following are summarised details of the most important elements of the Wiring Regulations that an electrician must test for, in order to confirm that the electrical installation meets the fundamental design requirements of BS 7671:2018 and is installed in conformance with the requirements of that British Standard.

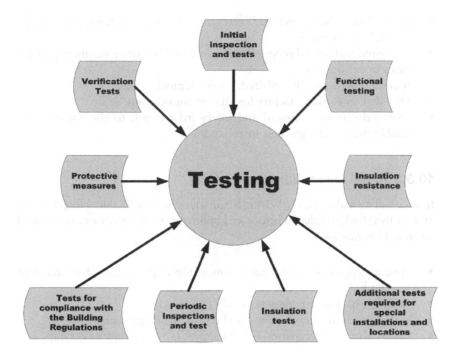

Figure 10.3 Testing.

10.3.1 Design requirements

Check to confirm that the number and type of circuits required for lighting, heating, power, control, signalling, communication and information technology etc. have taken consideration of:

- any special conditions;
- requirements for control, signalling, communication and information technology, etc.
- the daily and yearly variation of demand;
- the loads to be expected on the various circuits;
- the location and points of power demand.

10.3.2 Electricity distributor

Check to confirm that the electricity distributor has:

- ensured that their equipment on consumers' premises:
  - o is suitable for its purpose;
  - o is safe in its particular environment;
  - o clearly shows the polarity of the conductors;

- ensured that the cut-out and meter are mechanically protected and can be safely maintained;
- evaluated and agreed proposals for new installations or significant alterations to existing ones;
- maintained the supply within defined tolerance limits;
- provided an earthing facility for all new connections;
- provided certain technical and safety information to the consumer to enable them to design their installations.

10.3.3 Initial inspection and tests

In accordance with both the Wiring Regulations and the Building Regulations, all new installations (plus additions and/or alterations to existing circuits) need an initial verification to:

- ensure equipment and accessories meet the requirements of the relevant standard;
- comply with the requirements of BS 7671;
- comply with the requirements of the Building Regulations;
- ensure that the installation is not damaged so as to impair safety.

The contractor or other person responsible for the new work, or a person authorised to act on their behalf, shall record on the Electrical Installation Certificate or the Minor Electrical Installation Works Certificate any defects found, so far as is reasonably practicable, in the existing installation.

 The designer of the installation is responsible for recommending the interval to the first periodic inspection and test as detailed in Part 6 of BS 7671:2018.

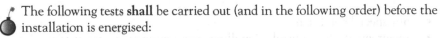 The following tests **shall** be carried out (and in the following order) before the installation is energised:

- a continuity test of all protective conductors (including main and supplementary equipotential bonding);
- a continuity test of all ring final circuit conductors;
- a measurement of the insulation resistance between live conductors and between each live conductor and Earth;
- a measurement of the insulation resistance of live parts from those of other circuits and those of other circuits and from Earth;
- a measurement of the insulation resistance of the main switchboard and each distribution circuit;
- a polarity test to verify that fuses, single-pole control and protective devices, lampholders and wiring meet requirements;
- confirmation that insulation for protection against direct and/or indirect contact meets requirements;

- confirmation that functional extra-low-voltage circuits meet all the test requirements for low-voltage circuits;
- verification that the separation of circuits is protected by SELV or PELV and that this electrical separation meets requirements;
- verification (by measurement) that the amount of protection against indirect contact provided by a non-conducting location meets requirements.

 The following tests **shall** then be carried out once the installation is energised:

- a measurement of the electrode resistance to Earth for earthing systems incorporating an Earth electrode;
- a measurement of Earth loop impedance;
- a measurement of prospective short-circuit and Earth fault;
- functional tests to verify the effectiveness of RCDs and test assemblies (e.g. switchgear, controlgear, drives, controls and interlocks) to show that they are properly mounted, adjusted and installed in accordance with the Regulations.

10.3.4 Functional testing

Where fault protection and/or additional protection is to be provided by an RCD, the effectiveness of any test facility incorporated in the device shall be verified.

Equipment such as:

- switchgear and controlgear assemblies, drives, controls and interlocks;
- systems for emergency switching off and emergency stopping;
- insulation monitoring;

shall be subjected to a functional test to show that it is properly mounted, adjusted and installed, and operates in accordance with the relevant requirements of the Wiring Regulations.

10.3.5 Insulation resistance

Check to confirm that:

- the insulation resistance (measured with all the installation's final circuits connected but with current-using equipment disconnected) between live (phase and neutral) conductors, and between each live conductor and Earth, is **not** less than that shown in Table 10.1;
- the separation of live parts from those of other circuits and from Earth is in accordance with the values show in Table 10.1, by measuring the insulation resistance.

Table 10.1 Minimum values of insulation resistance

Circuit nominal voltage (V)	Test voltage d.c. (V)	Minimum insulation resistance (MΩ)
SELV and PELV	250	0.25
Up to and including 500 V (with the exception of the above systems)	500	0.5
Above 500 V	1000	1.0

 Note: This test (more usually referred to as *meggering*) is, therefore, aimed at ensuring that the insulation of conductors, accessories and equipment is still capable of preventing dangerous leakage current between conductors and between conductors and Earth.

The resistance readings obtained should be greater than the minimum values shown in Table 10.1.

 Notes:

- Measurements shall be carried out with direct current.
- When the circuit includes electronic devices, only a measurement to protective Earth shall be made with the phase and neutral connected together.

 More stringent requirements are applicable for the wiring of fire alarm systems in buildings, see BS 5839-1.

Where a surge protective device (SPD) or other equipment is likely to influence the verification test, or be damaged, such equipment shall be disconnected before carrying out the insulation resistance test.

In locations exposed to a fire hazard, a measurement of the insulation resistance between the live conductors should be applied.

Where the circuit includes electronic devices which are likely to influence the results or be damaged, only a measurement between the live conductors connected together and the earthing arrangement shall be made.

10.3.5.1 Insulation resistance/impedance of floors and walls

In a non-conducting location at least three measurements shall be made in the same location, one of these measurements being approximately 1 m from any accessible extraneous-conductive-part in the location. The other two measurements shall be made at greater distances.

The above series of measurements shall be repeated for each relevant surface of the location.

 Further information on measurement of the insulation resistance/impedance of floors and walls can be found in Appendix 13 of BS 7671:2008.

Any insulation or insulating arrangement of extraneous-conductive-parts:

- when tested at 500 V d.c. shall be not less than 1 MΩ; and
- shall be able to withstand a test voltage of at least 2 kV a.c. rms; and
- shall not pass a leakage current exceeding 1 mA in normal conditions of use.

10.3.6 Site insulation

Confirm that:

- the insulation applied on site to protect against direct contact is capable of withstanding, without breakdown or flashover, an applied test voltage as specified in the British Standard for similar type-tested equipment;
- supplementary insulation applied to equipment during erection to protect against indirect contact is tested to ensure that the insulating enclosure;

10.3.7 Protective measures

The Wiring Regulations stipulate that a continuity test of all protective conductors (including main and supplementary equipotential bonding) shall be made to ensure that the correct degree of protection is being provided by the following protective measures, by confirming, checking and testing (Table 10.2):

Table 10.2 Protective measures

Type of protection	Description
Protection against overload current	That the protective device is capable of breaking any overload current flowing in the circuit conductors before the current can damage the insulation of the conductors.
Protection by Earth-free local equipotential bonding	That Earth-free local equipotential bonding prevents the appearance of a dangerous voltage between simultaneously accessible parts in the event of failure of the basic insulation.
Protection by electrical separation	That equipment used as a fixed source of supply has been manufactured so that the output is separated from the input and from the enclosure by insulation for protection against indirect contact.
	Note: This form of protection is intended for an individual circuit and is aimed at preventing shock current through contact with exposed-conductive-parts, which might be energised by a fault in the basic insulation of that circuit.

Table 10.2 *(continued)*

Type of protection	Description
Protection by extra-low-voltage systems (other than SELV)	That if an extra-low-voltage system complies with the requirements for SELV, it is not be connected to a live part or a protective conductor forming part of another system, and is not connected to:

- Earth;
- an exposed-conductive-part of another system;
- a protective conductor of any system; or
- an extraneous-conductive-part.

Protection against direct contact has been provided by either:

- insulation capable of withstanding 500 V a.c. rms for 60 seconds; or
- barriers or enclosures with a degree of protection of at least 1P2X or IPXXB.

This form of protection against direct contact is **not** required if the equipment is within a building in which main equipotential bonding is applied and the voltage does not exceed:

- 25 V a.c. rms or 60 V ripple-free d.c. when the equipment is normally only used in dry locations and large-area contact of live parts with the human body is not to be expected;
- 6 V a.c. rms or 15 V ripple-free d.c. in all other cases.

Note: When an extra-low-voltage circuit is used to supply equipment whose insulation does not comply with the minimum test voltage required for the primary circuit, the insulation of that equipment shall be reinforced to withstanding a voltage of 1500 V a.c. rms for 60 seconds.

Protection by insulation of live parts	That the insulation protection has been designed to prevent contact with live parts.

Whilst, generally speaking, this method is for protection against direct contact, it also provides a degree of protection against indirect contact

Protection by non-conducting location	That this form of protection prevents simultaneous contact with parts which may be at different potentials through failure of the basic insulation of live parts.

Whilst this sort of protection is not recognised in the Regulations for general use, it may be applied in special situations provided that they are under effective supervision.

Protection by non-conducting location shall **not** be used in installations and locations subject to increased risk of shock such as agricultural and horticultural premises, caravans, swimming pools etc.

Table 10.2 *(continued)*

Type of protection	Description
Protection by residual current devices (RCDs)	That: • parts of an TT system that are protected by a single RCD have been placed at the origin of the installation, unless that part between the origin and the device complies with the requirements for protection by using Class II equipment or an equivalent insulation; Where there is more than one origin, this requirement applies to each origin. • installations forming part of an IT system have been protected by an RCD supplied by the circuit concerned or make use of an insulation monitoring device.
Protection by SELV	That circuit conductors for each SELV system have been physically separated from those of any other system. Where this proves impracticable, SELV circuit conductors have been: • insulated for the highest voltage present; • enclosed in an insulating sheath additional to their basic insulation.
Protection by the use of Class II equipment or equivalent insulation	That this form of protection prevents a fault in the basic insulation causing a dangerous voltage to appear on the exposed metalwork of electrical equipment.

10.3.8 Protection by automatic disconnection of the supply

Where RCDs are also used for protection against fire, the conditions for protection by automatic disconnection of the supply shall be verified.

The verification of the effectiveness of the measures for fault protection by automatic disconnection of supply depends on the type of system used and is as follows:

10.3.8.1 TN systems

Compliance shall be verified by:

• measurement of the Earth fault loop impedance;
• verification of the characteristics and/or the effectiveness of the associated protective device.

10.3.8.2 TT systems

Compliance shall be verified by:

- measurement of the resistance of the Earth electrode for exposed-conductive-parts of the installation:
- verification of the characteristics and/or effectiveness of the associated protective device.

10.3.8.3 IT systems

Compliance shall be verified by calculation or measurement of the current (I_d) in case of a first fault at the live conductor.

Where conditions that are similar to the conditions of a TT system occur, in the event of a second fault in another circuit verification shall be made according to a TT system.

Where conditions that are similar to the conditions of a TN system occur, in the event of a second fault in another circuit verification shall be made as follows:

- For IT installations supplied from a local transformer, the Earth fault loop impedance is measured by inserting a connection with negligible impedance between a live conductor and Earth at the origin of the installation.
- For IT systems connected to a public grid, the Earth fault loop impedance is determined by verification of the continuity of the protective conductor and by measuring the loop impedance between two live conductors at the end of the circuit.

10.3.9 Protection against direct and indirect contact

The two methods for protecting against shock from both direct and indirect contact are:

- SELV (separated extra-low voltage) – where an extra-low-voltage electrical circuit that electrically separated from other circuits that carry higher currents;
- limitation of discharge of energy – where equipment is arranged so that current that can flow through the body (or livestock) is limited to a safe level (e.g. by using electric fences).

10.3.9.1 SELV

The system is **not** deemed to be SELV if any exposed-conductive-part of an extra-low-voltage system is capable of coming into contact with an exposed-conductive-part of any other system.

Note: In addition, a system which does **not** use a device such as an autotransformer, potentiometer, semiconductor device etc. to provide electrical separation is also **not** deemed to be a SELV system.

Protection by SELV is acceptable when:

- the nominal circuit voltage does not exceed extra-low voltage;
- the supply is from one of the following:

  o a safety isolating transformer complying with BS 3535;
  o a motor-generator with windings providing electrical separation equivalent to that of the safety isolating transformer specified above;
  o a battery or other form of electrochemical source;
  o a source independent of a higher-voltage circuit (e.g. an engine-driven generator);
  o electronic devices which (even in the case of an internal fault) restrict the voltage at the output terminals so that they do not exceed extra-low voltage.

Confirm and test that:

- a mobile source for SELV has been selected and erected in accordance with the requirements for protection by the use of Class II equipment or by an equivalent insulation where the user is protected by at least two layers of insulation;
- all live parts of a SELV system are:

  o not connected to Earth;
  o not connected to a live part or a protective conductor forming part of another system;

- circuit conductors for each SELV system are physically separated from those of any other system. Where this proves impracticable, SELV circuit conductors should be:

  o insulated for the highest voltage present; and (where this proves impracticable)
  o enclosed in an insulating sheath additional to their basic insulation;

- conductors of systems with a higher voltage than SELV are separated from the SELV conductors by an earthed metallic screen or an earthed metallic sheath;
- SELV circuit conductors that are contained in a multi-core cable with other circuits having different voltages are insulated, individually or collectively, for the highest voltage present in the cable or grouping;
- electrical separation between live parts of a SELV system (including relays, contactors and auxiliary switches) and any other system are maintained;

- exposed-conductive-parts of a SELV system are **not** connected to:

  o   Earth;
  o   an exposed-conductive-part of another system;
  o   a protective conductor of any other system;
  o   an extraneous-conductive-part;

Note: (Except where that particular electrical equipment cannot attain a voltage exceeding extra-low voltage);

- if the nominal voltage of a SELV system exceeds 25 V a.c. rms or 60 V ripple-free d.c., protection against direct contact is provided by one or more of the following:

  o   a barrier (or an enclosure) capable of providing suitable protection;
  o   insulation capable of withstanding a type-test voltage of 500 V a.c. rms for 60 seconds;

- the socket outlet of a SELV system:

  o   is incompatible with the plugs used for other systems in use in the same premises;
  o   does not have a protective conductor contact;

- luminaire-supporting couplers which have a protective conductor contact are not be installed in a SELV system.

Protection against both direct and indirect contact shall be judged to be provided when the equipment incorporates a means of limiting the amount of current which can pass through the body of a person or livestock to a value lower than that likely to cause danger.

10.3.10 Protection against direct contact

Electric shock caused by direct contact (i.e. when a bodily part directly touches live parts of electrical or electromechanical equipment or systems that are intended to be live) is particularly dangerous, as the full voltage of the supply can be developed across the body. On the whole, however, if an electrical installation has been designed and installed correctly, then there shouldn't be too much risk from direct contact – but carelessness (such as changing an electric light bulb without switching the mains off first) or overconfidence (such as working on a circuit with the power on) are the prime causes of injuries and death from electric shock.

The main protective methods against direct contact causing an electric shock are:

- barriers or an enclosure;
- insulation of live parts;

- by obstacles and placing out of reach;
- PELV and FELV.

 The use of an RCD cannot prevent direct contact, but may **supplement** other protective means.

10.3.10.1 Protection by a barrier or an enclosure provided during erection

Test to ensure that the degree of protection against direct contact provided by a barrier or an enclosure (provided during erection) is not less than IP2X or IPXXB or IP4X, as appropriate.

10.3.10.2 Protection by insulation of live parts

Complete a functional test to verify that protection by insulation of live parts **does** prevent contact with a live parts;

 Whilst, generally speaking, this basic form of insulation protection is for protection against direct contact, it also provides a degree of protection against indirect contact.

10.3.10.3 Protection by obstacles

Complete a functional test to verify that protection by obstacles prevents unintentional contact with a live part, but **not** intentional contact by deliberate circumvention of the obstacle.

 For some installations and locations where an increased risk of shock exists, this protective measure shall **not** be used.

10.3.10.4 PELV and FELV systems

Protective extra-low-voltage (PELV) and functional extra-low-voltage (FELV) systems shall provide protection against electric shock and meet the following requirements:

- barriers or enclosures with a degree of protection of at least 1P2X or IPXXB; or
- insulation capable of withstanding 500 V a.c. rms for 60 seconds;
- for extra-low-voltage systems that do not comply with the requirements for SELV in some respect, protection against direct contact shall be provided by one or more of the following:
  - o barriers or enclosures;
  - o insulation corresponding to the minimum voltage required for the primary circuit;

- when an extra-low-voltage circuit is used to supply equipment whose insulation does not comply with the minimum test voltage required for the primary circuit, the insulation of that equipment shall be reinforced to withstand a voltage of 1500 V a.c. rms for 60 seconds;
- if the primary circuit of the functional extra-low-voltage source is protected:

  o by automatic disconnection, then the exposed-conductive-parts of equipment in that functional extra-low-voltage system shall be connected to the protective conductor of the primary circuit;

  o by electrical separation, then the exposed-conductive-parts of equipment in that functional extra-low-voltage system shall be connected to the non-earthed protective conductor of the primary circuit;

- all socket outlets and luminaire-supporting couplers in a functional extra-low-voltage system shall use a plug which is dimensionally different from those used for any other system in use in the same premises.

10.3.10.5 Protection by placing out of reach

Check to ensure that:

- bare (or insulated) overhead lines being used for distribution between buildings and structures are installed in accordance with the Electricity Safety, Quality and Continuity Regulations 2002;
- bare live parts (other than overhead lines) are not be within arm's reach;
- bare live parts (other than an overhead line) are not within 2.5 m of:

  o an exposed-conductive-part;
  o an extraneous-conductive-part;
  o a bare live part of any other circuit;

- if a bulky or long conducting object is normally handled in these areas, the distances required shall be increased accordingly.

No additional protection against overvoltages of atmospheric origin is necessary for:

- installations that are supplied by low-voltage systems which do not contain overhead lines;
- installations that are supplied by low-voltage networks which contain overhead lines and their location is subject to less than 25 thunderstorm days per year;
- installations that contain overhead lines and their location is subject to less than 25 thunderstorm days per year;

 provided that they meet the required minimum equipment impulse withstand voltages shown in Table 10.3.

Table 10.3 Required minimum impulse to withstand voltage U$_w$(kV)

Required minimum impulse withstand voltage U$_w$(kV)				
Nominal voltage of the installation (V)	Category I Equipment with reduced impulse voltage)	Category II Equipment with normal impulse voltage)	Category III Equipment with high impulse voltage)	Category IV Equipment with very high impulse voltage)
230/240 277/480	1.5	2.5	4	6
400/690	2.5	4	6	8
1000	4	6	8	12

 Note: A suspended cable having insulated conductors with **earthed metallic coverings** is considered to be an *'underground cable'*.

Check to ensure that installations that are supplied by (or include) low-voltage overhead lines must incorporate protection against overvoltages of atmospheric origin or (if the location is subject to more than 25 thunderstorm days per year). This protection must be provided either by:

- a surge protective device with a protection level not exceeding Category II; or
- by other means providing an equivalent attenuation of overvoltages.

Where protective measures against indirect contact only have been dispensed with, confirm that:

- exposed-conductive-parts (including small isolated metal parts such as bolts, rivets, nameplates not exceeding 50 mm × 50 mm and cable clips) cannot be gripped or cannot be contacted by a major surface of the human body;
- inaccessible lengths of metal conduit do not exceed 150 mm^2;
- metal enclosures mechanically protecting equipment comply with the relevant British Standard;
- overhead line insulator brackets (and metal parts connected to them) are not within arm's reach;
- the steel reinforcement of steel reinforced concrete poles is not accessible;
- there is no risk of fixing screws used for non-metallic accessories coming into contact with live parts;
- unearthed street furniture that is supplied from an overhead line is inaccessible whilst in normal use.

10.3.11 Protection against indirect contact

Indirect contact (i.e. touching conductive parts which are not meant to be live, but which have become live due to a fault) is the other main cause of electric shock. Again, this is particularly dangerous, and the main protection against indirect contact is for the electrical installation to be correctly earthed and for the circuit to be fitted with some form of overcurrent cut-out device.

The main protective methods against indirect contact causing an electric shock are:

- automatic disconnection of supply;
- Earth-free local equipotential bonding;
- earthed equipotential bonding;
- earthing and protective conductors;
- earthing arrangements for combined protective and functional purposes;
- non-conducting location (absence of protective conductors);
- electrical separation;
- main equipotential bonding conductors;
- supplementary equipotential bonding conductors;
- use of Class II equipment.

10.3.11.1 Automatic disconnection of supply

The intention of this form of protection is to prevent a dangerous voltage occurring between simultaneously accessible conductive parts. For installations and locations with increased risk of shock (such as agricultural and horticultural buildings, saunas etc.) additional measures may be required. For example:

- automatic disconnection of supply by means of an RCD with a rated residual operating current not exceeding 30 mA;
- supplementary equipotential bonding;
- reduction of maximum fault clearance time.

Confirm and test that:

- installations which are part of a TN system meet the requirements for Earth fault loop impedance and for circuit protective conductor impedance, as specified in the Wiring Regulations regarding specified times for automatic disconnection of supplies;
- circuits supplying fixed equipment which are outside of the earthed equipotential zone and which have exposed-conductive-parts - that could be touched by a person who has direct contact with Earth - that the Earth fault loop impedance ensures that disconnection occurs within the time stated in Table 10.4 and as stated in BS 7671:2018;
- if the installation is part of a TT system, all socket outlet circuits are protected by an RCD;
- automatic disconnection using an RCD is not be applied to a circuit incorporating a PEN conductor;
- installations that provide protection against indirect contact by automatically disconnecting the supply have a circuit protective conductor run to (and terminated at) each point in the wiring and at each accessory; (except suspended lampholders which have no exposed-conductive-parts).

Table 10.4 Maximum disconnection times for TN systems

Installation nominal voltage	Maximum disconnection time (TN system)
50 V	0.8 sec
120V	0.4 sec
230V	0.2 sec
400V	0.2 sec
>400	0.1 s

10.3.11.2 Earth-free local equipotential bonding

Earth-free local equipotential bonding is effectively a Faraday cage, where all metal is bonded together (but **not** to Earth!) so as to prevent the appearance of a dangerous voltage occurring between simultaneously accessible parts in the event of failure of the basic insulation.

Confirm that:

- Earth-free local equipotential bonding has only been used in special situations which are Earth-free;
- a cautionary notice (warning that Earth-free local equipotential bonding is being used) has been fixed in a prominent position adjacent to every point of access to the location concerned.

 Note: For some installations and locations with an increased shock risk (e.g. agricultural and horticultural, saunas etc.), Earth-free local equipotential bonding shall **not** be used.

10.3.11.3 Earthing and protective conductors

A protective conductor may consist of one or more of the following:

- a single-core cable;
- a conductor in a cable;
- an insulated or bare conductor in a common enclosure with insulated live conductors;
- a fixed bare or insulated conductor;
- a metal covering (e.g. the sheath, screen or armouring of a cable);
- a metal conduit or other enclosure or electrically continuous support system for conductors;
- an extraneous-conductive-part.

Verify and test that:

- the thickness of tape or strip conductors is capable of withstanding mechanical damage and corrosion (see BS 7430);
- the connection of earthing conductors to the Earth electrode are:

  o soundly made;
  o electrically and mechanically satisfactory;
  o labelled in accordance with the Regulations;
  o suitably protected against corrosion;

- all installations have a main earthing terminal to connect the following to the earthing conductor:

  o the circuit protective conductors;
  o the main bonding conductors;
  o functional earthing conductors (if required);
  o lightning protection system bonding conductor (if any);

- earthing conductors are capable of being disconnected to enable the resistance of the earthing arrangement to be measured;
- all joints:

  o are capable of disconnection only by means of a tool;
  o are mechanically strong;
  o ensure the maintenance of electrical continuity.

 Note: for convenience (and if required) this may be combined with the main earthing terminal or bar.

- unless a protective conductor forms part of a multi-core cable (or cable trunking or a conduit is used as a protective conductor), the cross-sectional area, up to and including 6 mm², has been protected, throughout, by a covering at least equivalent in insulation to that of a single-core non-sheathed cable having a voltage rating of at least 450/750 V.

Where protective multiple earthing (PME) conditions apply, verify and test that:

- the main equipotential bonding conductor has been selected in accordance with the neutral conductor of the supply and Table 10.5;

 Note: The local distributor's network conditions may require a larger conductor.

- buried earthing conductors have a cross-sectional area not less than that stated in Table 10.6:

Table 10.5 Minimum cross-sectional area of the main equipotential bonding conductor in relation to the neutral

Copper equivalent cross-sectional area of the supply neutral conductor	Minimum copper equivalent cross-sectional area of the main equipotential bonding conductor
4 mm² up to 10 mm²	6 mm²
16 mm² up to 35 mm²	10 mm²
50 mm²	16 mm²
70 mm²	25 mm²

Table 10.6 Minimum cross-sectional areas of a buried earthing conductor

	Protected against mechanical damage	Not protected against mechanical damage
Protected against corrosion by a sheath		16 mm² copper 16 mm² coated steel
Not protected against corrosion	25 mm² copper 50 mm² steel	25 mm² copper 50 mm² steel

All protective conductors – particularly main equipotential and supplementary bonding conductors – must be tested for continuity using a low-resistance ohmmeter.

Verify and test that:

- the cross-sectional area of all protective conductors (less equipotential bonding conductors) is not less than

$$S = \frac{\sqrt{I^2 t}}{k}$$

- if the protective conductor:
  - o is not an integral part of a cable; or
  - o is not formed by conduit, ducting or trunking; or
  - o is not contained in an enclosure formed by a wiring system;

then the cross-sectional area shall be not less than:

  - o 2.5 mm² copper equivalent if protection against mechanical damage is provided; or
  - o 4 mm² copper equivalent if mechanical protection is not provided;

- protective conductors buried in the ground shall have a cross-sectional area not less than that stated in Table 10.6.

10.3.11.4 Earthing arrangements for combined protective and functional purposes

The following conductors may serve as a PEN conductor provided that the part of the installation concerned is not supplied through an RCD:

- a conductor of a cable not subject to flexing and with a cross-sectional area not less than 10 mm² (for copper) or 16 mm² for aluminium (this applies to a fixed installation by the way);
- the outer conductor of a concentric cable where that conductor has a cross-sectional area not less than 4 mm².

Verify and test (where necessary) that PEN conductors have only been used if:

- authorisation to use a PEN conductor has been obtained by the distributor; or
- the installation is supplied by a privately owned transformer or convertor and there is no metallic connection (except for the earthing connection) with the distributor's network; or
- the installation is supplied from a private generating plant.

Then you will need to check that:

- the outer conductor of a concentric cable is not be common to more than one circuit;
- the conductance of the outer conductor of a concentric cable (and the terminal link or bar):

  o for a single-core cable is not less than the internal conductor;
  o for a multi-core cable in a multiphase or multi-pole circuit is not be less than that of one internal conductor;
  o for a multi-core cable serving a number of points contained within one final circuit (or where the internal conductors are connected in parallel) is not be less than that of the internal conductors;

- the continuity of all joints in the outer conductor of a concentric cable (and at a termination of that joint) is supplemented by an additional conductor (additional, that is, to any means used for sealing and clamping the outer conductor);
- isolation devices or switching have not been inserted in the outer conductor of a concentric cable;
- PEN conductors of all cables have been insulated or have an insulating covering suitable for the highest voltage to which it may be subjected;
- if neutral and protective functions are provided by separate conductors, those conductors are not then be reconnected together beyond that point;

- separate terminals (or bars) have been provided for the protective and neutral conductors at the point of separation;
- PEN conductors have been connected to the terminals or bar intended for the protective earthing conductor and the neutral conductor.

 Note: Where earthing is required for protective as well as functional purposes, then the requirements for protective measures shall take precedence.

10.3.11.5 Protection by electrical separation

This form of protection is intended for an individual circuit and is aimed at preventing shock current through contact with exposed-conductive-parts which might be energised by a fault in the basic insulation of that circuit.
  Verify that:

- equipment used as a fixed source of supply is either:

  o selected and/or installed with double insulated electrical protection (Class II) or equivalent protection; or
  o manufactured so that the output is separated from the input and from the enclosure by insulation satisfying the conditions for Class II;

- for circuits supplying a single piece of equipment, no exposed-conductive-part of the separated circuit is connected to:

  o the protective conductor of the source;
  o any exposed-conductive-part of any other circuit;

- live parts of a separated circuit are not connected (at any point) to another circuit or to Earth;
- live parts of a separate circuit are electrically separated from all other circuits;

 Note: live parts of relays, contactors etc. included in a separated circuit (and between a separated circuit and other live parts of other circuits) shall be similarly electrically separated.

- mobile supply sources (fed from a fixed installation) are selected and/or installed with Class II or equivalent protection;
- protection by electrical separation has been applied to the supply of individual items of equipment by means of a transformer complying with BS 3535 (the secondary of which is not earthed) or a source affording equivalent safety;
- protection by electrical separation has been used to supply several items of equipment from a single separated source (but only for special situations);
- separated circuits, preferably, use a separate wiring system;

If this is not feasible, multi-core cables (without a metallic sheath) or insulated conductors (in an insulating conduit) may be used;

- source supplies are only supplying more than one item of equipment provided that;

  o all exposed-conductive-parts of the separated circuit are connected together by an insulated and non-earthed equipotential bonding conductor;

  o the non-earthed equipotential bonding conductor is not connected to a protective conductor, or to a exposed-conductive-part of any other circuit or to any extraneous-conductive-part;

  o all socket outlets are provided with a protective conductor contact that is connected to the equipotential bonding conductor;

  o all flexible equipment cables (other than reinforced or double insulated (Class H) equipment) have a protective conductor for use as an equipotential bonding conductor;

  o exposed-conductive-parts which are fed by conductors of different polarity (which are liable to a double fault occurring) are fitted with an associated protective device;

  o any exposed-conductive-part of a separated circuit cannot come in contact with an exposed-conductive-part of the source;

- the supply source to the circuit is either:

  o an isolating transformer complying with BS 3535; or
  o a motor-generator;

- the voltage of an electrically separated circuit does not exceed 500 V;
- all parts of a flexible cable (or cord) liable to mechanical damage are visible throughout its length;
- a warning notice (warning that protection by electrical separation is being used) is fixed in a prominent position adjacent to every point of access to the location concerned.

10.3.11.6 Main equipotential bonding conductors

All main equipotential (and supplementary) bonding conductors must be tested for continuity.

Confirm that main equipotential bonding conductors have (for each installation) been connected to the main earthing terminal of that installation. These can include the following:

- central heating and air-conditioning systems;
- exposed metallic structural parts of the building;
- gas installation pipes;
- water service pipes;

- other service pipes and ducting;
- the lightning protective system.

In accordance with the Building Regulatons, where an installation serves more than one building the above requirement shall be applied to each building.

10.3.11.7 Non-conducting location

This form of protection (as the name implies) consists of an area in which the floor, walls and ceiling are all insulated, and within which protective conductors and socket outlets do not have an earthing connection.

Test that the insulation of extraneous-conductive-parts:

- does not pass a leakage current exceeding I mA in normal use;
- is not less than 0.5 MΩ (when tested at 500 V d.c.);
- is able to withstand a test voltage of at least 2 kV a.c. rms;

Check that the:

- degree of protection against indirect contact provided by a non-conducting location is verified, by measuring the resistance of the location's floors and walls to the installation's main protective conductor at not less than three points on each relevant surface;

Note: One of these measurements should be not less than 1 m and not more than 1.2 m from any extraneous-conductive-part in the location. The other two measurements shall be made at greater distances.

- the insulation of extraneous-conductive-parts (to satisfy the requirements for protection be non-conducting location):
  - o are able to withstand a test voltage of at least 2 kV a.c. rms; and
  - o do not pass a leakage current exceeding I mA in normal use; and
  - o are not less than 0.5 MΩ when tested at 500 V d.c.

10.3.11.8 Supplementary equipotential bonding

Using a low-resistance ohmmeter (similar to that employed for testing main equipotential bonding conductors described above) test all supplementary equipotential bonding conductors for continuity, particularly in locations intended for livestock, to confirm that:

- any metallic grid that is laid in the floor for supplementary bonding is connected to the protective conductors of that installation;
- supplementary bonding connects all exposed- and extraneous-conductive-parts which can be touched by livestock.

10.3.11.9 Locations with increased risk of shock

For installations and locations with increased risk of shock (e.g. saunas, bathrooms and agricultural/horticultural premises etc.) certain additional measures may be required, such as:

- automatic disconnection of supply by means of an RCD with a rated residual operating current not exceeding 30 mA;
- supplementary equipotential bonding.

In these cases, test to confirm that for circuits supplying fixed equipment which are outside of the earthed equipotential zone (and which have exposed-conductive-parts that could be touched by a person who has direct contact directly with Earth) the Earth fault loop impedance ensures that disconnection occurs within the time stated in Table 10.4 and BS 7671:2018.

Test, measure and confirm that:

- automatic disconnection using an RCD has not been applied to a circuit incorporating a PEN conductor;
- if the installation is part of a TT system, all socket-outlet circuits have been protected by an RCD;
- installations that provide protection against indirect contact by automatically disconnecting the supply have a circuit protective conductor run to (and terminated at) each point in the wiring and at each accessory;

Except suspended lampholders, which have no exposed-conductive-parts.

- one or more of the following types of protective device have been used:

  o   an RCD;
  o   an overcurrent protective device;

- where an RCD is used in a TN-C-S system, a PEN conductor has not been used on the load side;
- the protective conductor to the PEN conductor is on the source side of the RCD;
- the maximum disconnection times to a circuit supplying socket outlets, and to other final circuits which supply portable equipment intended for manual movement during use, or hand-held Class I equipment, do not exceed those shown in Table 10.6;

Note: This requirement does not apply to a final circuit supplying an item of stationary equipment connected by means of a plug and socket outlet where precautions are already taken to prevent the use of the socket outlet for supplying hand-held equipment, nor to reduced low-voltage circuits.

- where a fuse is used to satisfy this disconnection requirement, maximum values of Earth fault loop impedance (Z_s) corresponding to a disconnection

Table 10.7 Maximum Earth fault loop impedance (Z_s) for fuses, for 0.4 s disconnection time with U_o of 230 V (see Regulation 413-02-10)

General purpose (gG) fuses to BS 88-2.1 and BS 88-6								
Rating (A)	6	10	16	20	25	32	40	50
Z_s (Ω)	8.89	5.33	2.82	1.85	1.5	1.09	0.86	0.63

time of 0.4 s are as stated in Table 10.7 for a nominal voltage to Earth (U_o) of 230 V;

- for a distribution circuit and a final circuit supplying only stationary equipment, the maximum disconnection time of 5 s is not exceeded:

 Notes:

1. The circuit loop impedances given in the table above should not be exceeded when the conductors are at their normal operating temperature. If the conductors are at a different temperature when tested, then the reading should be adjusted accordingly.
2. See appropriate part of BS 88 for types and rated currents of fuses other than those mentioned in Table 10.7.

10.3.11.10 Protection by separation of circuits

Ensure that:
- the separation of circuits shall be verified for protection by:
  - SELV;
  - PELV;
  - electrical separation; and
  - that the separation of live parts from those of other circuits and those of other circuits from Earth is verified by measuring that the insulation resistance is in accordance with values show in Table 10.8.
  - functional extra-low-voltage circuits meet all the test requirements for low-voltage circuits.

Table 10.8 Minimum values of insulation resistance

Circuit nominal voltage (V)	Test voltage d.c. (V)	Minimum insulation resistance (MΩ)
SELV and PELV	250	0.25
Up to and including 500 V (with the exception of the above systems)	500	0.5
Above 500 V	1000	1.0

10.3.11.11 Polarity

Complete a polarity test to verify that:

- circuits (other than BS EN 60238 E14 and E27 lampholders) which have an earthed neutral conductor centre-contact bayonet (and Edison screw lampholders) have their outer or screwed contacts connected to the neutral conductor;
- fuses and single-pole control and protective devices are only connected in the line conductor;
- wiring has been correctly connected to socket outlets and similar accessories.

10.3.11.12 Use of Class II equipment

Class II equipment (often referred to as double-insulated equipment) is typical of modern equipment intended to be connected to the fixed electrical installation (such as household appliances, portable tools and similar loads) and where all live parts are insulated so as to prevent a fault in the basic insulation causing a dangerous voltage to appear on the exposed metalwork of electrical equipment.

To verify compliance, confirm that:

- circuits supplying Class II equipment have a circuit protective conductor that is run to (and terminated at) each point in the wiring and at each accessory;

Except suspended lampholders which have no exposed-conductive-parts.

- the metalwork of exposed Class II equipment is mounted so that it is not in electrical contact with any part of the installation that is connected to a protective conductor;
- when Class II equipment is used as the sole means of protection against indirect contact, the installation or circuit concerned is under effective supervision whilst in normal use;

This form of protection shall **not** be used for circuits that include socket outlets or where a user can change items of equipment without authorisation.

10.3.12 Additional tests with the supply connected

Other than insulation tests, the following tests are to be completed with the supply connected:

- re-check of polarity;
- Earth electrode resistance;

- Earth fault loop impedance;
- prospective fault current.

10.3.12.1 Polarity

Repeat the polarity test shown in Section 10.3.11.11 to verify that:

- fuses and single-pole control and protective devices are only connected in the phase conductor;
- circuits (other than BS EN 60238 E14 and E27 lampholders) which have an earthed neutral conductor centre-contact bayonet (and Edison-screw lampholders) have their outer or screwed contacts connected to the neutral conductor;
- wiring has been correctly connected to socket outlets and similar accessories.

10.3.12.2 Earth electrode resistance

If the earthing system incorporates an Earth electrode as part of the installation, measure the electrode resistance to Earth (using test equipment similar to that shown in Annex 10A).

If the electrode under test is being used in conjunction with an RCD protecting an installation, test (prior to energising the remainder of the installation) between the phase conductor at the origin of the installation and the Earth electrode with the test link open.

 Note: The resulting impedance reading (i.e. the electrode resistance) should then be added to the resistance of the protective conductor for the protected circuits.

10.3.12.3 Earth fault loop impedance

This is an extremely important test to ensure that, under Earth fault conditions, overcurrent devices disconnect fast enough to reduce the risk of electric shock (see Figure 10.4). The Regulations stipulate that:

"If the protective measures employed require a knowledge of Earth fault loop impedance, then the relevant impedances must be measured."

Where a fuse is used, maximum values of Earth fault loop impedance (Z_s) corresponding to a disconnection time of 0.4 s are stated in Table 10.9 for a nominal voltage to Earth (U_o) of 230 V.

 Note: the circuit loop impedances given in the Table 10.9 should not be exceeded when the conductors are at their normal operating temperature.

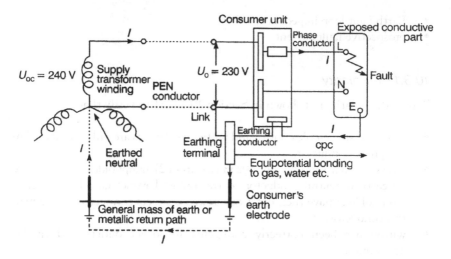

Figure 10.4 Testing Earth fault loop impedance.

If the conductors are at a different temperature when tested, then the reading should be adjusted accordingly.

Using test equipment (similar to that listed in Annex 10A) complete the following tests:

- Ensure that all main equipotential bonding is in place; connect the test equipment to the phase, neutral and Earth terminals at the remote end of the circuit under test. Press to test and record.

 Note: The circuit loop impedances given in the Table 10.9 should not be exceeded when the conductors are at their normal operating temperature. If the conductors are at a different temperature when tested, then the reading should be adjusted accordingly.

 Further information on measurement of Earth fault loop impedance can be found in Appendix 14 of BS 7671:2018.

Table 10.9 British Standards for fuse links

Circuit nominal voltage (V)	Test voltage d.c. (V)	Minimum insulation resistance (MΩ)
SELV and PELV	250	0.25
Up to and including 500V (with the exception of the above systems)	500	0.5
Above 500 V	1000	1.0

(Courtesy Brian Scaddon.)

10.3.12.4 Verification of voltage drop

When required, for compliance with the Regulations the voltage drop may be evaluated:

- by measuring the circuit impedance;
- by using calculations.

 Note: Verification of voltage drop is not normally required during initial verification.

10.3.12.5 Prospective fault current

Verify, test and ensure that:

- prospective fault currents (under both short-circuit and Earth fault conditions):
  - o have been assessed for each supply source;
  - o are calculated at every relevant point of the complete installation either by enquiry or by measurement;
  - o are measured at the origin and at other relevant points in the installation;
- protection of wiring systems against overcurrent takes into account minimum and maximum fault current conditions;
- fault current protective devices are provided:
  - o at the supply end of each parallel conductor where two conductors are in parallel;
  - o at the supply and load ends of each parallel conductor where more than two conductors are in parallel;
- fault current protective devices:
  - o have less than 3 m in length between the point where the value of current-carrying capacity is reduced and the position of the protective device;
  - o are installed so as to minimise the risk of fault current;
  - o are installed so as to minimise the risk of fire or danger to persons;
- fault current protective devices are placed (on the load side) at the point where the current-carrying capacity of the installation's conductors is likely to be lessened owing to:
  - o the method of installation;
  - o the cross-sectional area;

o   the type of cable or conductor used;
o   inherent environmental conditions;

- conductors are capable of carrying fault current without overheating.

Note: A single protective device may be used to protect conductors in parallel against the effects of fault current occurring.

Ensure that:

- fault current protection devices are capable of breaking;
- the breaking capacity rating of each device is not less than the prospective short-circuit current or Earth fault current at the point at which the device is installed;

Note: A lower breaking capacity is permitted if another protective device is installed on the supply side.

- the characteristics of each device used for overload current and/or for fault current protection have been co-ordinated so that the energy let through (i.e. by the fault current protective device) does not exceed the overload current protective device's limiting values;
- devices providing protection against both overload current and fault current are capable of breaking;

Note: Overload current protection devices may have a breaking capacity below the value of the prospective fault current at the point where the device is installed.

- circuit-breakers used as fault current protection devices:

  o   are capable of making any fault current up to and including the prospective fault current;
  o   break any fault current flowing before that current causes danger due to thermal or mechanical effects produced in circuit conductors or associated connections;
  o   break and make up any overcurrent up to and including the prospective fault current at the point where the device is installed;

- safety services with sources that are incapable of operating in parallel are protected against electric shock and fault current.

10.3.13 Insulation tests

When an electrical installation fails an insulation test, the installation must be corrected and the test made again. If the failure influences any previous tests that were made, then those tests must also be repeated.

10.3.13.1 Locations with a risk of fire due to the nature of processed and/or stored materials

Test that wiring systems (less those using mineral insulated cables and busbar trunking arrangements) have been protected against Earth insulation faults as follows:

- in TN and TT systems, by RCDs having a rated residual operating current (I_{An}) not exceeding 300 mA;
- in IT systems, by insulation monitoring devices with audible and visible signals.

 Notes:

1. Adequate supervision is required to facilitate manual disconnection as soon as appropriate.
2. The disconnection time of the overcurrent protective device, in the event of a second fault, shall not exceed 5 s.

10.3.13.2 Protection against electric shock – insulation tests

Confirm, measure and test that:

- circuit conductors for each SELV system are physically separated from those of any other system or (i.e. where this proves impracticable) confirm that SELV circuit conductors are:
  - o insulated for the highest voltage present;
  - o enclosed in an insulating sheath additional to their basic insulation;

- equipment is capable of withstanding all mechanical, chemical, electrical and thermal stresses normally encountered during service;

 Note: Paint, varnish, lacquer or similar products are **not** generally considered to provide adequate insulation for protection against direct contact in normal service.

- exposed-conductive-parts that might attain different potentials through failure of the basic insulation of live parts have been arranged so that a person will not come into simultaneous contact with two exposed-conductive-parts, or an exposed-conductive-part and any extraneous-conductive-part;

 This may be achieved if the location has an insulating floor and insulating walls and one or more of the following arrangements apply:

- o the distance between any separated exposed-conductive-parts (and between exposed-conductive-parts and extraneous-conductive-parts) is not less than 2.5 m (or 1.25 m if they are out of arm's reach);
- o protective obstacles (that are not connected to Earth or to exposed-conductive-parts and which are made out of insulating material) are used between exposed-conductive-parts and extraneous-conductive-parts;
- o the insulation is of acceptable electrical and mechanical strength;

- • if the nominal voltage of a SELV system exceeds 25 V a.c. rms or 60 V ripple-free d.c., then protection against direct contact has been provided by one (or more) of the following:

- o insulation capable of withstanding a type-test voltage of 500 V a.c. rms for 60 seconds; or
- o a barrier (or an enclosure) capable of providing protection to at least 1P2X or IPXXB;

- • in IT systems, an insulation monitoring device has been provided so as to indicate the occurrence of a first fault from a live part to an exposed-conductive-part or to Earth;
- • insulating enclosures:

- o are not pierced by conductive parts (other than circuit conductors) likely to transmit a potential;
- o do not contain any insulated screws of insulating material the future replacement of which by metallic screws could impair the insulation provided by the enclosure;
- o do not adversely affect the operation of the equipment protected;

 Note: Where the insulating enclosure has to be pierced by conductive parts (e.g. for operating handles of built-in equipment and for screws) protection against indirect contact shall not be impaired.

- • live parts are completely covered with insulation which:

- o can only be removed by destruction;
- o is capable of withstanding electrical, mechanical, thermal and chemical stresses normally encountered during service;

- • the basic insulation of operational electrical equipment is protected to at least 1P2X or IPXXB;
- • where insulation has been applied during the erection of the installation, the quality of the insulation has been verified.

 Where the risk of electric shock is increased by a reduction in body resistance and/or by contact with Earth potential, **confirm** that protection has been provided by insulation of live parts, protection by obstacles, barriers, enclosures or SELV.

10.3.13.3 Protection against electric shock – special installations or locations – verification tests

Where SELV or PELV is used (whatever the nominal voltage) in locations containing a bath, shower, hot air sauna and/or in a restrictive conductive location, confirm that protection against direct contact has been provided by:

- insulation capable of withstanding a type-test voltage of 500 V a.c. rms for 1 minute or
- barriers and/or enclosures providing protection to at least 1P2X or IPXXB.

 The above requirements do not apply to locations where freedom of movement is not physically constrained.

10.3.13.4 RCDs and RCBOs – verification tests

Test (on the load side of the RCD) that:

- general-purpose RCDs:
    - o do not open with a leakage current flowing equivalent to 50% of the rated tripping current;
    - o open in less than 200 ms with a leakage current flowing equivalent to 100% of the rated RCD tripping;
- general-purpose Residual Current Operated Circuit Breakers (RCCBs) conforming to BSEN 61008, or Residual Current Operated Circuit Breakers (RCBOs) without integral overcurrent protection conforming to BSEN 61009:
    - o do not open with a leakage current flowing equivalency to 50% of the rated tripping current of the RCD;
    - o open in less than 300 ms with a leakage current flowing equivalency to 100% of the RCD's tripping current;
- RCD-protected socket outlets to BS 7288:
    - o do not open with a leakage current flowing equivalent to 50% of the RCD's rated tripping current;
    - o open in less than 200 ms with a leakage current flowing equivalency to 100% of the RCD's rated tripping current.

10.3.13.5 Selection and erection of wiring systems

Test to confirm that:

- all electrical joints and connections meet stipulated requirements concerning conductance, insulation, mechanical strength and protection;

- cables that run in thermally insulated spaces are not actually covered by the thermal insulation;
- the current-carrying capacity of cables which are installed in thermally insulated walls or above a thermally insulated ceiling conforms with Appendix 4 to the Regulations;
- the current-carrying capacity of cables that are totally surrounded by thermal insulation for less than 0.5 m has been reduced according to the size of cable, length and thermal properties of the insulation;
- the insulation and/or sheath of cables connected to a bare conductor or busbar is capable of withstanding the maximum operating temperature of the bare conductor or busbar;
- wiring systems are capable of withstanding the highest and lowest local ambient temperature likely to be encountered, or are provided with additional insulation suitable for those temperatures;
- wiring systems have been selected and erected so as to minimise (i.e. during installation, use and maintenance) any damage to the insulation of cables, conductors and their terminations.

10.3.13.6 Site-applied insulation

Test to confirm that:

- insulation applied on site to protect against direct contact is capable of withstanding, without breakdown or flashover, an applied test voltage as specified in the British Standard for similar type-tested equipment;
- supplementary insulation applied to equipment during erection (i.e. to protect against indirect contact):
  - o protects to at least 1P2X or IPXXB; and
  - o is capable of withstanding, without breakdown or flashover, an applied test voltage as specified in the British Standard for similar type-tested equipment.

10.3.13.7 Supplies for safety services (IT systems)

In an IT system, confirm that continuous insulation monitoring has been provided to give audible and visible indications of a first fault.

10.3.14 Periodic inspections and tests

Periodic inspection and testing of all electrical installations must be carried out to confirm that the installation is in a satisfactory condition for continued service.

 Note: A generic list of examples of items requiring inspection together with examples of model forms for certification and reporting are given in Appendix 6 to BS 7671:2018.

The aim of periodic inspection and testing is to:

- confirm that the safety of the installation has not deteriorated or been damaged;
- ensure the continued safety of persons and livestock against the effects of electric shock and bums;
- identify installation defects and non-compliance with the requirements of the Regulations, which could give rise to danger;
- prevent property from being damaged by fire and heat caused by a defective installation.

Precautions shall be taken to ensure that inspection and testing does not cause:

- danger to persons or livestock;
- damage to property and equipment (even if the circuit is defective).

The frequency of periodic inspection and testing of installations will depend on:

- the type of installation, its use and operation;
- the quality the required maintenance; and
- external influences which the installation is (or might be) subjected to.

 Periodic inspection and testing of supervised installations may take the form of continuous monitoring and maintenance by skilled persons. Appropriate records shall be kept.

10.3.15 Verification tests

All completed installations (including additions and/or alteration to existing installations) shall be inspected and tested for conformance to the requirements of the Wiring and the Building Regulations).

10.3.15.1 Accessibility of connections

Confirm that all connections and joints are accessible for inspection, testing and maintenance, unless:

- they are in a compound-filled or encapsulated joint;
- the connection is between a cold tail and a heating element;
- the joint is made by welding, soldering, brazing or compression tool.

10.3.15.2 Appliances producing hot water or steam

Confirm that electric appliances producing hot water or steam have been protected against overheating.

10.3.15.3 Cables and conductors for low voltage

Confirm that flexible and non-flexible cables and flexible cords (and conductors used as an overhead line) operating at low voltage comply with the Relevant British or Harmonised Standard.

10.3.15.4 Electric surface heating systems

Confirm that the equipment, system design, installation and testing of all electric surface heating systems meet the requirements of BS 6351.

10.3.15.5 Emergency switching – verification tests

For any exterior installation, confirm that the switch is placed outside the building, adjacent to the equipment. Where this is not possible, a notice showing the position of the switch shall be placed adjacent to the equipment and a notice fixed near the switch shall indicate its use.

10.3.15.6 Forced air heating systems

Confirm (by inspection and test) that the electric heating elements of forced air heating systems (other than those of central-storage heaters):

- are incapable of being activated until the prescribed airflow has been established;
- deactivate when the airflow is reduced or stopped;
- do not have two, independent, temperature-limiting devices;
- have frames and enclosures which are constructed out of non-ignitable material.

10.3.15.7 Heating cables

Check that:

- heating conductors and cables that pass through (or are in close proximity to) to a fire hazard:
  o are enclosed in material with an ignitability characteristic 'P' as specified in BS 476;
  o are protected from any mechanical damage;
- heating cables that have been laid (directly) in soil, concrete, cement screed or other material used for road and building construction are:
  o capable of withstanding mechanical damage;
  o constructed of material that will be resistant to damage from dampness and/or corrosion;

- heating cable that have been laid (directly) in soil, a road or the structure of a building are installed so that it:
  - o is completely embedded in the substance it is intended to heat;
  - o is not damaged by movement (by it or the substance in which it is embedded);
  - o complies in all respects with the maker's instructions and recommendations;
- the maximum loading of floor-warming cable under operating conditions is no greater than the temperatures shown in Table 10.10.

Table 10.10 Maximum conductor operating temperatures for a floor-warming cable

Type of cable	Maximum conductor operating temperature (°C)
General-purpose PVC over conductor	70
Enamelled conductor, polychiorophene over enamel, PVC over all	70
Enamelled conductor, PVC over all	70
Enamelled conductor, PVC over enamel, lead-alloy 'E' sheath over all	70
Heat-resisting PVC over conductor	85
Nylon over conductor, heat-resisting PVC over all	85
Mineral insulation over conductor, copper sheath over all	Temperature dependent on type of seal employed, outer covering etc.
Silicone-treated woven-glass sleeve over conductor	180
Synthetic rubber or equivalent elastomeric insulation over conductor	85

10.3.15.8 Heating and ventilation systems

In locations where heating and ventilation systems containing heating elements are installed and where there is a risk of fire due to the nature of processed or stored materials, check to ensure that:

- the dust or fibre content and the temperature of the air does not present a fire hazard;
- temperature-limiting devices have a manual reset;
- heating appliances are fixed;
- heating appliances mounted close to combustible materials are protected by barriers to prevent the ignition of such materials;

- heat-storage appliances are incapable of igniting combustible dust and/or fibres;
- enclosures of equipment such as heaters and resistors do not attain higher surface temperatures than:
  - o 90°C under normal conditions; and
  - o 115°C under fault conditions.

10.3.15.9 Identification of conductors by letters and/or numbers

Test to confirm that all individual conductors and groups of conductors:

- have been identified by a label containing either letters or numbers that are clearly legible;
- have numerals that contrast, strongly, with the colour of the insulation; and that
- numerals 6 and 9 have been underlined to avoid confusion.

10.3.15.10 Plugs and socket outlets

Inspect and test to ensure that any plug and socket outlet used in single-phase a.c. or two-wire d.c. circuits that does **not** comply with BS 1363, BS 546, BS 196 or BS EN 60309-2 has either been designed especially for that purpose, or:

- the plug and socket outlet used for an electric clock has been specially designed for that purpose and the plug has a fuse not exceeding 3 A which complies with BS 646 or BS 1362;
- the plug and socket outlet used for an electric shaver is either part of the shaver supply unit that complies with BS 3535 or, in a room (other than a bathroom) that complies with BS 4573.

10.3.15.11 Water heaters

Confirm that water heaters (or boilers) having immersed and uninsulated heating elements are permanently connected to the electricity supply via a double-pole linked switch, which is either:

- separate from and within easy reach of the heater/boiler; or
- part of the boiler/heater.

10.3.15.12 Functional testing

The following are amongst the most important functional tests that should be completed:

- Verify the effectiveness of RCDs providing protection against indirect contact (or supplementary protection against direct contact) by a test simulating a typical fault condition.

- Functionally test assemblies (such as switchgear and controlgear assemblies, drives, controls and interlocks) to show that they are properly mounted, adjusted and installed in accordance with the Regulations.

10.3.16 Electrical connections

Of main concern are:

- connections between conductors and between a conductor and equipment;
- main earthing terminals or bars;
- final and distribution circuits.

Test and verify that:

- connections between conductors and between a conductor and equipment provide durable electrical continuity and adequate mechanical strength;
- the earthing conductor of main earthing terminals (or bars) is capable of being disconnected to enable the resistance of the earthing arrangements to be measured;

 For convenience (and if required) this may be combined with the main earthing terminal or bar.

- all joints:

  o are capable of disconnection only by means of a tool;
  o are mechanically strong;
  o ensure the maintenance of electrical continuity;

- the wiring of final and distribution circuits to equipment with a protective conductor current exceeding 10 mA have a protective connection complying with one or more of the following:

  o a single protective conductor with a cross-sectional area greater than 10 mm²;
  o a single (mechanically protected) copper protective conductor with a cross-sectional area greater than 4 mm²;
  o two individual protective conductors;
  o a BS 4444 Earth monitoring system that will automatically disconnect the supply to the equipment in the event of a continuity fault;

- connection (i.e. of the equipment) to the supply is by means of a double-wound transformer, which has its secondary winding connected to the protective conductor of the incoming supply and the exposed-conductive-parts of the above 10.1.1.

10.3.17 Tests for compliance with the building regulations

As shown in Table 10.11, there are four types of installation that have to be inspected and tested for compliance with the Building Regulations (see also Figure 10.5).

Table 10.11 Types of installation

Type of inspection	When is it used?	What should it contain?	Remarks
Electrical Installation Certificate	For the initial certification of a new installation or for the alteration or addition to an existing installation where new circuits have been introduced	A schedule of inspections and test results as required by Part 6 (of BS 7671) A certificate, including guidance for recipients (standard form from Appendix 6 of BS 7671)	The original certificate shall be given to the person ordering the work and a duplicate retained by the contractor
Minor Electrical Installation Works Certificate	For additions and alterations to an installation such as an extra socket outlet or lighting point to an existing circuit, the relocation of a light switch etc.	In accordance with the relevant provisions of Part 6 of BS 7671	This certificate may also be used for the replacement of equipment such as accessories or luminaires, but not for the replacement of distribution boards (or similar items) or the provision of a new circuit
Electrical Installation Report	For the inspection of an existing electrical installation	A schedule of inspections and a schedule of test results as required by Part 6 (of BS 7671)	For safety reasons, the electrical installation will need to be inspected at appropriate intervals by a competent person
Building Regulations Compliance Certificate	Confirmation that the work carried out complies with the Building Regulations	The basic details of the installation, the location, completion date and the name of the installer	A purchaser's solicitor may request this document when you come to sell your property. Looking further ahead, it may be required as one of the documents that will make up your Home Information Pack

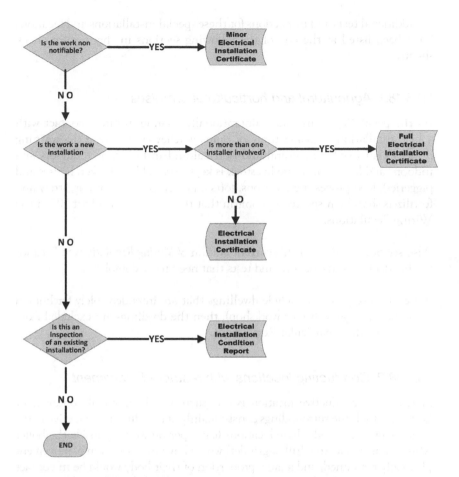

Figure 10.5 Choosing the correct Inspection Certificate.

 Part P of the Building Regulations applies only to fixed electrical installations that are intended to operate at low voltage or extra-low voltage which are not controlled by the Electricity Supply Regulations 1988 as amended, or the Electricity at Work Regulations 1989 as amended.

10.3.18 Additional tests required for special installations and locations

Part 7 of the Wiring Regulations contains additional requirements in respect of installations where the risk of electric shock is increased by a reduction in body resistance or by contact with Earth potential (e.g. locations containing a bath or shower, swimming pools, hot air sauna, construction installations, agricultural and horticultural premises, caravans, motor caravans and highway power supplies etc.).

Additional tests and inspections for these special installations and locations have been listed at the end of the following sections in the form of check sheets.

10.3.18.1 Agricultural and horticultural premises

As the possibility of animals unintentionally coming in direct contact with a live installation is greater than for humans (they cannot read the warning notices!), all fixed agricultural and horticultural installations (outdoors and indoors) and locations where livestock is kept (e.g. stables, chicken houses and piggeries), feed-processing locations, lofts and storage areas for hay, straw and fertilizers shall be inspected to confirm that they comply with Part 705 of the Wiring Regulations.

Also see Section 8.3.1 of this current edition of *Wiring Regulations in Brief* for further details of inspections and tests that need to be completed.

Note: If these premises include dwellings that are intended solely for human habitation (e.g. an office or workshop), then the dwellings are excluded from the scope of these particular Regulations.

10.3.18.2 Conducting locations with restricted movement

A restrictive conductive location is one such as a large metal container or boiler in which the surroundings consist mainly of metallic or conductive parts. People employed inside these locations (e.g. a person working inside the boiler whilst using an electric drill or grinder) would have their freedom of movement physically restrained, and a large proportion of their body would be in contact with the sides of that location and, therefore, prone to shock hazards.

The supplies to mobile equipment for use in such locations will have to be inspected to confirm that they comply with Section 706 of the Wiring Regulations.

Also see Section 8.3.2 of this current edition of *Wiring Regulations in Brief* for further details of inspections and tests that need to be completed.

10.3.18.3 Construction and demolition sites

Electrical installations at construction sites are primarily there so as to provide lighting and power for enabling work to proceed. As workmen will probably be working ankle deep in wet muddy conditions and using a selection of portable tools such as drills and grinders, they will be particularly susceptible to electric shock. To avoid this happening, PME earthing facility shall not be used for earthing an installation and all extraneous-conductive-parts must be reliably connected to the main earthing terminal.

Installations providing electricity supply for:

- new building construction;
- repairs, alterations, extensions or demolition of existing buildings;
- engineering construction;
- earthworks;

will have, therefore, to be inspected to confirm that they comply with Part 704 of the Wiring Regulations.

Also see Section 8.3.2 of this current edition of *Wiring Regulations in Brief* for further details of inspections and tests that need to be completed.

Note: These requirements do **not** apply to:

- construction site offices, cloakrooms, meeting rooms, canteens, restaurants, dormitories and toilets;
- installations covered by BS 6907.

10.3.18.4 Electrical installations in caravans, camping parks and similar locations

For electrical installations in in a caravan park or camping park, special consideration needs to be given to the:

- protection of people, due to the fact that the human body may be in contact with Earth potential;
- protection of wiring, due to tent pegs or ground anchors; and
- movement of heavy or high vehicles;

and these installations will have to be inspected to confirm that they comply with:

- Section 708 (for electrical installations within caravan parks, camping parks and similar locations);and/or
- Section 721 (for electrical installations inside caravans and motor caravans) of the Wiring Regulations.

Also see Section 8.3.3 of this current edition of *Wiring Regulations in Brief* for further details of inspections and tests that need to be completed.

10.3.18.4.1 Electrical installations in caravans and motor caravans

Caravans and motor caravans are designed as leisure accommodation vehicles which are either towed (e.g. by a car) or self-propelled to a caravan site. They will often contain a bath or a shower and special requirements for such installations will apply.

In addition to the normal dangers associated with fixed electrical installations, there is also the potential hazard of totally unskilled people moving the caravan/motor caravan, connecting and disconnecting the mains supply, and not ensuring that it is correctly earthed.

All electrical installations in caravans and motor caravans, therefore, need to be inspected to confirm that they comply with Part 721 of the Wiring Regulations.

Also see Section 8.3.3 of this current edition of *Wiring Regulations in Brief* for further details of inspections and tests that need to be completed on these particular installations.

Note: It should be remembered that the requirements of this section do **not** apply to:

- electrical circuits and equipment covered by the Road Vehicles Lighting Regulations 1989;
- installations covered by BS EN 1648-1 and BS EN 1648-2;
- internal electrical installations of mobile homes, fixed recreational vehicles, transportable sheds and temporary premises or structures.

10.3.18.5 Electric vehicle charging installations

With the increased number of all-electric or hybrid-electric cars currently being manufactured, the 2018 edition of BS 7671 now contains a new Section (722) to cover the requirements for circuits intended to supply electric vehicles for charging purposes.

Also see Section 8.3.4 of this current edition of *Wiring Regulations in Brief* for further details of inspections and tests that need to be completed.

10.3.18.6 Exhibitions, shows and stands

Whilst Chapter 51 of BS 7671:2018 still contains the requirements for external influences, the 18th edition of the Wiring Regulations has now been further developed in accordance with new ISO standards and the BS EN 60721 and BS EN 61000 series on environmental conditions.

Accordingly, the revised Regulations now require that all temporary electrical installations in exhibitions, shows and stands, including mobile and portable displays and equipment, must be inspected to confirm that they comply with Part 711 of the Wiring Regulations.

Also see Section 8.3.5 of this current edition of *Wiring Regulations in Brief* for further details of inspections and tests that need to be completed.

10.3.18.7 Floor and ceiling heating systems

This Section of BS 7671 has been completely revised since it was last published in 2008 and now applies to the installation of embedded electric floor and ceiling heating systems. For example, to enable surface heating these can be erected as either a thermal storage heating system or a direct heating system.

The requirements of this Section also apply to electric heating systems for de-icing, frost prevention and similar applications, and cover both indoor and outdoor systems such as walls, ceilings, floors, roofs, drainpipes, gutters, pipes, stairs, roadways, and non-hardened compacted areas (e.g. football fields, lawns).

In accordance with these revised Regulations, electric floor and ceiling heating systems which are erected as either thermal storage heating systems or direct heating systems must be inspected to confirm that they comply with Part 753 of the Wiring Regulations.

 Note: The regulations do not, however, apply to the installation of wall heating systems nor heating systems for industrial and commercial applications.

 Also see Section 8.3.6 of this current edition of *Wiring Regulations in Brief* for further details of inspections and tests that need to be completed.

10.3.18.8 Locations containing a bath or shower

When people use bathrooms and showers, most of the time they are naturally unclothed and wet and thus very vulnerable to electric shock due to their reduced body resistance (i.e. absence of shoes means less protection from shock, whilst water on their skin will tend to short-circuit its natural protection). Special measures are, therefore, required to ensure that the possibility of direct and/or indirect contact is reduced. Locations containing a fixed bath (bath tub or birthing pool), and showers and their surroundings, must, therefore, be inspected to confirm that they comply with Part 701 of the Wiring Regulations.

 Also see Section 8.3.7 of this current edition of *Wiring Regulations in Brief* for further details of inspections and tests that need to be completed.

 Locations containing baths or showers for medical treatment, or for disabled persons, may have special requirements.

10.3.18.9 Marinas and similar locations

For marinas, particular attention has to be given to the increased risk of electric shock owing to:

- contact of the body with Earth potential;
- presence of water;
- reduction in body resistance.

Circuits intended to supply pleasure craft or houseboats in marinas and similar locations must therefore be inspected to confirm that they comply with Section 709 of the Wiring Regulations.

 Also see Section 8.3.8 of this current edition of *Wiring Regulations in Brief* for further details of inspections and tests that need to be completed.

 Note: The Regulations do not apply to the supply for houseboats (if they are supplied directly from the public network) or to the internal electrical installations of pleasure craft or houseboats.

10.3.18.10 Medical locations

Section 710 of the Wiring Regulations is aimed at patient healthcare facilities and, although the requirements mainly refer to hospitals, private clinics, medical and dental practices, healthcare centres and dedicated medical rooms in the workplace, this Section also applies to electrical installations in locations designed for medical research and (where applicable) to veterinary clinics.

The International Electrotechnical Commission (IEC) and/or British Standards (BS) manufacturers' standards for medical electrical equipment consist of two types of testing: type testing and routine testing.

10.3.18.10.1 Type testing

Type testing is carried out by an approved test house (under tightly specified and tightly controlled environmental conditions) on a single representative sample of a piece of equipment for which certification of compliance with a standard is being sought. These tests are not intended for routine use – indeed, it has been documented that repetition of many of the tests would certainly cause deterioration in performance and safety of the equipment under test.

10.3.18.10.2 Routine testing

Routine testing, on the other hand, is intended to provide an indication of the inherent safety of the equipment without subjecting it to undue stress that would be liable to cause deterioration.

A full list of tests and inspections that need to be completed is contained in Section 7.3.10 of the Wiring Regulations. Further advice is also contained in Section 8.3.10 of this current edition of *Wiring Regulations in Brief.*

 Note: The requirements do not, however, apply to the actual medical electrical equipment itself, as this is fully covered in ISO 13485 (the Medical Devices Directive) and the BS EN 60601 series on Medical Electrical Equipment and Systems.

10.3.18.11 Mobile and transportable units

A vehicle and/or mobile (self-propelled or towed) or transportable structure (such as a container or cabin) in which all or part of an electrical installation is contained and which is provided with a temporary supply by means of, for example, a plug and socket outlet, shall be inspected to confirm that they comply with Part 717 of the Regulations.

Also see Section 8.3.11 of this current edition of *Wiring Regulations in Brief* for further details of inspections and tests that need to be completed.

10.3.18.12 Operating and maintenance gangways

This (comparatively small) Section of BS 7671:2018 centres on the operation and safe maintenance of switchgear and controlgear within areas that include gangways and where access is restricted to skilled or instructed persons. Access areas and the requirements for operating and maintenance gangways need to comply with Part 729 of the Wiring Regulations.

Also see Section 8.3.13 of this current edition of *Wiring Regulations in Brief* for further details of inspections and tests that need to be completed.

10.3.18.13 Rooms and cabins containing saunas

Saunas, similar to swimming pools, bathrooms and showers etc., are primarily used by people who are unclothed and wet and thus very vulnerable to electric shock due to their reduced body resistance (i.e. absence of shoes means less protection from shock, whilst water on their skin will tend to short-circuit its natural protection).

Special measures are, therefore, needed to ensure that the possibility of direct and/or indirect contact is reduced, and installations supplying electricity for locations in which hot air sauna heating equipment (in accordance with BSEN 60335-2-53) must be inspected to confirm that they comply with Part 703 of the Regulations.

Also see Section 8.3.9 of this current edition of *Wiring Regulations in Brief* for further details of inspections and tests that need to be completed.

10.3.18.14 Solar, photovoltaic power supply systems

Photovoltaics (PV) converts light directly into electricity by using photons from sunlight to knock electrons into a higher state of energy, thereby creating electricity. Solar cells are packaged in photovoltaic modules (often electrically connected in multiples as solar photovoltaic arrays) and convert energy from the sun into electricity.

Electrical installations of PV power supply systems (including subsystems with a.c. modules) need to be inspected to confirm that they comply with Part 712 of the Regulations.

 Also see Section 8.3.13 of this current edition of *Wiring Regulations in Brief* for further details of inspections and tests that need to be completed.

10.3.18.15 Swimming pools and other basins

Swimming pools are, by design, '*wet areas*', as people using them are normally wet, which will increase their vulnerability to electric shock. Special measures are, therefore, needed to ensure that all possibility of direct and/or indirect contact is reduced.

Requirements applicable to basins of swimming pools, paddling pools and other basins plus their surrounding zones must be inspected to confirm that they comply with Part 702 of the Regulations.

 Also see Section 8.3.16 of this current edition of *Wiring Regulations in Brief* for further details of inspections and tests that need to be completed.

10.3.18.16 Temporary electrical installations for structures, amusement devices and booths at fairgrounds, amusement parks and circuses

Temporarily erected mobile or transportable electrical machines, structures and electrical equipment are typically installed repeatedly and temporarily, at fairgrounds, amusement parks, circuses and/or similar places.

It is essential that an adequate number of socket outlets are installed to allow the user's requirements to be safely met, and their safe design, installation and operation shall be inspected to confirm that they comply with Part 740 of the Regulations.

 Also see Section 8.3.14 of this current edition of *Wiring Regulations in Brief* for further details of inspections and tests that need to be completed.

10.4 Identification and notices

For safety purposes, the Regulations require a number of notices and labels (see examples of these shown in Figure 10.6) to be used for electrical installations and these will need to be checked during initial and periodic inspections.

10.4.1 General

Verify that:

* labels (or other means of identification) indicate the purpose of each item of switchgear and controlgear;

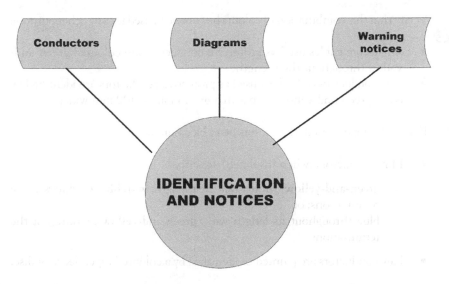

Figure 10.6 Identification and notices.

- wiring is marked and/or arranged so that it can be quickly identified for inspection, testing, repair or alteration of the installation;
- unambiguous marking has been provided at the interface between conductors identified in accordance with these Regulations and conductors identified to previous versions of the Regulations;
- **orange**-coloured conduits have been used to distinguish an electrical conduit from other services or other pipelines.

 Whilst Table 10.51 of BS 7671:2018 provides full details of conductor and conduit colours, Chapter 6 (Section 6.2.3.7) of this edition of *Wiring Regulations in Brief* deals with the main ones to remember.

10.4.2 Conductors

Verify that:

- conductor cable cores are identified by colour and/or lettering and/or numbering;
- binding and sleeves used for identifying protective conductors comply with BS 3858;
- neutral or mid-point conductors are coloured **blue**;
- protective conductor cable cores are identifiable at all terminations (and preferably throughout their length);
- protective conductors are a bi-colour combination of **green and yellow**, and that neither colour covers more than 70% of the surface being coloured;

Verify that this combination of colours has **not** been used for any other purpose;

- single-core cables used as protective conductors are coloured **green-and-yellow** throughout their length;
- bare conductors or busbars used as protective conductors are identified by equal **green and yellow** stripes that are 15 mm to 100 mm wide;

 If adhesive tape is used, then it has been bicoloured.

- PEN conductors (when insulated) are either:
  - o **green-and-yellow** throughout their length, with **blue** markings at the terminations; or
  - o **blue** throughout its length with, **green-and-yellow** markings at the terminations;
- bare conductors are painted or identified by a coloured tape, sleeve or disc.

ALL other conductors (including those used to identify conductor switchboard busbars and conductors) **shall** be coloured as shown in accordance with Table 10.12.

 The colour **green** shall **not** be used on its own.

Table 10.12 Identification of conductors

Function	Alphanumeric	Colour
a.c. power circuit (Note 1)		
Control circuits, ELV and other applications		
Functional earthing conductor		Cream
Mid-wire of three-wire circuit (Notes 2 and 3)	M	Blue
Negative (of negative earthed) circuit (Note 2)	M	Blue
Negative (of positive earthed) circuit	L-	Grey
Negative of three-wire circuit	L-	Grey
Negative of two-wire circuit	L-	Grey
Neutral of single- or three-phase circuit	N	Blue
Neutral or mid-wire (Note 4)	N or M	Blue
Outer negative of two-wire circuit derived from three-wire system	L-	Grey
Outer positive of two-wire circuit derived from three-wire system	L+	Brown
Phase 1 of three-phase a.c. circuit	L1	Brown
Phase 2 of three-phase a.c. circuit	L2	Black
Phase 3 of three-phase a.c. circuit	L3	Grey

Table 10.12   *(continued)*

Function	Alphanumeric	Colour
Phase conductor	L	Brown, black, red, orange, yellow, violet, grey, white, pink or turquoise
Phase of single-phase circuit	L	Brown
Positive (of negative earthed) circuit	L+	Brown
Positive (of positive earthed) circuit (Note 2)	M	Blue
Positive of three-wire circuit	L+	Brown
Positive of two-wire circuit	L+	Brown
Protective conductors		Green-and-yellow
Three-wire d.c. power circuit		
Two-wire earthed d.c. power circuit		
Two-wire unearthed d.c. power circuit		

Notes:

1. Power circuits include lighting circuits.
2. M identifies either the mid-wire of a three-wire d.c. circuit, or the earthed conductor of a two-wire earthed d.c. circuit.
3. Only the middle wire of three-wire circuits may be earthed.
4. An Earthed PELV conductor is **blue.**

Further information, concerning cable identification colours for extra-low-voltage and d.c. power circuits, is available from the IET website at http://www.iet.org.

10.4.2.1 Identification of conductors by letters and/or numbers

Where letters and/or numbers are used to identify conductors, check to confirm that:

- individual conductors and/or groups of conductors are identified by either letters or numbers that are clearly legible;
- all numerals contrast, strongly, with the colour of the insulation;
- numerals 6 and 9 are underlined;
- protective devices are arranged and identified so that the circuit protected is easily recognisable;
- protective conductors coloured **green-and-yellow** are not numbered other than for the purpose of circuit identification;
- the alphanumeric numbering system is in accordance with Table 10.12.

10.4.2.2 Omission od identification by colour or marking

Colour or marking is **not** required for:

- concentric conductors of cables;
- metal sheath or armour of cables (when used as a protective conductor);
- bare conductors (where permanent identification is impracticable);
- extraneous-conductive-parts used as a protective conductor;
- exposed-conductive-parts used as a protective conductor.

10.4.3 Diagrams

Verify that all available diagrams, charts, tables or schedules that have been used indicate:

- the type and composition of each circuit (points of utilisation served, number and size of conductors, type of wiring); and
- the method used for protection against indirect contact;
- the data (where appropriate) and characteristics of each protective device used for automatic disconnection;
- the identification (and location) of all protection, isolation and switching devices;
- the circuits or equipment items that are susceptible to a particular test.

 Verify that distribution board schedules **have been** provided within (or adjacent to) each distribution board.

 All symbols used shall comply with BS EN 60617.

10.4.4 Warning notices

Check to confirm that the following warning notices are appropriate to the situation, correctly positioned and contain the right information.

10.4.4.1 Earth-free local equipotential bonding

Where Earth-free local equipotential bonding has been used, confirm that an appropriate notice has been fixed in a prominent position adjacent to every point of access to the location concerned.

10.4.4.2 Earthing and bonding connections

Confirm that a permanent label with the words shown in Figure 10.7 has been permanently fixed at or near:

- the connection point of every earthing conductor to an Earth electrode;

Figure 10.7 Safety notice.

- the connection point of every bonding conductor to an extraneous-conductive-part;
- the main Earth terminal (when separated from the main switchgear).

Where protection is by Earth-free local equipotential bonding or electrical separation, confirm that the warning notice reads as per Figure 10.8.

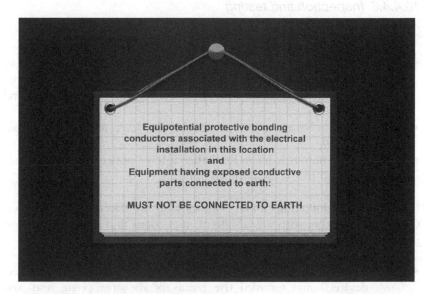

Figure 10.8 Earthing, bonding and electrical separation notice.

10.4.4.3 Electrical installations in caravans, motor caravans and caravan parks

Confirm that:

- a notice has been fixed on (or near) the electrical inlet recess of the cara-van or motor caravan installation confirming that:

  o   the supply to the caravan is suitable for its intended purpose;
  o   the RCDs (if fitted) are suitably installed;

that a notice, worded as shown in BS 7671:2018 Figure 10.721, has been perma-nently fixed near the main isolating switch.

10.4.4.4 Emergency switching

For any exterior installation making use of an emergency switching device, confirm that the switch is placed outside the building, adjacent to the equip-ment, with a notice fixed near the switch to indicate its use.

10.4.4.5 Final circuit distribution boards

Confirm that distribution boards for socket-outlet final circuits have a notice that clearly indicates circuits which have a high protective conductor current, and that this information is positioned so as to be plainly visible to a person employed in modifying or extending the circuit.

10.4.4.6 Inspection and testing

Confirm that:

- a notice has been fixed in a prominent position at or near the origin of every installation upon completion of any work (e.g. initial verification, alterations and additions to an installation and periodic inspection and testing);
- the notice has been inscribed in indelible characters and reads as shown in Figure 10.9;
- if an installation includes an RCD then it has a notice (fixed in a promi-nent position) that reads as shown in Figure 10.10.

Confirm that, following initial verification, an Electrical Installation Certificate (together with a schedule of inspections and a schedule of test results), has been given to the person ordering the work and that:

- the schedule of test results identified every circuit, its related protec-tive device(s) and recorded the results of the appropriate tests and measurements;

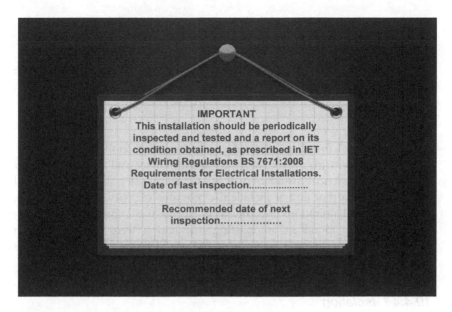

Figure 10.9 Inspection and testing notice.

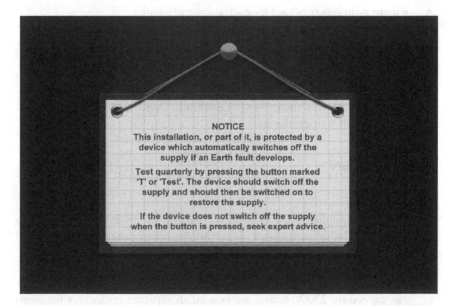

Figure 10.10 Earth fault inspection and testing notice.

- the certificate took account of the respective responsibilities for the safety of that installation and the relevant schedules;
- defects or omissions revealed during inspection and testing of the installation work covered by the certificate were rectified before the certificate is issued.

Confirm that:

- an Electrical Installation Certificate (containing details of the installation together with a record of the inspections made and the test results) or a Minor Electrical Installation Works Certificate (for all minor electrical installations that do not include the provision of a new circuit) was provided for all alterations or additions to electrical circuits; and that
- any defects found in an existing installation were recorded on an Electrical Installation Certificate or a Minor Electrical Installation Works Certificate.

10.4.4.7 Isolation

Confirm that a notice is fixed in each position where there are live parts which are not capable of being isolated by a single device.

 The location of each isolator shall be indicated, unless there is no possibility of confusion.

If an installation is supplied from more than one source, confirm that:

- a main switch is provided for each source of supply;
- a notice has been placed warning operators that more than one switch needs to be operated.

Unless an interlocking arrangement is provided, confirm that a notice has been provided warning people that they will need to use the appropriate isolating devices if an equipment item or enclosure that contains live parts cannot be isolated by a single device.

10.4.4.8 Voltage

Verify that:

- items of equipment (or enclosures) within which a nominal voltage exceeding 230 V exists (and where the presence of such a voltage would **not** normally be expected) are arranged so that before access is gained to a live part, a warning of the maximum voltage present is clearly visible;
- where terminals (or other fixed live parts between which a nominal voltage exceeding 230 V exists) are housed in separate enclosures (or items of equipment which, although separated, can be reached simultaneously

Figure 10.11 Non-standard colours.

by one person), a notice has been secured in a position such that anyone, before gaining access to such live parts, is warned of the maximum voltage which exists between those parts;

• means of access to all live parts of switchgear and other fixed live parts where different nominal voltages exist are marked to indicate the voltages present.

10.4.4.9 Warning notice – non-standard colours

If an installation that was wired according to a previous version of the Regulations is partially altered or rewired according to the current requirements of BS 7671:2018, verify that a warning notice (see Figure 10.11) has been placed at (or near) the appropriate distribution board.

10.5 What type of certificates and reports are there?

There are three types of certificates associated with electrical installations (as shown in Figure 10.12). These are:

Electrical Installation Certificate	For of the design, construction, inspection and testing of the work
Minor Electrical Installation Works Certificate	For the design, construction, inspection and testing of minor work
Electrical Installation Condition Report	For the regular inspection and testing of an electrical installation

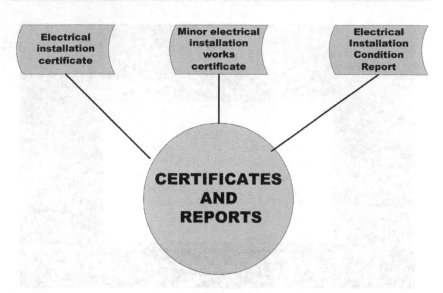

Figure 10.12 Types of certificates and reports.

The template of these certificates and information concerning them can be downloaded from https://electrical.theiet.org/bs-7671/model-forms/ (or from a multitude of other websites) and, provided your version of the certificate is based on the models shown in Appendix 6 of BS 7671:2018, it can be modified to suit your own particular company's quality-management documentation and include your company logo, address etc.

All certificates need to be made out and signed (or otherwise authenticated) by a competent person or persons and may be produced in any durable medium, including written and electronic media.

10.5.1 Electrical Installation Certificate

Upon completion of the verification of a new installation or changes to an existing installation, an Electrical Installation Certificate (based on the model shown in Figure 10.13) shall be provided, together with a schedule of inspections and a schedule of test results, and this shall be given to the person ordering the work.

The schedule of test results shall identify every circuit, including its related protective device(s), and shall require that the results of the appropriate tests and measurements are recorded.

The Electrical Installation Certificate contains details of the installation, together with a record of the inspections made and the test results. It is **only** used for the initial certification of a new installation or for an alteration or addition to an existing installation where new circuits have been introduced.

ELECTRICAL INSTALLATION CERTIFICATE

STINGRAY

(REQUIREMENTS FOR ELECTRICAL INSTALLATIONS – BS 7671:2018)

DETAILS OF THE CLIENT	☐
....	

INSTALLATION ADDRESS	☐
....	

DESCRIPTION AND EXTENT OF THE INSTALLATION	☐
Description of installation:	New installation
Extent of installation covered by this Certificate:	Addition to an existing installation
	Alteration to an existing installation
(Use continuation sheet if necessary) See continuation sheet No:	

FOR DESIGN

I/We being the person(s) responsible for the design of the electrical installation (as indicated by my/our signatures below), particulars of which are described above, having exercised reasonable skill and care when carrying out the design and additionally where this certificate applies to an addition or alteration, the safety of the existing installation is not impaired, hereby CERTIFY that the design work for which I/we have been responsible is to the best of my/our knowledge and belief in accordance with BS 7671:2018, amended to .... (date) except for the departures, if any, detailed as follows:

Details of departures from BS 7671 (Regulations 120.3, 133.1.3 and 133.5):

Details of permitted exceptions (Regulation 411.3.3). Where applicable, a suitable risk assessment(s) must be attached to this Certificate.

Risk assessment attached

The extent of liability of the signatory or signatories is limited to the work described above as the subject of this Certificate.

For the DESIGN of the installation: **(Where there is mutual responsibility for the design)

Signature: ....   Date: ....   Name (IN BLOCK LETTERS): .... Designer No 1

Signature: ....   Date: ....   Name (IN BLOCK LETTERS): .... Designer No 2**

FOR CONSTRUCTION

I being the person responsible for the construction of the electrical installation (as indicated by my signature below), particulars of which are described above, having exercised reasonable skill and care when carrying out the construction hereby CERTIFY that the construction work for which I have been responsible is to the best of my knowledge and belief in accordance with BS 7671:2018, amended to ....(date) except for the departures, if any, detailed as follows:

Details of departures from BS 7671 (Regulations 120.3 and 133.5):

The extent of liability of the signatory is limited to the work described above as the subject of this Certificate.

For CONSTRUCTION of the installation:

Signature: ....   Date: ....   Name (IN BLOCK LETTERS): .... Constructor

FOR INSPECTION & TESTING

I being the person responsible for the inspection and testing of the electrical installation (as indicated by my signature below), particulars of which are described above, having exercised reasonable skill and care when carrying out the inspection and testing hereby CERTIFY that the work for which I have been responsible is to the best of my knowledge and belief in accordance with BS 7671:2018, amended to ....(date) except for the departures, if any, detailed as follows:

Details of departures from BS 7671 (Regulations 120.3 and 133.5):

The extent of liability of the signatory is limited to the work described above as the subject of this Certificate.

For INSPECTION AND TESTING of the installation:

Signature: ....   Date: ....   Name (IN BLOCK LETTERS): .... Inspector

NEXT INSPECTION

I/We the designer(s), recommend that this installation is further inspected and tested after an interval of not more than .... years/months.

Figure 10.13 Electrical Installation Certificate.

Notes:

1. The Electrical Installation Certificate is **not** to be used for a periodic inspection (for which an Electrical Installation Condition Report form should be used).
2. For an alteration or addition which does not involve the introduction of a new circuit(s), a Minor Electrical Installation Works Certificate should be used.
3. The original certificate shall be given to the person ordering the work and a duplicate retained by the contractor.
4. The Electrical Installation Certificate is **only** valid if accompanied by the Schedule of Inspections and the Schedule(s) of Test Results.

5. The Electrical Installation Certificate **shall** be compiled and signed by a competent person(s).

The Electrical Installation Report is 14 pages long and (speaking from experience) most electricians find that having to complete this lengthy form is (to put it mildly) a bit of a bore! However, it is a very important document, and one that is repeatedly relied on for completing other inspections (such as the Minor Works Certificate or the Installation Condition Report).

10.5.2 Minor Electrical Installation Works Certificate

A Minor Electrical Installation Works Certificate (see model in Figure 10.14) is to be used **only** for minor electrical work which does not include the provision of a new circuit, in accordance with BS 7671:2018.

A Minor Electrical Installation Works Certificate (also referred to as the Minor Works Certificate) is used for additions and alterations to an installation, such as an extra socket outlet or lighting point to an existing circuit, the relocation of a light switch etc. This certificate may also be used for the replacement of equipment such as accessories or luminaires, but **not** for the replacement of distribution boards (or similar items) or the provision of a new circuit.

The object of this certificate is to confirm that a minor electrical installation alteration or improvement has been designed, constructed and tested in accordance with BS 7671:2018.

It is a legal requirement to which everybody must comply and includes:

- full details of al minor works completed (e.g. departures from BS 7671:2018, location and description etc.);
- details of the modified circuit (e.g. system type, earthing, wiring system used, protective measures etc.);
- confirmation that all necessary inspections and tests have been completed (e.g. insulation resistance, adequate protective bonding, RCD operation etc.).

MINOR ELECTRICAL INSTALLATION WORKS CERTIFICATE

STINGRAY

(REQUIREMENTS FOR ELECTRICAL INSTALLATIONS – BS 7671:2018)

To be used only for minor electrical work which does not include the provision of a new circuit

PART 1: Description of the minor works

1. Details of the Client .... Date minor works completed ....

2. Installation location/address ....

3. Description of the minor works ....

4. Details of departures, if any, from BS 7671:2018 for the circuit altered or extended (Regulation 120.3, 133.1.3 and 133.5):
   Where applicable, a suitable risk assessment(s) must be attached to the Certificate
   .... Rrisk assessment attached

5. Comments on (including any defects observed in) the existing installation (Regulation 644.1.2):

....

PART 2: Presence and adequacy of installation earthing and bonding arrangements (Regulation 132.16)

1. System earthing arrangement: TN-S   TN-C-S   TT

2. Earth fault loop impedance at distribution board (Z_{db}) supplying the final circuit .... Ω

3. Presence of adequate main protective conductors:

   Earthing conductor
   Main protective bonding conductor(s) to: Water   Gas   Oil   Structural steel   Other.....................

PART 3: Circuit details

DB Reference No.: .... DB Location and type: ....

Circuit No.: .... Circuit description: ....

Circuit overcurrent protective device: BS (EN) .... Type .... Rating .... A

Conductor sizes: Live .... mm² cpc .... mm²

PART 4: Test results for the circuit altered or extended (where relevant and practicable)

Protective conductor continuity: $R_1 + R_2$ .... Ω   or R_2 .... Ω

Continuity of ring final circuit conductors: L/L............... Ω   N/N .... Ω   cpc/cpc .... Ω

Insulation resistance: Live - Live .... MΩ   Live - Earth .... MΩ

Polarity satisfactory: Maximum measured earth fault loop impedance: Z_s .... Ω

RCD operation: Rated residual operating current ($I_{\Delta n}$) .... mA

        Disconnection time .... ms

        Satisfactory test button operation

PART 5: Declaration

I certify that the work covered by this certificate does not impair the safety of the existing installation and the work has been designed, constructed, inspected and tested in accordance with BS 7671:2018 (IET Wiring Regulations) amended to .... (date) and that to the best of my knowledge and belief, at the time of my inspection, complied with BS 7671 except as detailed in Part 1 above.

Name: ....

For and on behalf of: ....

Address: ....   Signature: ....

....   Position: ....

....   Date: ....

Figure 10.14 Minor Electrical Installation Works Certificate.

10.5.3 Electrical Installation Condition Report

An Electrical Installation Condition Report (see model in Figure 10.15) is used for reporting on the condition of an existing installation and will include schedules of both the inspection and the test results.

ELECTRICAL INSTALLATION CONDITION REPORT StingRay

SECTION A. DETAILS OF THE PERSON ORDERING THE REPORT
Name ....
Address ....

SECTION B. REASON FOR PRODUCING THIS REPORT ....

Date(s) on which inspection and testing was carried out ....
SECTION C. DETAILS OF THE INSTALLATION WHICH IS THE SUBJECT OF THIS REPORT
Occupier ....
Address ....

Description of premises
Domestic ☐   Commercial ☐   Industrial ☐   Other (include brief description) ☐ ....
Estimated age of wiring system ....years
Evidence of additions / alterations   Yes ☐   No ☐   Not apparent ☐   If yes, estimate age ....years
Installation records available? (Regulation 651.1)   Yes ☐   No ☐   Date of last inspection .... (date)
SECTION D. EXTENT AND LIMITATIONS OF INSPECTION AND TESTING
Extent of the electrical installation covered by this report
....

Agreed limitations including the reasons (see Regulation 653.2) ....

Agreed with: ....
Operational limitations including the reasons (see page no.............) ....

The inspection and testing detailed in this report and accompanying schedules have been carried out in accordance with BS 7671:2018 (IET Wiring Regulations) as amended to ....
It should be noted that cables concealed within trunking and conduits, under floors, in roof spaces, and generally within the fabric of the building or underground, have **not** been inspected unless specifically agreed between the client and inspector prior to the inspection. An inspection should be made within an accessible roof space housing other electrical equipment.
SECTION E. SUMMARY OF THE CONDITION OF THE INSTALLATION
General condition of the installation (in terms of electrical safety) ....
....

Overall assessment of the installation in terms of its suitability for continued use
SATISFACTORY / UNSATISFACTORY* (Delete as appropriate)
*An unsatisfactory assessment indicates that dangerous (code C1) and/or potentially dangerous (code C2) conditions have been identified.
SECTION F. RECOMMENDATIONS
Where the overall assessment of the suitability of the installation for continued use above is stated as UNSATISFACTORY, I / we recommend that any observations classified as 'Danger present' (code C1) or 'Potentially dangerous' (code C2) are acted upon as a matter of urgency.
Investigation without delay is recommended for observations identified as 'Further investigation required' (code FI).
Observations classified as 'Improvement recommended' (code C3) should be given due consideration.

Subject to the necessary remedial action being taken, I / we recommend that the installation is further inspected and tested by ....(date)
SECTION G. DECLARATION
I/We, being the person(s) responsible for the inspection and testing of the electrical installation (as indicated by my/our signatures below), particulars of which are described above, having exercised reasonable skill and care when carrying out the inspection and testing, hereby declare that the information in this report, including the observations and the attached schedules, provides an accurate assessment of the condition of the electrical installation taking into account the stated extent and limitations in section D of this report.

Inspected and tested by:	Report authorised for issue by:
Name (Capitals)	Name (Capitals) ....
Signature	Signature ....
For/on behalf of	For/on behalf of ....
Position	Position ....
Address	Address ....
Date	Date ....

SECTION H. SCHEDULE(S)
...........schedule(s) of inspection and ....schedule(s) of test results are attached.
The attached schedule(s) are part of this document and this report is valid only when they are attached to it.

Figure 10.15 Electrical Installation Condition Report.

An installation which was designed to an earlier edition of the Regulations and which does not fully comply with the current edition is not necessarily unsafe for continued use, or requiring upgrading. Only damage, deterioration, defects, dangerous conditions and non-compliance with the requirements of the Regulations – which may give rise to danger – should be recorded.

10.6 Test requirements specific for compliance to the Building Regulations

10.6.1 Mandatory requirements

10.6.1.1 Part P – electrical safety

Confirm that:

- reasonable provision has been made in the design, installation, inspection and testing of electrical installations to protect persons from fire or injury;
- sufficient information has been provided so that persons wishing to operate, maintain or alter an electrical installation can do so with reasonable safety.

10.6.1.2 Part M – access and facilities for disabled people

Confirm that (in addition to the requirements of the Disability Discrimination Act 1995) precautions have been taken to ensure that:

- new non-domestic buildings and/or dwellings (e.g. houses and flats used for student living accommodation etc.);
- extensions to existing non-domestic buildings;
- non-domestic buildings that have been subject to a material change of use (e.g. so that they become a hotel, boarding house, institution, public building or shop);

are capable of allowing people, regardless of their disability, age or gender to:

- gain access to buildings;
- gain access within buildings;
- be able to use the facilities of the buildings (both as visitors and as people who live or work in them);
- use sanitary conveniences in the principal storey of any new dwelling.

 Note: From 1 October 2010, the Equality Act replaced most of the Disability Discrimination Act (DDA). However, the Disability Equality Duty in the DDA continues to apply.

10.6.1.3 Part L – conservation of fuel and power

Confirm that:

- energy efficiency measures have been provided which ensure that lighting systems utilise energy-efficient lamps with:

  o   manual switching controls; or
  o   automatic switching (in the case of external lighting fixed to the building); or
  o   both manual and automatic switching controls;

so that the lighting systems can be operated effectively with regard the conservation of fuel and power.

Confirm that building occupiers have been supplied with sufficient information (including results of performance tests carried out during the works) to show how the heating and hot water services can be operated and maintained.

10.6.2 Inspection and test

Verify that all electrical installations have been inspected and tested during and at the end of installation, and before they are taken into service, and that they:

- are reasonably safe and that they comply with BS 7671:2018;
- meet the relevant equipment and installation standards.

Confirm that all components that are part of an electrical installation have been inspected (during installation as well as on completion) to verify that the components have been:

- been selected and installed in accordance with the requirements of BS 7671:2018;
- been made in compliance with appropriate British or Harmonised European Standards;
- been evaluated against external influences (such as the presence of moisture);
- been checked to see that they have not been visibly damaged and are not defective so as to be unsafe;
- been tested to check satisfactory performance with respect to continuity of conductors, insulation resistance, separation of circuits, polarity, earthing and bonding arrangements, Earth fault loop impedance and functionality of all protective devices including RCDs;
- had their test results recorded;
- had their test results compared with the relevant performance criteria to confirm compliance.

 Note: inspections and testing of DIY work should **also** meet the above requirements.

10.6.3 Consumer units

Ensure that accessible consumer units have been fitted with a child-proof cover or are installed in a lockable cupboard.

10.6.4 Design

Confirm that electrical installations have been designed and installed so that they:

- are suitably enclosed (and appropriately separated) to provide mechanical and thermal protection;
- do not present an electric shock or fire hazard to people;
- meet the requirements of the Building Regulations;
- provide adequate protection for persons against the risks of electric shock, burn or fire injuries;
- provide adequate protection against mechanical and thermal damage.

10.6.5 Earthing

Inspect and confirm that:

- electrical installations have been properly earthed;
- lighting circuits include a circuit protective conductor;
- socket outlets which have a rating of 32 A or less and which may be used to supply portable equipment for use outdoors are protected by an RCD;
- the distributor has provided an earthing facility for all new connections;
- new or replacement, non-metallic light fittings, switches or other components do not require earthing (e.g. non-metallic varieties) unless new circuit protective (earthing) conductors have been provided;
- if there are any socket outlets that will accept unearthed (e.g. two-pin) plugs supply equipment that needs to be earthed is not used.

10.6.6 Electricity distributors' responsibilities

Prior to starting works, confirm that the electricity distributor has:

- accepted responsibility for ensuring that the supply is mechanically protected and can be safely maintained;
- evaluated and agreed the proposal for a new (or significantly altered) installation;
- installed the cut-out and meter in a safe location.

Confirm that the distributor has:

- provided an earthing facility for new connections;
- maintained the supply within defined tolerance limits;

- provided certain technical and safety information to the consumer to enable them to design their installation(s);
- ensured that their equipment on consumers' premises:

  o   is suitable for its purpose;
  o   is safe in its particular environment;
  o   clearly shows the polarity of the conductors.

10.6.7 Electrical installations

Verify (by inspection and test) that during installation, at the end of installation and before they are taken into service all electrical installations:

- have been designed and installed (suitably enclosed and appropriately separated) to provide mechanical and thermal protection;
- provide adequate protection for persons against the risks of electric shock, burn or fire injuries;
- meet the requirements of the Building Regulations.

Confirm that all electrical installation work:

- has been carried out professionally;
- complies with the Electricity at Work Regulations 1989 as amended;
- has been carried out by persons who are competent to prevent danger and injury while doing it, or who are appropriately supervised.

10.6.8 Extensions, material alterations and material changes of use

Where any electrical installation work is classified as an extension, a material alteration or a material change of use, confirm that:

- the existing fixed electrical installation in the building is capable of supporting the amount of additions and alterations that will be required;
- the earthing and bonding systems are satisfactory and meet the requirements;
- the mains supply equipment is suitable and can carry the additional loads envisaged;
- any additions and alterations to the circuits which feed them comply with the requirements of the Regulations;
- the protective measures meet the requirements of the Regulations;
- the rating and the condition of existing equipment (belonging to both the consumer and the electricity distributor) is sufficient.

10.6.9 Wiring and wiring systems

Confirm that:

- cables concealed in floors and walls (if required):
  - o have an earthed metal covering, or
  - o are enclosed in steel conduit; or
  - o have some form of additional mechanical protection;
- cables to an outside building (e.g. garage or shed), if run underground, have been routed and positioned so as to give protection against electric shock and fire as a result of mechanical damage to a cable;
- heat-resisting flexible cables (if required) have been supplied for the final connections to certain equipment (see maker's instructions).

10.6.9.1 Continuity of conductors

The continuity of conductors and connections to exposed-conductive-parts and extraneous-conductive-parts, if any, shall be verified by a measurement of resistance on:

- protective conductors, including protective bonding conductors; and
- in the case of ring final circuits, live conductors.

 Note: It is recommended that this continuity test is carried out with a supply having a no-load voltage between 4 V and 24 V, d.c. or a.c., and a short-circuit current of not less than 200 mA.

10.6.9.2 Continuity of ring final circuit conductors

A test shall be made to verify the continuity of each conductor, including the protective conductor, of every ring final circuit.

10.6.9.3 Equipotential bonding conductors

Confirm that:

- main equipotential bonding conductors for water service pipes, gas installation pipes and oil supply pipes, and certain other 'earthy' metalwork, have been provided;
- where there is an increased risk of electric shock (e.g. such as in bathrooms and shower rooms), supplementary equipotential bonding conductors have been installed;
- the minimum size of supplementary equipotential bonding conductors (without mechanical protection) is 4 mm^2.

10.6.10 Socket outlets

Confirm by inspection that:

- older types of socket outlet that have been designed with non-fused plugs, are not connected to a ring circuit;
- RCD protection has been provided for all socket outlets which have a rating of 32 A or less and which may be used to supply portable equipment for use outdoors;
- switched socket outlets indicate whether they are ON;
- socket outlets that will accept unearthed (two-pin) plugs are not used to supply equipment that needs to be earthed;
- the following requirements (see Table 10.13) for wall sockets have been met:

Table 10.13 Building Regulations requirements for wall sockets

Type of wall	Requirement
Cavity masonry	The position of sockets has been staggered on opposite sides of the separating wall
	Deep sockets and chases have not been used in a separating wall
	Deep sockets and chases in a separating wall have not been placed back to back
Framed walls with absorbent material	Sockets have: • been positioned on opposite sides of a separating wall • not been connected back to back • been staggered a minimum of 150 mm edge to edge
Solid masonry	Deep sockets and chases have not been used in separating walls
	The position of sockets has been staggered on opposite sides of the separating wall
Timber-framed	Power points: • that have been set in the linings have a similar thickness of cladding behind the socket box. • have not been placed back to back across the wall

Confirm (by inspection and testing) that all socket outlets used for lighting:

- are wall-mounted;
- are easily reachable;
- have been installed between 450 mm and 1200 mm from the finished floor level;
- are located no nearer than 350 mm to room corners;
- if of a switched variety, indicate whether they are ON;

10.6.11 Switches

Ensure that all controls and switches:

- are easy to operate, visible and free from obstruction;
- are located between 750 mm and 1200 mm above the floor;
- do not require the simultaneous use of both hands (unless necessary for safety reasons) to operate;

and that:

- mains and circuit isolator switches clearly indicate whether they ON or OFF;
- individual switches on panels and on multiple socket outlets have been well separated;
- front plates contrast visually with their backgrounds;
- controls that need close vision (e.g. thermostats) have been located between 1200 mm and 1400 mm above the floor;
- where possible, light switches with large push pads shave been used in preference to pull-cords;
- the colours **red** and **green** have **not** been used in combination as indicators of ON and OFF for switches and controls;
- all switches used for lighting:

  o are easily reachable;
  o have been installed between 450 mm and 1200 mm from the finished floor level.

Confirm that light switches:

- have large push pads (in preference to pull-cords);
- align horizontally with door handles;
- are between 900 and 1100 mm from the entrance door opening;
- are located between 750 mm and 1200 mm above the floor;
- are **not** coloured **red** and **green** (i.e. as a combination) as indicators for ON and OFF.

10.6.12 Telephone points and TV sockets

Confirm that all telephone points and TV sockets are located between 400 mm and 1000 mm above the floor (or between 400 mm and 1200 mm above the floor for permanently wired appliances).

10.6.13 Equipment and components

10.6.13.1 Emergency alarms

Test and inspect to ensure that:

- emergency assistance alarm systems have:

  - o ' visual and audible indicators to confirm that an emergency call has been received;
  - o a reset control that is reachable from a wheelchair, WC, or shower/ changing seat;
  - o a signal that is distinguishable visually and audibly from the fire alarm;

- emergency alarm pull cords are:

  - o coloured **red**;
  - o located as close to a wall as possible;
  - o have two **red** 50 mm diameter bangles;

- front plates contrast visually with their backgrounds;
- the colours **red** and **green** have **not** been used (in combination) to indicate ON and OFF for switches and controls.

10.6.13.2 Fire alarms

Verify (by test and inspection that fire-detection and fire-warning systems have been properly designed, installed and maintained, and that:

- all buildings have arrangements for detecting fire;
- all buildings have been fitted with a suitable (electrically operated) fire warning system (in compliance with BS 5839) or have means of raising an alarm in case of fire (e.g. rotary gongs, handbells or by shouting **'FIRE'**);
- fire alarms emit both audio and visual signals to warn occupants with hearing or visual impairments;
- the fire warning signal is distinct from other signals which may be in general use;
- in premises that are used by the general public (e.g. large shops and places of assembly) a staff alarm system (complying with BS 5839) has been used.

10.6.13.3 Heat emitters

Check that heat emitters:

- are either screened or have their exposed surfaces kept at a temperature below 43°C;
- that are located in toilets and bathrooms do not restrict:

  - o the minimum clear wheelchair manoeuvring space;
  - o the space beside a WC used to transfer from the wheelchair to the WC.

10.6.13.4 Portable equipment for use outdoors

Verify that RCDs have been provided for all socket outlets which have a rating of 32 A or less and which may be used to supply portable equipment for use outdoors.

10.6.13.5 Power-operated entrance doors

Confirm that all power-operated doors have been provided with:

- safety features to prevent injury to people who are struck or trapped (such as a pressure-sensitive door edge which operates the power switch);
- a readily identifiable (and accessible) stop switch;
- a manual or automatic opening device in the event of a power failure, where and when necessary for health or safety.

Confirm that:

- all doors to accessible entrances have been provided with a power-operated door-opening and closing system if a force greater than 20 N is required to open or shut a door;
- once open, all doors to accessible entrances are wide enough to allow unrestricted passage for a variety of users, including wheelchair users, people carrying luggage, people with assistance dogs, and parents with pushchairs and small children;
- power-operated entrance doors:

  o have a sliding, swinging or folding action controlled manually (by a push pad, card swipe, coded entry, or remote control) or automatically controlled by a motion sensor or proximity sensor such as contact mat;
  o open towards people approaching the doors;
  o provide visual and audible warnings that they are operating (or about to operate);
  o incorporate automatic sensors to ensure that they open early enough (and stay open long enough) to permit safe entry and exit;
  o incorporate a safety stop that is activated if the doors begin to close when a person is passing through;
  o revert to manual control (or fail safe) in the open position in the event of a power failure;
  o when open, do not project into any adjacent access route;

- power-operated doors' manual controls:

  o are located between 750 mm and 1000 mm above floor level;
  o are operable with a closed fist;
  o are set back 1400 mm from the leading edge of the door when fully open;
  o are clearly distinguishable against the background;
  o contrast visually with the background.

10.6.14 Thermostats

Check that all controls that need close vision (e.g. thermostats) are located between 1200 mm and 1400 mm above the floor.

10.6.14.1 Smoke alarms – dwellings

Confirm by test and inspection that smoke alarms have been positioned:

- in the circulation space within 7.5 m of the door to every habitable room;
- in the circulation spaces between sleeping spaces and places where fires are most likely to start (e.g. kitchens and living rooms);
- on every storey of a house (including bungalows).

Confirm by test and inspection that:

- kitchen areas that are not separated from the stairway or circulation space by a door have been equipped with an additional heat detector in the kitchen, that is interlinked to the other alarms;
- if more than one smoke alarm has been installed in a dwelling, they have been linked so that if a unit detects smoke it will operate the alarm signal of all the smoke detectors.

Verify by inspection that smoke alarms:

- have ideally been mounted between 25 mm and 600 mm below the ceiling (25–150 mm in the case of heat detectors) and at least 300 mm from walls and light fittings;
- have not been fixed over a stair shaft or any other opening between floors;
- have not been fitted:
  - o in places that get very hot (such as a boiler room);
  - o in a very cold area (such as an unheated porch);
  - o in bathrooms, showers, cooking areas or garages, or any other place where steam, condensation or fumes could give false alarms;
  - o next to or directly above heaters or air-conditioning outlets;
  - o on surfaces which are normally much warmer or colder than the rest of the space.

Test, inspect and confirm that the power supply for a smoke alarm system:

- has been derived from the dwelling's mains electricity supply via a single independent circuit at the dwelling's main distribution board (consumer unit);
- includes a stand-by power supply that will operate during mains failure;
- is (preferably) not protected by an RCD.

10.6.15 Lighting

10.6.15.1 External lighting fixed to the building

Confirm (by test and inspection) that all external lighting (including lighting in porches, but not lighting in garages and carports):

- automatically extinguishes when there is enough daylight, and when not required at night;
- has sockets that can **only** be used with lamps having an efficacy greater than 40 lumens per circuit Watt (such as fluorescent or compact fluorescent lamp types, and **not** GLS tungsten lamps with bayonet-cap or Edison-screw bases).

10.6.15.2 Fittings, switches and other components

Confirm that new or replacement, non-metallic light fittings, switches or other components that require earthing (e.g. non-metallic varieties) have been provided with new circuit-protective (earthing) conductors.

10.6.15.3 Fixed lighting

Ensure that in locations where lighting can be expected to have most use, fixed lighting (e.g. fluorescent tubes and compact fluorescent lamps – but **not** GLS tungsten lamps with bayonet-cap or Edison-screw bases) with a luminous efficacy greater than 40 lumens per circuit-Watt have been made available.

10.6.15.4 Lighting circuits

Verify that all lighting circuits include a circuit-protective conductor.

10.6.16 Lecture/conference facilities

In lecture hall and conference facilities, confirm that artificial lighting has been designed to:

- give good colour rendering of all surfaces;
- be compatible with other electronic and radio-frequency installations.

10.6.17 Cellars or basements

Ensure that LPG storage vessels and LPG-fired appliances that are fitted with automatic ignition devices or pilot lights have not been installed in cellars or basements.

10.7 What about test equipment?

Although BS 7671 lays great emphasis on the requirements for inspection and testing in Chapter 6 of the Regulations, the only reference (as far as I can tell!) to actual test equipment concerning insulation monitoring devices is a statement which says that:

- "an insulation monitoring device shall be so designed or installed that it can only be possible to modify the setting with the use of a key or a tool;
- in an IT system, an insulation monitoring device shall be provided so as to indicate the occurrence of a first fault from a live part to an exposed-conductive-part, or to Earth;
- installations forming part of an IT system can make use of an insulation monitoring device."

Of course, the actual choice of test equipment that the electrician chooses to use will normally be based on personal preference and experience (see Figure 10.16). Nevertheless, it is essential that any piece of test equipment (including software) that is used when installing or inspecting electrical installations for compliance with the Regulations can be relied on to produce accurate results.

Figure 10.16 A selection of test equipment.

ISO 9001:2015 (i.e. the internationally recognised standard for quality management) specifies the requirements for the control of test equipment (although they actually refer to them as 'measuring and monitoring devices' – see Figure 10.17).

Proof	The controls that an organisation has in place to ensure that equipment (including software) used for conformance to specified requirements is properly maintained.
Likely documentation	Equipment records of maintenance and calibration. Work instructions.

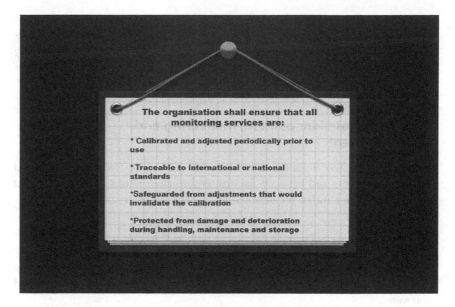

The organisation shall ensure that all monitoring services are:

* Calibrated and adjusted periodically prior to use

* Traceable to international or national standards

*Safeguarded from adjustments that would invalidate the calibration

*Protected from damage and deterioration during handling, maintenance and storage

Figure 10.17 Mandatory requirements from the Building Regulations.

Although the majority of electricians probably work on an individual basis and the requirement to operate as an accredited and registered ISO 9001:2015 company doesn't really apply, following the recommendations of this standard can only help to improve the quality of any organisation – no matter what its size.

In general, therefore:

- all measuring and test equipment that is used by an electrician needs to be well maintained, in good condition and capable of safe and effective operation within a specified tolerance of accuracy;
- all measuring and test equipment should be regularly inspected and/or calibrated to ensure that it is capable of accurate operation (and where necessary by comparison with external sources traceable back to national standards);
- any electrostatic protection equipment that is utilised when handling sensitive components should be regularly checked to ensure that it remains fully functional;
- the control of measuring and test equipment (whether owned by the electrician, on loan, hired or provided by the customer) should always include a check that the equipment:
  - o is exactly what is required;
  - o has been initially calibrated before use;
  - o operates within the required tolerances;
  - o is regularly recalibrated; and that
  - o facilities exist (either within the organisation or via a third party) to adjust, repair or recalibrate as necessary.

If the measuring and test equipment is used to verify process outputs against a specified requirement, then the equipment needs to be maintained and calibrated against national and international standards, and the results of any calibrations carried out **must** be retained and the validity of previous results reassessed if they are subsequently found to be out of calibration.

10.7.1 Control of inspection, measuring and test equipment

Measuring and test equipment items should always be stored correctly and satisfactorily protected between uses (to ensure their bias and precision), and should be verified and/or recalibrated at appropriate intervals.

10.7.2 Computers

Special attention **must** be paid to computers if they are used in controlling processes and particularly to the maintenance and accreditation of any related software.

10.7.3 Software

Software used for measuring, monitoring and/or testing specified requirements should be validated prior to use.

10.7.4 Calibration

Without exception, all measuring instruments can be subject to damage, deterioration or just general wear and tear when they are in regular use. The electrician should, therefore, take account of this fact and ensure that all of his test equipment is regularly calibrated against a known working standard.

The accuracy of the instrument will depend very much on what items it is going to be used to test and the frequency of use of the test instrument, and the electrician will have to decide on the maximum tolerance of accuracy for each item of test equipment.

Of course, calibrating against a 'working standard' is pretty pointless if that particular standard cannot be relied upon, and so the workshop standard **must also** be calibrated, on a regular basis, at either a recognised calibration centre or at the National Physical Laboratory, against one of the national standards.

The electrician will, therefore, have to make allowances for:

- the calibration and adjustment of all measuring and test equipment that can affect the product quality of their inspection and/or test;
- the documentation and maintenance of calibration procedures and records;

- the regular inspection of all measuring or test equipment items to ensure that they are capable of the accuracy and precision that is required;
- environmental conditions which may affect the suitability of the calibrations, inspections, measurements and tests to be completed.

If the instrument is found to be outside of its tolerance of accuracy, any items previously tested with the instrument must be regarded as suspect. In these circumstances, it would be wise to review the test results obtained from the individual instrument. This could be achieved by compensating for the extent of inaccuracy to decide if the acceptability of the item would be reversed.

10.7.4.1 Calibration methods

There are various possibilities, such as:

- sending all working equipment to an external calibration laboratory;
- sending one of each item (i.e. a *'workshop standard'*) to a calibration laboratory, then sub-calibrating each working item against the workshop standard;
- testing by attributes – take a known *'faulty'* product and a known *'good'* product, and then test each one to ensure that the test equipment can identify the faulty and good product correctly.

10.7.4.2 Calibration frequency

The calibration frequency depends on how much the instrument is used, its ability to retain its accuracy and how critical the items being tested are.

Infrequently used instruments are often only calibrated prior to their use whilst frequently used items would normally be checked and recalibrated at regular intervals depending, again, on product criticality, cost, availability etc.

Normally 12 months is considered to be about the maximum calibration interval.

10.7.4.3 Calibration ideals

- Each instrument should be uniquely identified, allowing it to be traced.
- The calibration results should be clearly indicated on the instrument.
- The calibration results should be retained for reference.
- The instrument should be labelled to show the next *'calibration-due'* date, in order to easily avoid its use outside of the period of confidence.
- Any means of adjusting the calibration should be sealed, allowing easy identification if it has been tampered with (e.g. a label across the joint of the casing).

Note: Examples of test equipment normally used by electricians are shown in Annex 10A.

Annex 10A – Examples of test equipment used to test electrical installations

The following are examples of instruments that are required to test electrical installations for compliance with the requirements of BS 7671.

 Quite a lot of test equipment manufacturers now produce dual or multifunctional instruments, so it is quite common to find an instrument that is capable of performing a number of different types of tests – for example, continuity and insulation resistance, loop impedance and prospective fault current. It is, therefore, wise to carry out a little research before purchasing!

Continuity tester

All protective and bonding conductors must be tested to ensure that they are electrically safe and correctly connected. Low-resistance ohmmeters and simple multimeters are normally used for continuity testing. Ideally they should have a no-load voltage of between 4 V and 24 V, be capable of producing an a.c. or d.c. short-circuit voltage of not less than 200 mA and have a resolution of at least 0.01 mΩ.

Insulation resistance tester

A low resistance between phase and neutral conductors, or from live conductors to Earth, will cause a leakage current which will cause weakening of the insulation, as well as involving a waste of energy which would increase the running costs of the installation. To overcome this problem, the resistance between poles or to Earth must never be less than 0.5 mΩ for the usual supply voltages (see Figure 10.18).

Loop impedance tester

Loop testing is a quick, convenient, and highly specific method of testing an electrical circuit for its ability to engage protective devices (circuit-breakers and fuses etc.) by simulating a fault from live to Earth or from live to neutral (short-circuit). The tester first measures the unloaded voltage, then connects a known resistance between the conductors, thereby simulating a fault. The voltage drop is measured across the known resistor, in series with the loop, and the proportion of the supply voltage that appears across the resistor will be dependent on the impedance of the loop.

RCD tester

The standard method for protecting electrical installations is to make sure that an Earth fault results in a fault current that is high enough to operate the protective device quickly so that fatal shock is prevented. However, there are cases

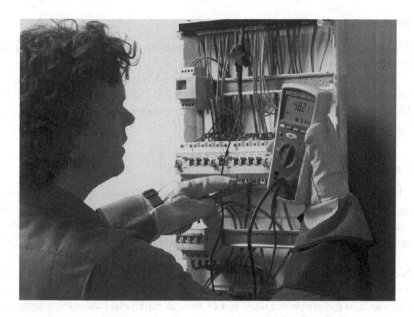

Figure 10.18 Insulation tester.

where the impedance of the Earth-fault loop, or the impedance of the fault itself, are too high to enable enough fault current to flow. In such a case, either:

- the current will continue to flow to Earth, perhaps generating enough heat to start a fire; or
- metalwork which can be touched may be at a high potential relative to Earth, resulting in severe shock danger.

Either or both of these possibilities can be removed by the installation of a residual current device (RCD).

 Note: RCDs are also, sometimes, referred to as:

RCCB   residual current operated circuit-breaker;
SRCD   socket outlet incorporating an RCD;
PRCD   portable RCD, usually an RCD incorporated into a plug;
RCBO   an RCCD which includes overcurrent protection;
SRCBO   a socket outlet incorporating an RCBO and RCD tester, allowing a selection of out-of-balance currents to flow through the RCD and cause its operation.

 The RCD tester should **not** be operated for longer than 2 s.

Prospective fault current tester

A prospective fault current (PFC) tester is used to measure the prospective phase neutral fault current.

Test lamp or voltage indicator

These types of tester (often referred to as a *tetrascope* or *neon screwdriver*) are frequently used by electricians.

These compact screwdriver multi-testers are normally water and impact resistant, with a.c. voltage test, contact test 70–250 V a.c., non-contact 100–1000 V a.c., polarity test 1.5 V d.c.–36 V d.c., continuity check 0–5 Ω and auto power on/off.

Earth electrode resistance

The Earth electrode (when used), is the means of making contact with the general mass of Earth and should be regularly tested to ensure that good contact is made. In all cases, the aim is to ensure that the electrode resistance is not so high that the voltage from earthed metalwork to Earth exceeds 50 V.

Note: Acceptable electrodes are rods, pipes, mats, tapes, wires, plates and structural steelwork buried or driven into the ground. The pipes of other services such as gas and water must **not** be used as Earth electrodes (although they must be bonded to Earth).

Author's end note

That completes what I consider to be the essential parts of the Wiring Regulations that you should be reminded of. However, for full details of all the official requirements and recommendations for electrical installations, you will need to obtain a copy of BS 7671:2018.

The IET website (https://electrical.theiet.org/bs-7671) also provides in-depth information on the latest changes made to this standard, its relationship with the Building Regulations and a whole host of other useful information.

'Mr Google' can also provide some very useful on-site hints as well.

To support the content of this book, a group of useful Annexes follows this chapter. These Annexes are:

Annex A – Symbols used in electrical installations;
Annex B – List of electrical and electromechanical symbols;
Annex C – SI units;
Annex D – The IP Code;
Annex E – Acronyms and abbreviations;
Annex F – Standards;
Annex G – Books by the same author (including a series of books on the Building Regulations as well as on ISO 9001:2018 concerning Quality Management Systems).

Annex A

Symbols used in electrical installations

Socket outlet		Microphone		Operating device (coil)	
Switched socket outlet		Loudspeaker		Make contact - normally open	
Switch		Antenna		Break contact - normally closed	
Two-way switch, single-pole		Machine *function M = Motor G = Generator		Manually operated switch	
Intermediate switch				Three-phase winding - Delta	
Pull switch, single-pole		Generator		Three-phase winding - Star	
Lighting outlet position		Indicating instrument *function V = Voltmeter A = Ammeter		Changer, Converter	
Fluorescent luminaire		Integrating instrument or Energy meter *function Wh = Watt-hour VArh = Volt ampere reactive hour		Rectifier	
Wall mounted luminaire				Invertor	
Emergency lighting luminaire (or special circuit)				Primary cell - longer line positive, shorter line negative	
Self-contained emergency lighting luminaire		Load *details		Battery	
Push button		Motor starter *indicates type		Transformer - general symbol	
Clock		Class II appliance			
Bell		Class III appliance			
Buzzer		Safety isolating transformer			
Horn		Isolating transformer			
Telephone handset		Fuse link, rated current in amperes			

10^9	giga	G
10^6	mega	M
10^3	kilo	k
10^{-3}	milli	m
10^{-6}	micro	μ
10^{-9}	nano	n

Annex B

List of electrical and electromechanical symbols

SYMBOL	DESCRIPTION
β°	tube oscillating angle
°C	degrees Celsius
Ω	ohm
μg	microgram
μg/m³	micrograms per cubic metre
μm	micrometre
μs	microsecond
a	amplitude
A	ampere
A/m	amperes per metre
am	attometre
atm	standard atmosphere
C	coulomb
cd	candela
cd/m²	candelas per square metre
dB	decibels
dB(A)	decibel amp
dBm	decibel metre
dm³	cubic decimetre
dm³/mm	cubic decimetres/millimetre – flow
Em	exametre
eV	electronvolt
f	frequency
F	farad
fm	femtometre
ft	foot
g	gram
G	gauss
G	shock
g²/Hz	accelerated spectral density
GHz	gigahertz – frequency
Gm	gigametre
g/m³	grams per cubic metre
gn	peak acceleration
Gs	setting value of a characteristic quantity
h	hour
H	henry
ha	hectare

SYMBOL	DESCRIPTION
hp	horsepower
hr(s)	hour(s) – alternative to h
Hz	Hertz
I	amp
I^2R	power
in	inch
J	joule
k	constant of the relay
K	kelvin
kA	kiloamp
kA/μs	kiloamps per microsecond
kg	kilogram
kg/m^3	kilograms per cubic metre
kgf	kilogram force
kHz	kilohertz
kPa	kilo Pascal – pressure
ks	kilosecond
kV	kilovolt
kW	kilowatt
kW/m^2	kilowatts per square metre – irradiance
l	litre
lb	pound
lb/in	pounds per square inch
m	metre
m/s	metres per second
m/s^2	metres per second per second – amplitude
m^2	square metre
m^3	cubic metre
mbar	millibar – pressure
MHz	megahertz
min	minute
mm	millimetre
Mm	megametre
mm/h	millimetres per hour
mm/m^2	millimetres/square metre – exposure
mol	mole
ms	millisecond
mV	millivolt
MVA	megavolt amp
N	newton
N/m^2	newtons per square metre
NaCl	sodium chloride
nF	nano Farad
nm	nanometre
pH	alkalinity/acidity value
pm	picometre
Pm	petametre
R	intensity of dropfield in mm/h
R	resistance
rad/s	radians per second
s	second
S	siemens

SYMBOL	DESCRIPTION
t	tonne
T	time
T	tesla
Tm	terametre
û	amplitude of voltage surge
Un	nominal voltage
V	volt
V/μs	volts per microsecond
V/km	volts per kilometre
Vm	volts per metre
W	watt
Wb	weber
W/m^2	watts per square metre – irradiance
yd	yard
ym	yocotmetre
Ym	Yottametre
zm	zeptometre
Zm	zettametre

Annex C
SI units for existing technology

The revised SI (an acronym for the International System of Units, which is informally known as the metric system) rests on a foundation of seven values, known as the constants, whose values are the same everywhere in the universe. In the revised SI, these constants completely define the seven base SI units, from the second to the candela.

Table C1 Basic SI units

SI nomenclature	Abbreviation	Quantity
metre	m	length
kilogram	kg	mass
second	s	time
ampere	A	electrical current
kelvin	K	temperatures
mole	mol	amount of substance
candela	cd	luminous intensity

Kilogram

Of the seven units, only the kilogram (kg) is represented by a physical object, namely a cylinder of platinum-iridium kept at the International Bureau of Weights and Measures at Sèvres, near Paris, with a duplicate at the US Bureau of Standards.

Metre

The metre (m), on the other hand, "is the length of the path travelled by light in a vacuum during a time interval of 1/299,792,458 of a second".

Second

The second is defined by taking the fixed numerical value of the caesium frequency Δv, the unperturbed ground-state hyperfine transition frequency of the caesium 133 atom, to be 9 192 631 770 when expressed in the unit Hz, which is equal to s^{-1}. The wording of the definition was updated in 2019.

Ampere

The ampere (A) is "that constant current which, if maintained in two straight parallel conductors of infinite length, of negligible circular cross section and placed 1 m apart in vacuum, would produce between these conductors a force equal to 2×10^{-7} newtons per metre length".

Kelvin

The unit of temperature is the kelvin (K), which is a thermodynamic measurement as opposed to one based on the properties of real material. Its origin is at absolute zero and there is a fixed point where the pressure and temperature of water, water vapour and ice are in equilibrium, which is defined as 273.16 K.

Mole

The mole (mol) is "that quantity of substance of a system which contains as many elementary entities as there are atoms in 0.012 kg of carbon-12". For definition purposes the entities **must** be specified, (e.g. atoms, electrons, ions or any other particles or groups of such particles).

Candela

Finally there is the candela (cd), the unit of light intensity. This is defined as 'the luminous intensity, in the perpendicular direction, of a surface of $1/600,000$ m^2 of a black body at the temperature of freezing platinum under a pressure of 101,325 N/m^2'.

Small-number SI prefixes

Within the SI units there is a distinction between a quantity and a unit. Length is a quantity, but a metre (abbreviated to m) is a unit.

Table C2 Small number SI units

Measurement	Symbol	Equivalent to
millimetre	mm	0.001 m or 10^{-3} m
micrometre	μm	0.000 001 m or 10^{-6} m
nanometre	nm	0.000 000 001 m or 10^{-9} m

(Continued)

Table C2 Small number SI units (*Continued*)

Measurement	Symbol	Equivalent to
picometre	pm	0.000 000 000 001 m or 10^{-12} m
femtometre	fm	0.000 000 000 000 001 m or 10^{-15} m
attometre	am	0.000 000 000 000 000 001 m or 10^{-18} m
zeptometre	zm	0.000 000 000 000 000 000 001 m or 10^{-21} m
yoctometre	ym	0.000 000 000 000 000 000 000 001 m or 10^{-24} m

Large-number SI prefixes

Table C3 Large number SI prefixes

Measurement	Symbol	Equivalent to
megametre	Mm	1,000,000 m or 10^6 m
gigametre	Gm	1,000,000,000 m or 10^9 m
terametre	Tm	1,000,000,000,000 m or 10^{12} m
petametre	Pm	1,000,000,000,000,000 m or 10^{15} m
exametre	Em	1,000,000,000,000,000,000 m or 10^{18} m
zettametre	Zm	1,000,000,000,000,000,000,000 m or 10^{21} m
yottametre	Ym	1,000,000,000,000,000,000,000,000 m or 10^{24} m

Deprecated prefixes

Some non-SI fractions and multiples are occasionally used (see below), but they are not encouraged.

Table C4 Deprecated prefixes

Fractions	Prefix	Abbreviation	Multiple	Prefix	Abbreviation
10^{-1}	deci	d	10	deka	da
10^{-2}	centi	c	10^2	hecto	h

Derived units

Some units, derived from the basic SI units, have been given special names, many of which originate from a person's name (e.g. Siemens).

Table C5 Derived units

Quantity	Name of unit	Abbreviation (symbol)	Expression in terms of other SI units
energy	joule	J	Nm
force	newton	N	-
power	watt	W	J/s
electric charge	coulomb	C	As
potential difference (voltage)	volt	V	W/A
electrical resistance (or reactance or impedance)	ohm	Ω	V/A
electrical capacitance	farad	F	C/V
magnetic flux	weber	Wb	Vs
inductance (note that the plural of henry is henrys)	henry	H	Wb/A
magnetic flux density	tesla	T	Wb/m^2
admittance (electrical conductance)	siemens	S	A/V (= Ω^{-1})
frequency	hertz	Hz	cycles per second (or events per second)

Units without special names

Other derived units, without special names, are listed below.

Table C6 Units without special names

Quantity	Unit	Abbreviation
area	square metres	m^2
volume	cubic metres	m^3
density	kilograms per cubic metre	kg/m^3
velocity	metres per second	m/s
angular velocity (angular frequency)	radians per second	rad/s
acceleration	metres per second per second	m/s^2
pressure	newtons per square metre	N/m^2
electric field strength	volts per metre	V/m
magnetic field strength	amperes per metre	A/m
luminance	candelas per square metre	cd/m^2

Tolerated units

Some non-SI units are tolerated in conjunction with SI units.

Table C7 Tolerated units

Quantity	Unit	Abbreviation (symbol)	Definition
area	hectare	ha	10^4 m^2
volume	litre	l	10^{-3} m^3
pressure	standard atmosphere	atm	101,325 Pa
mass	tonne	t	10^3 kg (Mg)
energy	electronvolt	eV	1.6021 x 10^{19} J
magnetic	gauss	G	10^{-4} T

Obsolete units

For historical interest, (as well as for completeness), the following table gives a list of obsolete units.

Table C8 Obsolete units

Quantity	Unit	Abbreviation (symbol)	Definition
length	inch	in	0.0254m
	foot	ft	0.3048m
	yard	yd	0.9144m
	mile	mi	1.60394km
mass	pound	lb	0.4539237kg
force	dyne	dyn	10^{-5}N
	poundal	pdl	0.138255N
	pound force	lbf	4.44822N
	kilogram force	kgf	9.80665N
pressure	atmosphere	atm	101.325kN/m^2
	torr	torr	133.322N/m^2
	pounds per square inch	lb/in	6894.76N/m^2
energy	erg	erg	10^{-7}J
power	horsepower	hp	745.700W

Table of the SI units, symbols and abbreviations

Table C9 SI units and symbols

SI unit name	SI unit symbol	Quantity measured
ampere	A	Electric current
ampere per metre	A/m	Magnetic field strength
ampere per square metre	A/m^2	Current density
becquerel	Bq s^{-1}	Activity – of radionuclide
candela	cd	Luminous intensity
candela per square metre	cd/m^2	Luminance
coulomb	C	Electric charge, quantity of electricity
coulomb per cubic metre	C/m^3	Electric charge density
coulomb per kilogram	C/kg	Exposure (X rays and gamma rays)
coulomb per square metre	C/m^2	Electric flux density
cubic metre	m^3	Volume
cubic metre per kilogram	m^3/kg	Specific volume
degree Celsius	°C	Celsius temperature
farad	F C/V	Capacitance
farad per metre	F/m	Permittivity
gray	Gy	Absorbed dose, specific energy imparted, absorbed dose index
gray per second	Gy/s	Absorbed dose rate
henry	H Wb/A	Inductance
henry per metre	H/m	Permeability
hertz	Hzs^{-1}	Frequency
joule	J N·m	Energy, work, quantity of heat
joule per cubic metre	J/m^3	Energy density
joule per kelvin	J/K	Heat capacity, entropy
joule per kilogram	J/kg	Specific energy
joule per kilogram kelvin	J/(kg·K)	Specific heat capacity
joule per mole	J/mol	Molar energy

(Continued)

Table C9 SI units and symbols (*Continued*)

SI unit name	SI unit symbol	Quantity measured
joule per mole kelvin	J/(mol·K)	Molar heat capacity, molar entropy
kelvin	K	Absolute temperature, sometimes referred to as thermodynamic temperature
kilogram	kg	Mass
kilogram per cubic metre	kg/m^3	Density, mass density
lumen	lm	Luminous flux
lux	lx lm/m^2	Illuminance
metre	m	Length
metre per second	m/s	Speed, velocity
metre per second squared	m/s^2	Acceleration
mole	mol	Amount of substance
mole per cubic metre	mol/m^3	Concentration
newton	N	Force
newton metre	N·m	Moment of force
newton per metre	N/m	Surface tension
ohm	Ω V/A	Electric resistance
pascal	Pa N/m^2	Pressure, stress
pascal second	Pa·s	Dynamic viscosity
radian	rad	Plane angle
radian per second	rad/s	Angular velocity
radian per second squared	rad/s^2	Angular acceleration
second	s	Time or time interval
siemens	S A/V	Electric conductance (1/electric resistance)
sievert	Sv	Dose equivalent (index)
square metre	m^2	Area
steradian	sr	Solid angle
tesla	T Wb/m^2	Magnetic flux density
volt	V W/A	Electrical potential or potential difference, electromotive force
volt per metre	V/m	Electric field strength
watt	W J/s	Power
watt per metre kelvin	W/(m·K)	Thermal conductivity

(*Continued*)

Table C9 SI units and symbols (*Continued*)

SI unit name	SI unit symbol	Quantity measured
watt per square metre	W/m^2	Power density, heat flux density, irradiance
watt per square metre steradian	W $\cdot$ m$^2 \cdot$ sr^1	Radiance
watt per steradian	W/sr	Radiant intensity
weber	Wb	Magnetic flux

The table above gives some of the most commonly used SI symbols, units and abbreviations which are seen in scientific and electromechanical engineering applications.

Annex D
The IP code

The IP code (or 'Ingress Protection') ratings are an internationally recognised protection classification system which was created by the International Electrotechnical Commission (IEC) and published as a British, European and International Standard, namely BS EN 60529:1992.

This particular standard describes, classifies and rates the degree of protection provided by mechanical casings and electrical enclosures against intrusion, dust, moisture and accidental contact by electrical enclosures and mechanical casings.

It is particularly useful in the electrical engineering and manufacturing world, because it gives us a clue as to the types of equipment we can use in difficult environments.

Code breakdown

The code itself comprises the letters 'IP' followed by two single digits. The first number indicates the level of protection the enclosure gives against solid particles. Liquid ingress protection is denoted by the second number.

The following table shows what each digit or part of the IP code represents.

Table D1 IP coding description

IP indication	First digit: solid particle protection	Second digit: liquid ingress protection	Third digit: mechanical impact resistance	Additional letter: other protections
IP	Single numeral: 0–6 or 'X'	Single numeral: 0–9 or 'X'	Single numeral: 0–9	Single letter
Mandatory	Mandatory	Mandatory	No longer used	Optional

First digit	Protection against solid particles	Second digit	Protection against liquid ingress
1	No protection	0	No protection
2	Protects against solid objects over 50 mm, like an object being accidently touched by a hand	1	Protection from falling drops of water (equivalent to 1 mm of rainfall per minute) or condensation
3	Protection against solid objects over 2.5 mm, such as tools and thick wires	2	Protection from vertically falling water drips, which will not harm the equipment if it's tilted at an angle up to 15° (equivalent to 3 mm of rainfall per minute)
4	Protection against objects larger than 1 mm, such as wires, screws or nails etc.	3	Protection against spraying water at a pressure of 80–100 kPa
5	Prevents dust entering in quantities large enough to cause problems	4	Protection against splashing water at a pressure of 80–100 kPa
6	Totally protected against dust	5	Protection against water jets projected from a nozzle
		6	Protection against powerful water jets, heavy seas or water ingress when the object is immersed in up to 1 m of water
		6K	Protection against powerful water jets with increased pressure
		7	Protection when temporarily immersed in water up to 1 m in depth
		8	Protection against continuous immersion in water more than 1 m in depth
		9K	Protection against close-range high pressure, high-temperature spray-downs

Example: IP68 = 6 (dust-tight) + 8 (protection from continuous immersion in water).

Additional letters

Further letters can be attached so as to provide additional information related to the protection of the device.

Letters A–D detail specific items that cannot penetrate an item. The final four letters simply give a little more information about the test conditions that led to the IP rating assigned.

A = back of hand
B = finger
C = tool
D = wire
F = oil resistant
H = high-voltage device
M = device moving during water test
S = device standing still during water test
W = weather conditions

IPX coding

The letter X is used to show that a manufacturer has avoided specifying a digit for a specific reason.

Example: All plug sockets are IP2X, showing that they prevent people from sticking their fingers in the holes.

Annex E

Acronyms and abbreviations

a.c.	alternating current
ACS	assembly for construction sites
ADP	automatic data processing
ADS	automatic disconnection of supply
AFDD	arc fault detection devices
BEC	British Electrotechnical Committee
BRE	Building Research Establishment Ltd
BS	British Standards
BSI	British Standards Institution
CAD	computer aided design
CB	circuit-breaker
CBN	common bonding network
CE	Conformity Europe
CECC	CENELEC Electronic Components Committee
CEN	Comité Européen de Normalisation
CENELEC	Comité Européen de Normalisation Électrotechnique
CHP	combined heat and power generation
CNE	combined neutral and earth
CO	carbon monoxide
CPC	circuit protective conductor
CPR	Construction Products Regulation
CPS	control and protective switching device
d.c.	direct current
DCL	device for connecting a luminaire
DDA	Disability Discrimination Act
DIY	do it yourself
EBADS	equipotential bonding and automatic disconnection of supplies
ECA	Electrical Contractors Association
EEBAD or EEBADS	Earthed Equipotential Bonding and Automatic Disconnection of Supply
EEC	European Economic Commission
EIC	Electrical Installation Certificate
EICR	Electrical Installation Condition Report

ELECSA	Electrical or Renewable Technology Systems Installation in Homes'
ELV	extra-low voltage
EMC	electromagnetic compatibility
emf	electromotive force
EMI	electromagnetic interference
EN	European Normalisation
ESQCR	Electricity Safety, Quality and Continuity Regulations
EU	European Union
EV	electric vehicle
EVSE	electric vehicle supply equipment
EWR	Electricity at Work Regulations
FE	functional Earth
FENSA	fenestration self-assessment scheme
FELV	functional extra-low voltage
HBES	home and building electronic systems
HD	Harmonised Directive
HELA	Health and Safety Executive/Local Authorities
HEMP	high-altitude electromagnetic pulse
HSE	Health & Safety Executive
HV	high voltage
I/O	input/output
ICM	insulation current monitoring device
IEC	International Electrotechnical Commission
IEE	Institution of Electrical Engineers (who later changed their name to the Institution of Engineering and Technology – IET)
IET	Institution of Engineering and Technology
IFLS	insulation fault location system
IIE	Institution on of Incorporated Engineers
ILU	integrated logistic unit
IMD	insulation monitoring device
Inspd	nominal discharge current
IP2X	protection against approach by hands
IPC	implant point of coupling
IPXXB	the degree of protection provided once the cover is taken off a distribution board or consumer unit
ISM	industrial, scientific and medical
ISO	International Standards Organisation
IT	information technology
ITCZ	international conveyance zone
ITE	information technology equipment
Limp	impulse current
LPS	lightning protection system
LSC	luminaire supporting coupler
LUR	logical user requirement

LV	low voltage
MDD	Medical Devices Directive
ME	medical electrical
MED-IMD	medical insulation monitoring device
MEIC	Minor Electrical Installation Works Certificate
MET	main earthing terminal
MKS	metre-kilogram-second
MMI	man machine interface
MTBF	mean time between failures
N	neutral
NAPIT	National Association of Professional Inspectors and Testers
NICEIC	National Inspection Council for Electrical Installation Counselling
NSO	National Standards Organisation
OCPD	overcurrent protective device
OJT	on-the-job-training
OPSI	Office of Public Sector Information
PAT	portable appliance testing
PCB	printed circuit board
PE	protective Earth
PELV	protective extra-low voltage
PEN	combined protective and neutral conductors
PHEV	plug-in hybrid electric vehicle
PME	protective multiple earthing
PRCD	portable RCD, usually an RCD incorporated into a plug
PV	photovoltaic power supply system
PVC	polyvinyl chloride
RAH	relative air humidity
RAM	reliability, availability and maintainability
RCBO	residual current operated circuit breaker without integral overcurrent protection
RCCB	residual current operated circuit breaker with integral overcurrent protection
RCCD	residual current operated circuit breaker
RCD	residual current device
RCM	residual current monitor
RF	radio frequency
RH	relative humidity
SELV	separated extra-low voltage
SI	Statutory Instrument
SI	Système International d'Unités
SPD	surge protective device
SRCBO	socket outlet incorporating an RCBO and RCD tester
SRCD	socket outlet incorporating an RCD

SSEG	small-scale embedded generators
STE	Society of Telegraph Engineers
TDS	time delay switches
TLV	threshold limit values
TOM	temporary overvoltages
TQM	total quality management
UPS	uninterruptible power supply
VDU	visual display unit
VSD	variable speed drive
WAUILF	workplace applied uniform indicated low frequency (application)
YFR	yearly forecast rationale

Annex F
Standards

F.1 British and international standards currently used with the Wiring Regulations

By the time you read this edition of *Wiring Regulations in Brief*, it is quite possible that some of the Standards and Directives listed in this book **and** the IET Wiring Regulations will have been reviewed and updated and, although the ones listed in this annex are still relevant, the latest edition of these documents should **always** be taken into account.

For this reason, the reader is recommended to always have a quick check via Google (or some other search engine) to make sure that they are using the most up-to-date standard, regulation or recommendation.

BS or EN number	Title
BS 31	Steel conduit and fittings for electrical wiring (now superseded)
BS 67:1987	Specification for ceiling roses
BS 88-1:1967	Specification for cartridge fuses for voltages up to 660 V. Performance and dimensions of fuse links (now superseded)
BS 196	Specification for protected-type non-reversible plugs, socket outlets cable-couplers and appliance-couplers with earthing contacts for single phase a.c. circuits up to 250 V (now superseded)
BS 476 (series)	Fire tests on building materials and structures
BS 546: 1988	Plugs, socket outlets and socket-outlet adaptors
BS 559:2009	Design, construction and installation of signs
BS 646:2013	Cartridge fuse links (rated up to 5 A) for a.c. and d.c. service
BS 731 (series)	Specification. Flexible steel conduit for cable protection and flexible steel tubing to enclose flexible drives (now superseded)
BS 951:2009	Electrical earthing. Clamps for earthing and bonding
BS 1361:1971	Specification for cartridge fuses for a.c. circuits in domestic and similar premises (now superseded)

F.1 British and international (*continued*)

BS or EN number	Title
BS 1362:1992	Specification for general-purpose fuse links for domestic and similar purposes (primarily for use in plugs)
BS 1363 (series)	13 A plugs, socket outlets, connection units and adaptors
BS 3036:1992	Semi-enclosed electric fuses (ratings up to 100 A and 240 V to Earth)
BS 3535	Isolating transformers and safety isolating transformers. General requirements (now superseded)
BS 3858:2014	Binding and identification sleeves for use on electric cables and wires
BS 4177:2015	Specification for cooker control units
BS 4293:1983	Specification for residual current-operated circuit-breakers (now superseded)
BS 4444:1995	Guide to electrical Earth monitoring and protective conductor proving
BS 4573:2016	Specification for two-pin reversible plugs and shaver socket outlets
BS 4607 (series)	Non-metallic conduits and fittings for electrical installations. Specification for fittings and components of insulating material
BS 4662:2009	Boxes for flush mounting of electrical accessories. Requirements and test methods and dimensions
BS 4727 (series)	Glossary of electrotechnical power, telecommunications, electronics, lighting and colour terms
BS 5266 (series)	Emergency lighting
BS 5467:2016	Electric cables. Thermosetting insulated, armoured cables for voltages of 600/1000 V and 1900/3300 V for fixed installations
BS 5499 (series)	Graphical symbols and signs. Safety signs, including fire safety signs
BS 5655 (series)	Lifts and service lifts
BS 5733:2014	Specification for general requirements for electrical accessories
BS 5803-5:1985	Thermal insulation for use in pitched roof spaces in dwellings. Specification for installation of man-made mineral fibre and cellulose fibre insulation
BS 5839 (series)	Fire detection and fire alarm systems for buildings
BS 6004:2012	Electric cables. PVC insulated, non-armoured cables for voltages up to and including 450/750 V, for electric power, lighting and internal wiring
BS 61535:2006	Installation couplers intended for permanent connection in fixed installations

F.1 British and international (*continued*)

BS or EN number	Title
BS 6231:2006	Electric cables. Single-core PVC-insulated flexible cables of rated voltage 600/1000 V for switchgear and controlgear wiring
BS 6346:2005	Electric cables. PVC-insulated, armoured cables for voltages of 600/1000 V and 1900/3300 V
BS 6351(series)	Electric surface heating
BS 6500:2000	Electric cables. Flexible cords rated up to 300/500 V, for use with appliances and equipment intended for domestic, office and similar environments
BS 6701:2016	Telecommunications equipment and telecommunications cabling. Specification for installation, operation and maintenance
BS 6724:2016	Electric cables. Thermosetting insulated, armoured cables for voltages of 600/1000 V and 1900/3300 V, having low emission of smoke and corrosive gases when affected by fire
BS 6907 (series)	Electrical installations for open-cast mines and quarries
BS 6972:2012	Specification for general requirements for luminaire supporting couplers for domestic, light industrial and commercial use
BS 6991:2012	Specification for 6/10 A, two-pole weather-resistant couplers for household, commercial and light industrial equipment
BS 7001:1988	Specification for interchangeability and safety of a standardised luminaire supporting coupler
BS 7211:2012	Electric cables. Thermosetting insulated, non-armoured cables for voltages up to and including 450/750 V, for electric power, lighting and internal wiring, and having low emission of smoke and corrosive gases when affected by fire
BS 7375:1996	Code of practice for distribution of electricity on construction and building sites
BS 7430:2015	Code of practice for earthing
BS 7454:2010	Method for calculation of thermally permissible short-circuit currents, taking into account non-adiabatic heating effects
BS 7629-1:2015)	Specification for 300/500 V fire resistant electric cables having low emission of smoke and corrosive gases when affected by fire. Multi-core cables
BS 7697:2010	Nominal voltages for low voltage public electricity supply systems
BS 7698-12:1998	Reciprocating internal combustion engine driven alternating current generating sets. Emergency power supply to safety devices

F.1 British and international (*continued*)

BS or EN number	Title
BS 7769 (series)	Electric cables. Calculation of the current rating
BS 7846:2015	Electric cables. 600/1000 V armoured fire-resistant cables having thermosetting insulation and low emission of smoke and corrosive gases when affected by fire
BS 7889:2012	Electric cables. Thermosetting insulated, unarmoured cables for a voltage of 600/1000 V
BS 7909:2011	Code of practice for design and installation of temporary distribution systems delivering a.c. electrical supplies for lighting, technical services and other entertainment-related purposes
BS 7919:2006	Electric cables. Flexible cables rated up to 450/750V, for use with appliances and equipment intended for industrial and similar environments
BS 8436:2011	Electric cables. 300/500 V screened electric cables having low emission of smoke and corrosive gases when affected by fire, for use in walls, partitions and building voids. Multi-core cables
BS 8450:2006	Code of practice for installation of electrical and electronic equipment in ships
BS 8519:2010	Selection and installation of fire-resistant power and control cable systems for life safety and fire-fighting applications
BS 9999:2017	Fire safety in the design, management and use of buildings
BS AU 149a:1987	Specification for electrical connections between towing vehicles and trailers with 6 V or 12 V electrical equipment: type 12 N (normal)
BS AU 177a:1987	Specification for electrical connections between towing vehicles and trailers with 6 V or 12 V electrical equipment: type 12 S (supplementary)
BS EN 81 (series)	Safety rules for the construction and installation of lifts.
BS EN 115 (series)	Safety of escalators and moving walks
BS EN ISO 13297	Small craft. Electrical systems. Alternating current installations
BS EN 13636:2004	Cathodic protection of buried metallic tanks and related piping
BS EN 15112:2006	External cathodic protection of well casing
BS EN 1648 (series)	Leisure accommodation vehicles
BS EN 1838:2013	Lighting applications. Emergency lighting
BS EN 50081 (series)	Electromagnetic compatibility (now superseded)
BS EN 50085 (series)	Cable trunking and cable ducting systems for electrical installations
BS EN 50086 (series)	Specification for conduit systems for cable management

F.1 British and international (*continued*)

BS or EN number	Title
BS EN 50107 (series)	Signs and luminous-discharge-tube installations operating from a no-load rated output voltage exceeding 1 kV but not exceeding 10 kV
BS EN 50171:2001	Central power supply systems
BS EN 50174 (series)	Information technology – cabling installation
BS EN 50200:2015	Method of test for resistance to fire of unprotected small cables for use in emergency circuits
BS EN 50265-2-1	Common test methods for cables under fire conditions
BS EN 50266 (series)	Common test methods for cables under fire conditions. Test for vertical flame spread of vertically mounted bunched wires or cables (now superseded)
BS EN 50362:2003	Method of test for resistance to fire of larger unprotected power and control cables for use in emergency circuits
BS EN 50438:2013	Requirements for the connection of micro-generators in parallel with public low-voltage distribution networks
BS EN 50565-1: 2014	Guide to use for cables with a rated voltage not exceeding 450/750 V
BS EN 60038:2011	CENELEC standard voltages
BS EN 60079 (series)	Electrical apparatus for explosive gas atmospheres
BS EN 60092-507:2014	Electrical installations in ships – pleasure craft
BS EN 60146-2:2000	Semiconductor convertors. General requirements and line commutated convertors. Self-commutated semiconductor converters including direct d.c. converters
BS EN 60204 (series)	Safety of machinery. Electrical equipment of machines
BS EN 60228:2005	Conductors of insulated cables
BS EN 60238:2011	Edison screw lampholders. BS EN 60238:1999 remains current
BS EN 60269 (series)	Low-voltage fuses
BS EN 60309 (series)	Plugs, socket outlets and couplers for industrial purposes
BS EN 60320-1:2015	Appliance couplers for household and similar general purposes. General requirements
BS EN 60332-1-2:2016	Tests on electric and optical-fibre cables under fire conditions. Test for vertical flame propagation for a single insulated wire or cable. Procedure for 1 kW pre-mixed flame
BS EN 60335 (series)	Household and similar electrical appliances. Safety. General requirements
BS EN 60423:2007	Conduit systems for cable management. Outside diameters of conduits for electrical installations and threads for conduits and fittings

F.1 British and international (*continued*)

BS or EN number	Title
BS EN 60445:2010	Basic and safety principles for man–machine interface, marking and identification. Identification of equipment terminals and of terminations of certain designated conductors, including general rules for an alphanumeric system
BS EN 60529:2013	Specification for degrees of protection provided by enclosures (IP code)
BS EN 60570:2003	Electrical supply track systems for luminaires. Replaces BS EN 60570:1997 and BS EN 60570-2-1:1995 which remain current
BS EN 60598 (series)	Luminaires
BS EN 60601 (series)	Medical electrical equipment
BS EN 60617	Graphical symbols for diagrams (now superseded)
BS EN 60623:2017	Secondary cells and batteries containing alkaline or other non-acid electrolytes. Vented nickel-cadmium prismatic rechargeable single cells
BS EN 60664-1:2007	Insulation coordination for equipment within low-voltage systems. Principles, requirements and tests
BS EN 60669 (series)	Switches for household and similar fixed electrical installations
BS EN 60670 (series)	Boxes and enclosures for electrical accessories for household and similar fixed electrical installations
BS EN 60684	Flexible insulating sleeving
BS EN 60702-1:2015	Mineral-insulated cables and their terminations with a rated voltage not exceeding 750 V. Cables
BS EN 60721 (series)	Classification of environmental conditions
BS EN 60896 (series)	Stationary lead-acid batteries
BS EN 60898 series)	Circuit-breakers for a.c. operation
BS EN 60947 (series)	Low-voltage switchgear and controlgear
BS EN 60998 (series)	Connecting devices for low-voltage circuits for household and similar purposes
BS EN 6100-1	Glossary of building and civil engineering terms
BS IEC 61000 (series)	Electromagnetic compatibility (EMC). Testing and measurement techniques. IEMI immunity test methods for equipment and systems
BS EN 61008-1:2015	Residual current operated circuit-breakers without integral overcurrent protection for household and similar uses (RCCBs). General rules
BS EN 61011 (series)	Electric fence energisers (now superseded)
BS EN 61034-2:2013	Measurement of smoke density of cables burning under defined conditions. Test procedure and requirements

F.1 British and international (*continued*)

BS or EN number	Title
BS EN 61095:2009	Specification for electromechanical contactors for household and similar purposes
BS EN 61140:2016	Protection against electric shock. Common aspects for installation and equipment
BS EN 61184:2016	Bayonet lampholders
BS EN 61215:2005	Crystalline silicon terrestrial photovoltaic (PV) modules. Design qualification and type approval
BS EN 61241 (series)	Electrical apparatus for use in the presence of combustible dust
BS EN 61347 (series)	Lamp controlgear
BS EN 61386 (series)	Conduit systems for cable management
BS EN 61439-3:2012	Low-voltage switchgear and controlgear assemblies. Distribution boards intended to be operated by ordinary persons
BS EN 61534 (series)	Powertrack systems
BS EN 61537:2007	Cable tray systems and cable ladder systems for cable management. BS EN 61537:2002 remains current
BS EN 61557 (series)	Electrical safety in low-voltage distribution systems up to 1000 V a.c. and 1500 V d.c. Equipment for testing, measuring or monitoring of protective measures. General requirements
BS EN 61558 (series)	Safety of power transformers, power supply units and similar. General requirements and tests BS EN 61558-1:1998 remains current
BS EN 61643-11:2012	Low-voltage surge protective devices. Surge protective devices connected to low-voltage power systems. Requirements and test methods
BS EN 62020:1999	Electrical accessories. Residual current monitors for household and similar uses (RCMs)
BS EN 62040	Uninterruptible power systems (UPS)
BS EN 62196-2:2017	Plugs, socket outlets, vehicle connectors and vehicle inlets. Conductive charging of electric vehicles. Dimensional compatibility and interchangeability requirements for a.c. pin and contact-tube accessories
BS EN 62208:2011	Empty enclosures for low-voltage switchgear and controlgear assemblies. General requirements
BS EN 62262:2002	Degrees of protection provided by enclosures for electrical equipment against external mechanical impacts (IK code)
BS EN 62305 (series)	Protection against lightning
BS EN 62423:2012	Type F and type B residual current operated circuit-breakers with and without integral overcurrent protection for household and similar uses

F.1 British and international (*continued*)

BS or EN number	Title
BS EN ISO 11446:2012	Road vehicles. Connectors for the electrical connection of towing and towed vehicles. 13-pole connectors for vehicles with 12 V nominal supply voltage
BS EN 61995-1	Devices for the connection of luminaires for household and similar purposes. General requirements

F.2 British Standards associated with the Wiring Regulations – listing by subject

BS or EN number	Title
BS 1363 (series)	13 A plugs, socket outlets, connection units and adaptors
BS EN 60320-1:2015	Appliance couplers for household and similar general purposes. General requirements
BS EN 60445:2010	Basic and safety principles for man–machine interface, marking and identification. Identification of equipment terminals and of terminations of certain designated conductors, including general rules for an alphanumeric system
BS EN 61184:2016	Bayonet lampholders
BS 3858:2014	Binding and identification sleeves for use on electric cables and wires
BS EN 60670 (series)	Boxes and enclosures for electrical accessories for household and similar fixed electrical installations
BS 4662:2009	Boxes for flush mounting of electrical accessories. Requirements and test methods and dimensions
BS EN 61537:2007	Cable tray systems and cable ladder systems for cable management. BS EN 61537:2002 remains current
BS EN 50085 (series)	Cable trunking and cable ducting systems for electrical installations
BS EN 13636:2004	Cathodic protection of buried metallic tanks and related piping
BS EN 60038:2011	CENELEC standard voltages
BS EN 50171:2001	Central power supply systems
BS EN 60898 series)	Circuit breakers for a.c. operation
BS EN 60721 (series)	Classification of environmental conditions

F.2 British Standards (*continued*)

BS or EN number	Title
BS 7909:2011	Code of practice for design and installation of temporary distribution systems delivering a.c. electrical supplies for lighting, technical services and other entertainment-related purposes
BS 7375:1996	Code of practice for distribution of electricity on construction and building sites
BS 7430:2015	Code of practice for earthing
BS 8450:2006	Code of practice for installation of electrical and electronic equipment in ships
BS EN 50266 (series)	Common test methods for cables under fire conditions. Test for vertical flame spread of vertically mounted bunched wires or cables (now superseded)
BS EN 60228:2005	Conductors of insulated cables
BS EN 61386 (series)	Conduit systems for cable management
BS EN 60423:2007	Conduit systems for cable management. Outside diameters of conduits for electrical installations and threads for conduits and fittings
BS EN 60998 (series)	Connecting devices for low-voltage circuits for household and similar purposes
BS EN 61215:2005	Crystalline silicon terrestrial photovoltaic (PV) modules. Design qualification and type approval
BS EN 62262:2002	Degrees of protection provided by enclosures for electrical equipment against external mechanical impacts (IK code)
BS EN 61995-1	Devices for the connection of luminaires for household and similar purposes. General requirements
BS EN 60238:2011	Edison screw lampholders. BS EN 60238:1999 remains current
BS 8436:2011	Electric cables. 300/500 V screened electric cables having low emission of smoke and corrosive gases when affected by fire, for use in walls, partitions and building voids. Multi-core cables
BS 7846:2015	Electric cables. 600/1000 V armoured fire-resistant cables having thermosetting insulation and low emission of smoke and corrosive gases when affected by fire
BS 7769 (series)	Electric cables. Calculation of the current rating
BS 7919:2006	Electric cables. Flexible cables rated up to 450/750V, for use with appliances and equipment intended for industrial and similar environments
BS 6500:2000	Electric cables. Flexible cords rated up to 300/500 V, for use with appliances and equipment intended for domestic, office and similar environments
BS 6346:2005	Electric cables. PVC-insulated, armoured cables for voltages of 600/1000 V and 1900/3300 V

F.2 British Standards (*continued*)

BS or EN number	Title
BS 6004:2012	Electric cables. PVC-insulated, non-armoured cables for voltages up to and including 450/750 V, for electric power, lighting and internal wiring
BS 6220:1999	Electric cables. Single-core PVC-insulated flexible cables of rated voltage 600/1000 V for switchgear and controlgear wiring
BS 6231:2006	Electric cables. Single-core PVC-insulated flexible cables of rated voltage 600/1000 V for switchgear and controlgear wiring
BS 5467:2016	Electric cables. Thermosetting insulated, armoured cables for voltages of 600/1000 V and 1900/3300 V for fixed installations
BS 6724:2016	Electric cables. Thermosetting insulated, armoured cables for voltages of 600/1000 V and 1900/3300 V, having low emission of smoke and corrosive gases when affected by fire
BS 7211:2012	Electric cables. Thermosetting insulated, non-armoured cables for voltages up to and including 450/750 V, for electric power, lighting and internal wiring, and having low emission of smoke and corrosive gases when affected by fire.
BS 7889:2012	Electric cables. Thermosetting insulated, unarmoured cables for a voltage of 600/1000 V
BS EN 61011 (series)	Electric fence energisers (now superseded)
BS 6351(series)	Electric surface heating
BS EN 62020:1999	Electrical accessories. Residual current monitors for household and similar uses (RCMs)
BS EN 60079 (series)	Electrical apparatus for explosive gas atmospheres
BS EN 61241 (series)	Electrical apparatus for use in the presence of combustible dust (now superseded)
BS 951:2009	Electrical earthing. Clamps for earthing and bonding. Specification
BS 6907 (series)	Electrical installations for open-cast mines and quarries
BS EN 60092-507:2014	Electrical installations in ships – pleasure craft
BS EN 61557 (series)	Electrical safety in low voltage distribution systems up to 1000 V a.c. and 1500 V d.c. Equipment for testing, measuring or monitoring of protective measures. General requirements
BS EN 60570:2003	Electrical supply track systems for luminaires. Replaces BS EN 60570:1997 and BS EN 60570-2-1:1995 which remain current
BS IEC 61000 (series)	Electromagnetic compatibility (EMC). Testing and measurement techniques. IEMI immunity test methods for equipment and systems

F.2 British Standards (*continued*)

BS or EN number	Title
BS EN 50081 (series)	Electromagnetic compatibility (now superseded)
BS 5266 (series)	Emergency lighting
BS EN 62208:2011	Empty enclosures for low-voltage switchgear and controlgear assemblies. General requirements
BS EN 15112:2006	External cathodic protection of well casing
BS 5839 (series)	Fire-detection and fire alarm systems for buildings
BS 9999:2017	Fire safety in the design, management and use of buildings
BS 476 (series)	Fire tests on building materials and structures
BS EN 60684	Flexible insulating sleeving
BS EN 6100-1	Glossary of building and civil engineering terms
BS 4727 (series)	Glossary of Electrotechnical power, telecommunications, electronics, lighting and colour terms
BS 5499 (series)	Graphical symbols and signs. Safety signs, including fire safety signs
BS EN 60617	Graphical symbols for diagrams (now superseded)
BS 4444:1995	Guide to electrical Earth monitoring and protective conductor proving
BS EN 50565-1: 2014	Guide to use for cables with a rated voltage not exceeding 450/750 V
BS EN 60335 (series)	Household and similar electrical appliances. Safety. General requirements
BS EN 50174 (series)	Information technology – cabling installation
BS 61535:2006	Installation couplers intended for permanent connection in fixed installations
BS EN 60664-1:2007	Insulation coordination for equipment within low-voltage systems. Principles, requirements and tests
BS 3535	Isolating transformers and safety isolating transformers. General requirements (now superseded)
BS EN 61347 (series)	Lamp controlgear
BS EN 1648 (series)	Leisure accommodation vehicles
BS 5655 (series)	Lifts and service lifts
BS EN 1838:2013	Lighting applications. Emergency lighting
BS EN 60269 (series)	Low-voltage fuses
BS EN 61643-11:2012	Low-voltage surge protective devices. Surge protective devices connected to low-voltage power systems. Requirements and test methods
BS EN 60947 (series)	Low-voltage switchgear and control gear
BS EN 61439-3:2012	Low-voltage switchgear and controlgear assemblies. Distribution boards intended to be operated by ordinary persons

F.2 British Standards (*continued*)

BS or EN number	Title
BS EN 60598 (series)	Luminaires
BS EN 61034-2:2013	Measurement of smoke density of cables burning under defined conditions. Test procedure and requirements
BS EN 60601 (series)	Medical electrical equipment.
BS 7454:2010	Method for calculation of thermally permissible short-circuit currents, taking into account non-adiabatic heating effects
BS EN 50362:2003	Method of test for resistance to fire of larger unprotected power and control cables for use in emergency circuits
BS EN 50200:2015	Method of test for resistance to fire of unprotected small cables for use in emergency circuits
BS EN 60702-1:2015	Mineral-insulated cables and their terminations with a rated voltage not exceeding 750 V. Cables
BS 7697:2010	Nominal voltages for low voltage public electricity supply systems
BS 4607 (series)	Non-metallic conduits and fittings for electrical installations. Specification for fittings and components of insulating material
BS EN 60309 (series)	Plugs, socket outlets and couplers for industrial purposes
BS EN 62196-2:2017	Plugs, socket outlets, vehicle connectors and vehicle inlets. Conductive charging of electric vehicles. Dimensional compatibility and interchangeability requirements for a.c. pin and contact-tube accessories
BS EN 61534 (series)	Powertrack systems
BS EN 61140:2016	Protection against electric shock. Common aspects for installation and equipment
BS EN 62305 (series)	Protection against lightning
BS 7698-12:1998	Reciprocating internal combustion engine driven alternating current generating sets. Emergency power supply to safety devices
BS EN 50438:2013	Requirements for the connection of micro-generators in parallel with public low-voltage distribution networks
BS EN 61008-1:2015	Residual current operated circuit-breakers without integral overcurrent protection for household and similar uses (RCCBs). General rules
BS EN ISO 11446:2012	Road vehicles. Connectors for the electrical connection of towing and towed vehicles. 13-pole connectors for vehicles with 12 V nominal supply voltage
BS EN 115 (series)	Safety of escalators and moving walks

F.2 British Standards (*continued*)

BS or EN number	Title
BS EN 60204 (series)	Safety of machinery. Electrical equipment of machines
BS EN 61558 (series)	Safety of power transformers, power supply units and similar. General requirements and tests BS EN 61558-1:1998 remains current
BS EN 81 (series)	Safety rules for the construction and installation of lifts
BS EN 60623:2017	Secondary cells and batteries containing alkaline or other non-acid electrolytes. Vented nickel-cadmium prismatic rechargeable single cells
BS 8519:2010	Selection and installation of fire-resistant power and control cable systems for life safety and fire-fighting applications
BS EN 60146-2:2000	Semiconductor convertors. General requirements and line commutated convertors. Self-commutated semiconductor converters including direct d.c. converters
BS 3036:1992	Semi-enclosed electric fuses (ratings up to 100 A and 240 V to earth)
BS EN 50107 (series)	Signs and luminous-discharge-tube installations operating from a no-load rated output voltage exceeding 1 kV but not exceeding 10 kV
BS EN ISO 13297	Small craft. Electrical systems. Alternating current installations
BS 4573:2016	Specification for two-pin reversible plugs and shaver socket-outlets
BS 7629-1:2015)	Specification for 300/500 V fire resistant electric cables having low emission of smoke and corrosive gases when affected by fire. Multi-core cables
BS 6991:2012	Specification for 6/10 A, two-pole weather-resistant couplers for household, commercial and light industrial equipment
BS 1361:1971	Specification for cartridge fuses for a.c. circuits in domestic and similar premises (now superseded)
BS 88-1:1967	Specification for cartridge fuses for voltages up to 660 volts. Performance and dimensions of fuse links (now superseded)
BS 67:1987	Specification for ceiling roses
BS EN 50086 (series)	Specification for conduit systems for cable management
BS 4177:2015	Specification for cooker control units
BS EN 60529:2013	Specification for degrees of protection provided by enclosures (IP code)
BS 559:2009	Specification for design, construction and installation of signs

F.2 British Standards (*continued*)

BS or EN number	Title
BS AU 149a:1987	Specification for electrical connections between towing vehicles and trailers with 6 V or 12 V electrical equipment: type 12 N (normal)
BS AU 177a:1987	Specification for electrical connections between towing vehicles and trailers with 6 V or 12 V electrical equipment: type 12 S (supplementary)
BS EN 61095:2009	Specification for electromechanical contactors for household and similar purposes
BS 1362:1992	Specification for general-purpose fuse links for domestic, and similar purposes (primarily for use in plugs)
BS 5733:2014	Specification for general requirements for electrical accessories
BS 6972:2012	Specification for general requirements for luminaire supporting couplers for domestic, light industrial and commercial use
BS 7001:1988	Specification for interchangeability and safety of a standardised luminaire supporting coupler
BS 196	Specification for protected-type non-reversible plugs, socket outlets cable-couplers and appliance-couplers with earthing contacts for single-phase a.c. circuits up to 250 V (now superseded)
BS 4293:1983	Specification for residual current-operated circuit-breakers (now superseded)
BS 646:2013	Specification. Cartridge fuse links (rated up to 5 A) for a.c. and d.c. service
BS 731 (series)	Specification. Flexible steel conduit for cable protection and flexible steel tubing to enclose flexible drives (now superseded)
BS 546: 1988	Specification. Two-pole and earthing-pin plugs, socket outlets and socket-outlet adaptors
BS EN 60896 (series)	Stationary lead-acid batteries
BS 31	Steel conduit and fittings for electrical wiring (now superseded)
BS EN 60669 (series)	Switches for household and similar fixed electrical installations
BS 6701:2016	Telecommunications equipment and telecommunications cabling. Specification for installation, operation and maintenance
BS EN 60332-1-2:2016	Tests on electric and optical-fibre cables under fire conditions. Test for vertical flame propagation for a single insulated wire or cable. Procedure for 1 kW pre-mixed flame
BS 5803-5:1985	Thermal insulation for use in pitched roof spaces in dwellings. Specification for installation of man-made mineral fibre and cellulose fibre insulation

F.2 British Standards (*continued*)

BS or EN number	Title
BS EN 62423:2012	Type F and type B residual current operated circuit-breakers with and without integral overcurrent protection for household and similar uses
BS EN 62040	Uninterruptible power systems (UPS)

F.3 Other standards to which reference is made in the Regulations

F.3.1 IEC and ISO

IEC 60038-am 2 Ed 6	IEC standard voltages
IEC 60364	Low-voltage electrical installations
IEC 60364-5-51	Electrical installations of buildings – Part 5-51: Selection and erection of electrical equipment – Common rules
IEC 60449-am 1 Ed 1	Voltage bands for electrical installations of buildings
IEC 60502-1 Ed 2	Power cables with extruded insulation and their accessories for rated voltages from 1 kV (Urn = 1.2 kV) up to 30 kV (Urn = 36 kV) – Part 1: Cables for rated voltages of 1 kV (Urn = 1.2 kV) and 3 kV (Urn = 3.6 kV)
IEC 60755-am 2	General requirements for residual current operated protective devices
IEC 60884 Ed 3.1	Plugs and socket-outlets for household and similar purposes. Part 1: General requirements
IEC 60906	IEC system of plugs and socket outlets for household and similar purposes
IEC 61201:1992	Extra-low voltage (ELV). Limit values. Also known as PD 6536
IEC 61386	Conduit systems for cable management (BS EN 61386 series)
IEC 61386-24 Ed 1	Particular requirements – Conduit systems buried underground
IEC 61662 TR2 Ed 1	Assessment of the risk of damage due to lightning
IEC 61936-1 Ed 1	Power installations exceeding 1 kV a.c. - Part 1: Common rules
IEC 61995-1 Ed 1	Devices for the connection of luminaires for household and similar purposes – Part 1: General requirements
IEC/TS 62081 Ed 1	Arc welding equipment. Installation and use
ISO 8820	Road vehicles. Fuse links

F.3.2 CENELEC Harmonised Documents

BS 7671 Requirements for Electrical Installations takes account of the technical substance of agreements reached in CENELEC. In particular, the technical intent of the following CENELEC Harmonisation Documents is included.

F.3.2.1 CENELEC Harmonised Documents – listed by subject

Agricultural and horticultural premises	HD 60364-7-705:2007
Application of measures for protection against overcurrent	HD 384.4.473 AI:1980
Caravan parks, camping parks and similar locations	HD 60364-7-708:2009
Conducting locations with restricted movement	HD 60364-7-706:2007
Construction and demolition site installations	HD 60364-7-704:2007
Devices for protection against overvoltage	HD 60364-5-534:2008
Earthing arrangements	HD 60364-5-54:2007
Electrical installations in caravans and motor caravans	HD 60364-7-721:2009
Exhibitions, shows and stands	HD 384.7.711:2003
External influences	HD 60364-5-51:2006
Extra-low-voltage lighting installations	HD 60364-7-715:2005
Fundamental principles, assessment of general characteristics and definitions	HD 60364-1:2008
Identification of cores in cables and flexible cords	HD 308 S2:2001
Initial verification	HD 60364-6:2007
Locations containing a bath or shower	HD 60364-7-701:2007
Low voltage generating sets	HD 384.5.551:1997
Marinas and similar locations	HD 60364-7-709:2009
Measures against electromagnetic disturbances	FprHD 60364-4-444:200X
Measures against overcurrent	HD 60364-4-43:2008
Medical locations	FprHD 60364-7-710:2010
Mobile or transportable units	FprHD 60364-7-717:2009
Operating and maintenance gangways	HD 60364-7-729:2009
Outdoor lighting installations	HD 60364-5-559:2005
Protection against electric shock	HD 60364-4-41:2007
Protection against fire	HD 60364-4-42:2001
Protection against fire where particular risks or danger exist	HD 384.4.482 51:1997
Protection against overcurrent	HD 60364-4-43:2008
Protection against overvoltages	HD 60364-4-443:2006

Protection against thermal effects	HD 60364-4-42;2001
Protection of low-voltage installations against temporary overvoltages	HD 60364-4-442:1997
Rooms and cabins containing sauna heaters	HD 384.7.703:2005
Sauna heaters	HD 60364-7-703:2005
Selection and erection of equipment – common rules	prHD 60364-5-51:2003
Solar photovoltaic (PV) power supply systems	HD 60364-7-712:2005
Swimming pools and other basins	FprHD 60364-7-702:2009
Temporary electrical installations for structures, amusement devices	HD 60364-7-740:2006

F.3.2.1 CENELEC Harmonised Documents – listed by Directive

FprHD 60364-4-444:200X	Measures against electromagnetic disturbances
FprHD 60364-7-702:2009	Swimming pools and other basins
FprHD 60364-7-710:2010	Medical locations
FprHD 60364-7-717:2009	Mobile or transportable units
HD 308 S2:2001	Identification of cores in cables and flexible cords
HD 384.4.41 S2/AI:2002	Protection against electric shock
HD 384.4.42 S1 A2:1994	Protection against thermal effects
HD 384.4.43 S2:2001	Protection against overcurrent
HD 384.4.443 S1:2000	Protection against overvoltages
HD 384.4.473 AI:1980	Application of measures for protection against overcurrent
HD 384.4.482 51:1997	Protection against fire where particular risks or danger exist
HD 384.5.551:1997	Low voltage generating sets
HD 384.6.61 S2:2003	Initial verification
HD 384.7.702 S2:2002	Swimming pools and other basins
HD 384.7.703:2005	Rooms and cabins containing sauna heaters
HD 384.7.708:2005	Caravan parks, camping parks and similar locations
HD 384.7.711:2003	Exhibitions, shows and stands
HD 384.7.714 51:2000	Outdoor lighting installations
HD 60364-1:2008	Fundamental principles, assessment of general characteristics and definitions
HD 60364-4-41:2007	Protection against electric shock
HD 60364-4-42:2001	Protection against fire
HD 60364-4-42;2001	Protection against thermal effects

HD 60364-4-43:2008	Measures against overcurrent
HD 60364-4-443:2006	Protection against overvoltages
HD 60364-5-51:2006	Selection and erection – common rules
HD 60364-5-51:2006	External influences
HD 60364-5-534:2008	Devices for protection against overvoltage
HD 60364-5-54:2007	Earthing arrangements
HD 60364-5-559:2005	Outdoor lighting installations
HD 60364-6:2007	Initial verification
HD 60364-7-701:2007	Locations containing a bath or shower
HD 60364-7-703:2005	Sauna heaters
HD 60364-7-704:2007	Construction and demolition site installations
HD 60364-7-705:2007	Agricultural and horticultural premises
HD 60364-7-706:2007	Conducting locations with restricted movement
HD 60364-7-708:2009	Caravan parks, camping parks and similar locations
HD 60364-7-709:2009	Marinas and similar locations
HD 60364-7-712:2005	Solar photovoltaic (PV) power supply systems
HD 60364-7-715:2005	Extra-low-voltage lighting installations
HD 60364-7-717:2004	Mobile or transportable units
HD 60364-7-721:2009	Electrical installations in caravans and motor caravans
HD 60364-7-729:2009	Operating and maintenance gangways
HD 60364-7-740:2006	Temporary electrical installations for structures, amusement devices
IEC 60364-4-44:2008	Introduction to voltage and electro disturbances
prHD 60364-5-51:2003	Selection and erection of equipment – common rules
prHD 60364-5-54:2004	Earthing arrangements, protective conductors and protective bonding conductors
prHD 60364-7-709:2007	Marinas and similar locations
prHD 60364-7-721:2007	Electrical installations in caravans and motor caravans
prHD 60364-7-740:2006	Temporary electrical installations for structures, amusement devices and booths at fairgrounds, amusement parks and circuses

Where the above Harmonised Documents contain UK special national conditions, those conditions have been incorporated within BS 7671. If BS 7671 is being applied in other countries, the above HDs should be consulted to confirm the status of a particular regulation.

 BS 7671 will continue to be amended from time to time to take account of the publication of new or amended CENELEC standards.

Annex G

Other books associated with the Wiring Regulations

Title	Extracts from book reviews	ISBN
Building Regulations in Brief (9th edition) 	This ninth edition of the most popular and trusted guide reflects all the latest amendments to the Building Regulations, planning permission and the Approved Documents in England and Wales. This includes coverage of the new Approved Document Q on security, and a second part to Approved Document M which divides the regulations for *'dwellings' and 'buildings other than dwellings'*. A new chapter has been added to incorporate these changes and to make the book more user friendly. Giving practical information throughout on how to work with (and within) the Regulations, this book enables compliance in the simplest and most cost-effective manner possible. The no-nonsense approach of *Building Regulations in Brief* cuts through any confusion and explains the meaning of the Regulations. Consequently, it has become a favourite for anyone in the building industry or studying, as well as those planning to have work carried out on their home.	 Paperback: ISBN 9781138285163 Hardback: ISBN 9781138309364 e-Book: ISBN 9781315269153

(continued)

Title	Extracts from book reviews	ISBN
Building Regulations Pocket Book (1st edition) 	This handy guide provides you with all the information you need to comply with the UK Building Regulations and Approved Documents. On site, in the van, in the office, wherever you are, this is the book you'll refer to time and time again to double check the regulations on your current job. The *Building Regulations Pocket Book* is the most reliable and portable guide to compliance with the Building Regulations. Part 1 provides an overview of the Building Act. Part 2 offers a handy guide to the do's and don'ts of gaining the local council's approval for planning permission and Building Regulations Approval. Part 3 presents an overview of the requirements of the Approved Documents associated with the Building Regulations. Part 4 is an easy-to-read explanation of the essential requirements of the Building Regulations that any architect, builder or DIYer needs to know to keep their work safe and compliant on both domestic or non-domestic jobs. This book is essential reading for all building contractors and sub-contractors, site engineers, building engineers, building control officers, building surveyors, architects, construction site managers and DIYers. Homeowners will also find it useful to understand what they are responsible for when they have work done on their home (ignorance of the regulations is no defence when it comes to compliance!).	 Paperback: ISBN 9780815368380 Hardback: ISBN 9780815368373 e-Book: ISBN 9781351254809

(continued)

Title	Extracts from book reviews	ISBN

Scottish Building Standards in Brief (1st edition)

Scottish Building Standards in Brief takes the highly successful formula of Ray Tricker's *Building Regulations in Brief* and applies it to the requirements of the Building (Scotland) Regulations 2004. With the same no-nonsense and simple-to-follow guidance, but written specifically for the Scottish Building Standards, it's the ideal book for builders, architects, designers and DIY enthusiasts working in Scotland.

The book explains the meaning of the regulations, their history, current status, requirements, associated documentation and how local authorities view their importance, and emphasises the benefits and requirements of each one.

There is no easier or clearer guide to help you to comply with the Scottish Building Standards in the simplest and most cost-effective manner possible.

 **Routledge** Taylor & Francis Group

Paperback:
ISBN 9780750685580
Hardback:
ISBN 9781138162365
e-Book:
ISBN 9780080942513

Water Regulations In Brief (1st edition)

Water Regulations in Brief is a unique reference book, providing all the information needed to comply with the regulations, in an easy-to-use, full-colour format.

Crucially, unlike other titles on this subject, this book doesn't just cover the Water Regulations, it also clearly shows how they link in with the Building Regulations, water bylaws and the Wiring Regulations, providing the only available complete reference to the requirements for water fittings and water systems. Structured in the same logical, time-saving way as the author's other bestselling ... *in Brief* books, *Water Regulations in Brief* will be a welcome change to anyone tired of wading through complex, jargon-heavy publications in search of the information they need to get the job done.

Routledge Taylor & Francis Group

Paperback:
ISBN 9781856176286
Hardback:
ISBN 9781138408661
e-Book:
ISBN 9780080950945

(continued)

Title	Extracts from book reviews	ISBN
Quality Management Systems: A Practical Guide to Standards Implementation (1st edition) 	This book provides a clear, easy-to-digest overview of Quality Management Systems (QMS). Critically, it offers the reader an explanation of the International Standards Organization's (ISO) requirement that in future all new and existing Management Systems Standards will need to have the same high-level structure, commonly referred to as Annex SL, with identical core text, as well as common terms and definitions. In addition to explaining what Annex SL entails, this book provides the reader with a guide to the principles, requirements and interoperability of Quality Management System standards, how to complete internal and external management reviews, third-party audits and evaluations, as well as how to become an ISO certified organisation once your QMS is fully established. As a simple and straightforward explanation of QMS standards and their current requirements, this is a perfect guide for practitioners who need a comprehensive overview to put theory into practice, as well as for undergraduate and postgraduate students studying quality management as part of broader operations and management courses.	 Paperback: ISBN 9780367223533 Hardback: ISBN 9780367223519 e-Book: ISBN 9780429274473

(continued)

Title	Extracts from book reviews	ISBN
ISO 9001:2015 In Brief (4th edition) 	*ISO 9001: 2015 In Brief* provides an introduction to quality management systems for students, newcomers and busy executives, with a user-friendly, simplified explanation of the history, requirements and benefits of the new standard. This short, easy-to-understand reference tool also helps organisations to quickly set up an ISO 9001:2015-compliant Quality Management System for themselves at minimal expense and without high consultancy fees. Now in its fourth edition, *ISO 9001:2015 In Brief* consists of a number of chapters covering topics like: What is Quality? What is a Quality Management Who produces Quality Standards? What is ISO 9001:2015?. What other standards are based on ISO 9001:2015? What to do once your QMS is established – Process improvement tools, internal auditing and the road to ISO 9001:2015 certification The book is then supported by: Annex A – A summary of the requirements of ISO 9001:2015, including an overview of the content of the various clauses and sub-clauses, the likely documentation required and how these would affect an organisation.	 Paperback: ISBN 9781138025868 Hardback: ISBN 9781138025851 e-Book: ISBN 9781315774831

(continued)

Title	Extracts from book reviews	ISBN
ISO 9001:2015 for Small Businesses (6th edition) 	Small businesses face many challenges today, including the increasing demand by larger companies for ISO 9001 compliance, a challenging task for any organisation and in particular for a small business without quality assurance experts on its payroll. Ray Tricker has already guided hundreds of businesses through to ISO accreditation, and this sixth edition of his life-saving ISO guide provides all you need to meet the new 2015 standards. *ISO 9001:2015 for Small Businesses* helps you understand what the new standard is all about and how to achieve compliance in a cost-effective way. Covering all the major changes to the standards, this book provides direct, accessible and straightforward guidance. This edition includes: down-to-earth explanations to help you determine what you need to enable you to work in compliance with and/or achieve certification to ISO 9001:2015; a detailed description of the structure of ISO 9001:2015 and its compliance with Annex SL; coverage of the new requirements for Risk Management and Risk Analysis; an example of a complete, generic Quality Management System consisting of a Quality Manual plus a whole host of Quality Processes, Quality Procedures and Word Instructions; and access to a free software copy of these generic QMS files to give you a starting point from which to develop your own documentation.	 Paperback: ISBN 9781138025868 Hardback: ISBN 9781138025820 e-Book: ISBN 9781315774855

(continued)

Title	Extracts from book reviews	ISBN

ISO 9001:2015 Audit Procedures (4th edition)

Revised and fully updated, *ISO 9001:2015 Audit Procedures* describes the methods for completing management reviews and quality audits and describes the changes made to the standards for 2015 and how they are likely to impact on your own audit procedures.

Now in its fourth edition, this text includes essential material on process models, generic processes and detailed coverage of auditor questionnaires. Part II includes a series of useful checklists to assist auditors in compiling their own systems and individual audit check sheets. The whole text is also supported with a glossary of terms as well as explanations of acronyms and abbreviations used in quality.

ISO 9001:2015 Audit Procedures is for auditors of small businesses looking to complete a quality audit review for the 2015 standards. This book will also prove invaluable to all professional auditors completing internal, external and third party audits.

The book also includes access to a free software copy of these generic ISO 9001:2015 audit files, to give you a starting point from which to develop your own audit procedures.

Routledge
Taylor & Francis Group

Paperback ISBN
9781138025899
Hardback ISBN
9781138025882
e-Book ISBN
9781315774817

Index

9780367431983